BEYOND *Capricorn*

BEYOND *Capricorn*

HOW PORTUGUESE ADVENTURERS
SECRETLY DISCOVERED AND MAPPED AUSTRALIA AND NEW
ZEALAND 250 YEARS BEFORE CAPTAIN COOK

PETER TRICKETT

East Street Publications
www.eaststreet.com.au

Published by East Street Publications, 11 Gibson Street, Bowden,
Adelaide, SA 5007

First published 2007
Reprinted 2007

Front and back cover illustrations: Portuguese carrack off the coast of Queensland, early sixteenth century – montage from the Terra Java map of the Vallard Atlas, 1547. Reproduced with kind permission of the Huntington Library, San Marino, California.

National Library of Australia Cataloguing-in-Publication data:

Trickett, Peter, 1934- .

Beyond Capricorn : how Portuguese adventurers secretly
discovered and mapped Australia and New Zealand 250 years before Captain Cook.

Bibliography.
Includes index.

ISBN 9780975114599.
ISBN 0 9751145 9 X.

1. Australia - Discovery and exploration - Portuguese. I. Title.

919.4041

Cover design Nada Backovic
Book designed and typeset by David Bradbury
Printed by Griffin Press, South Australia

www.eaststreet.com.au

To Myrna

Contents

List of maps and illustrations

Sketch Maps

Maps

Illustrations

List of Plates

Colour plates between pages 180–181

Black and white plates between pages 196–197

Acknowledgements

I have acknowledged in more detail elsewhere the invaluable assistance given by Dr Inacio Rebello de Andrade in making sense of the numerous Portuguese place names appearing on the Vallard map coastlines of Australia. Another person to whom I owe a particular debt of gratitude in this regard is Vera Morais, tutor in Portuguese at the Australian National University's Centre for Continuing Education—in particular for her help in providing a tentative Portuguese 'translation' of the intriguing New Zealand Maori fishermen's chant. Others whose assistance I gratefully acknowledge include Father Anscar McPhee, of the Kalumburu Mission on the Kimberley coast; Khamis bin HJ Abas, Curator, Melaka Museums Corporation, for help in identifying Malay cannons; Dan Fitzhenry for details of his proposed Gabo Island shipwreck investigation; Ben Cruise for information about the oral traditions of his ancestors, the Aboriginal people of the Twofold Bay region; and Jody White, curator at the Eden Killer Whale Museum. José Barata, in Portugal, provided valuable new information concerning Christopher de Mendonça's fleet as it was preparing to embark on its quest for the 'Island of Gold'.

I would also like to express my thanks to William P Frank, Curator, Hispanic, Cartographic & Western Manuscripts at the Huntington Library, Los Angeles, for his assistance with access to copies of the library's Vallard maps. My grateful appreciation also to many staff members at the National Library of Australia, especially the Maps Section, for their unfailing helpfulness and courtesy, and to my old friend Ray Hukins for help in tracking down the 'Spanish Helmet' at Te Papa Tongarewa, the Museum of New Zealand. And last but far from least, my sincere thanks go to Jane Macduff, publisher at East Street Publications, for all her tireless work in helping to refine and improve my original manuscript.

Introduction

History is full of surprises: we all too often underestimate the capabilities of people living in earlier times. Until the latter part of the twentieth century hardly anyone was willing to entertain the idea that the Vikings could have reached North America centuries before Columbus—even though just such a discovery appeared to be related in detail in the Viking sagas. But in fact, as an independently minded Norwegian archaeologist, Helge Ingstad, demonstrated to the world in the 1960s, conclusive archaeological evidence of the Vikings' presence in North America a thousand years ago had been there all along in the fields of Newfoundland, simply waiting to be unearthed.

So it is with the Portuguese discovery of Australia. The map evidence that Portuguese explorers sailed to and charted the coasts of Australia 250 years before Captain Cook is there for anyone to see—and has been for centuries—yet most Australian academics dismiss it out of hand as either a fabrication or a fantasy. As for supporting archaeological evidence of the Portuguese discovery, this too, has in recent years come to light; but in contrast to the now universal acceptance of the early Viking presence in America, the comparable evidence of a Portuguese presence on Australia's coasts is still met in orthodox academic circles with blank denial.

This book owes its origins to a combination of chance and coincidence. Its genesis was a display of antique maps which I happened to come across at the National Library of Australia, in Canberra. One exhibit in particular caught my attention. It was a world map by a German theologian, Heinrich Bunting, published in Magdeburg in 1581. The coastline of a land called *India Meridionalis* which Bunting had placed in the southern Indian Ocean looked strangely familiar: surely this could be none other than the west coast of Australia—yet how could that possibly be, if, as the caption claimed, this map predated the first recorded European knowledge of Australia?

And indeed there could be no denying the fact that the Bunting map was printed at least two decades before Dutch captains first began sailing by the western coasts of Australia en route to their East Indies base at Batavia, now Jakarta. So how could a German map-maker have known about the shape of this apparent Australian coast and, more to the point, who could have charted it, and with such seemingly accurate detail?

These were no more than thoughts in the back of my mind when, thumbing through the offerings on the 'remainders' table at a Canberra bookshop a year or two later I came across a set of reproductions of a collection of sixteenth century world maps originating in France. As examples of early cartography they were clearly of outstanding quality. What particularly attracted my attention were two maps which depicted a strange-looking southern continent with the name 'Terra Java'. Could this somehow hold a key to the puzzle of how the Bunting map had originated?

An initial examination of the names on these 'French' maps soon left little doubt that they were in reality of Portuguese origin. But how to discover what lay behind them? The history of Portugal's 'Golden Age' in the sixteenth century was recorded in great detail by a succession of officially sanctioned chroniclers, so this seemed the obvious place to begin. And by an unexpected stroke of good fortune there was no need to contemplate travelling to Lisbon to read the chroniclers' works. Copies of almost every single one of them were to be found in the collection of the National Library of Australia, located on my doorstep in Canberra. Many of these vital reference books were available thanks to the library's acquisition in 1966 of the Braga Collection, comprising some 6000 books, pictures and other documents painstakingly collected by Jose Maria Braga, a businessman, teacher and scholar living in the then Portuguese colony of Macao.

Some of these historical works are reprints published by the government of Portugal between the 1920s and 1940s. But who would have thought that the library would also be the owner of a 300-year-old copy of Manuel de Faria e Sousa's comprehensive three-volume work, *Asia Portuguesa*, presenting a detailed, blow-by-blow account of the fortunes of Portugal's far-flung overseas empire up to the time of the author's death in 1649. No twentieth century reprint this, but a set of gold-embossed, leather-bound chronicles published in Lisbon between 1674 and 1703, almost a century that is, before Captain Cook first set eyes on the coast of Australia! The entire set was still in pristine condition, complete with line illustrations and an ornate coat of arms on the title page.

In terms of antiquity, though, even Faria e Sousa's works were easily eclipsed by another leather-bound volume, a study of Portugal's Asian exploits written in Latin by an Italian Jesuit, Giovanni Pietro Maffei. Entitled *Historiarum Indicarum* this weighty volume's chapters all begin with elaborate letters of the alphabet illuminated in monastic style. It was published in Rome in 1585—three years, that is, before the Spanish Armada and at a time when William Shakespeare was just beginning his career as a playwright. Notable among many other rare works furnished by the National Library's collection was a beautifully preserved volume of the Portuguese chronicler Diogo do Couto's *Decadas da Asia* (Decades of Asia), printed in 1736.

In scope, the many Portuguese chronicles range over the whole of that nation's seaborne empire, covering a vast canvas extending from West Africa to Brazil and from Mozambique to India, the Persian Gulf, South-East Asia, China and the Spice Islands. Voyages in the vicinity of Australia were but a tiny part of the whole.

For me the one big problem was that the chronicles were all written (naturally enough) in Portuguese, and the Old Portuguese of Shakespeare's day at that. And at that time I knew scarcely a word of Portuguese, either ancient or modern!

At first I was able to get assistance in my transcription efforts from a Brazilian postgraduate student studying at the Australian National University, but when at the end of the first year he had to move to Sydney I was thrown back on my own devices. Fortunately, the university at about that time happened to begin offering an evening course in Portuguese. With this and the help of a Portuguese-English dictionary I managed to struggle along, gradually getting the hang of the chroniclers' sometimes convoluted phraseology.

It was the beginning of a long journey of discovery. At first I thought the project would be fairly straightforward but that proved to be a naive idea. As when the proverbial stone is thrown into a calm pond, the ripples kept getting wider and wider. At times it seemed almost like a detective story: new clues constantly appeared and demanded to be followed up. What had the Portuguese really been up to? One avenue of investigation led into the field of Australian Aboriginal tradition and rock art; another to Normandy, in France. Portuguese battles against the Sultan of Malaya intruded into the scene, as did a mysterious voyage out into the vastness of the Pacific.

In the central task of trying to make sense of the many place names adorning what were evidently the coasts of Australia in the early maps I was working on, I received invaluable assistance from Dr Inacio Rebello de Andrade, former Portuguese Ambassador to Australia who now makes his home in Canberra. His insights into the meaning of these sometimes cryptic place names proved crucial in helping to confirm beyond doubt that it was indeed Portuguese mariners who had made the remarkable voyages to which these maps were witness.

I am not an academic historian but perhaps this is just as well, since had I been, I probably would not have dared take this project on. One soon discovers that in orthodox academic circles, advocacy of a Portuguese discovery of Australia is little short of heresy. The subject is not even deemed worthy of serious discussion. Crucial discoveries supported by rigorous scientific research which confirm a Portuguese maritime presence of Australia's coasts in the sixteenth century have, as a rule, been either ignored or denigrated by the vast majority of Australian historians and maritime archaeologists. What it amounts to is a culture of denial. Even the National Library of Australia, despite possessing in its own archives a copy of an early sixteenth century Portuguese-origin map catalogued 'First Map of Australia' still doggedly maintains in its own publications that it was not until the seventeenth century, when the Dutch appeared on the scene, that any mapping of Australia actually began.

Unfortunately, most of Portugal's important archives and maps were lost in the great earthquake, tsunami and fire which ravaged Lisbon in 1755. The Casa da India, where the state archives were housed, was among the many buildings totally destroyed. Detractors have claimed that this loss is 'convenient' for those pursuing the story of Portuguese voyages to Australia—which they maintain never took place. In fact the reverse is the truth. But for this disaster, some written account of the remarkable voyage or voyages which resulted in the charting of the coastlines of Australia might well have come to light.

As for my personal standpoint, having discovered so much about unknown feats of courage and seamanship rivalling those of the great Captain Cook, in an absurdly quixotic sort of way I started to feel that I somehow owed it to the memory of these courageous mariners to make known the story of how they risked (and sometimes lost) their lives in incredibly daring voyages of exploration at the furthest ends of the earth.

In writing this book I have been conscious of the danger that point-by-point identification of Australian coastal features in the context of Portuguese voyages might at times strain the patience of some readers. For better or worse, however, this detailed process of identification is unavoidable, because in the final analysis it is the only way in which the accuracy of the maps can ultimately be proved. In the absence of first-hand accounts of the Portuguese expeditions I have tried to bring to life the prosaic details of names and places by introducing, where appropriate, eyewitness descriptions of the same places on the Australian coasts by later voyagers such as Cook, Dampier, Flinders and the French captain, Nicolas Baudin.

Some of the remnants of the great Portuguese trading empire are only a short plane ride from Australia's northernmost city, Darwin, yet they are almost unknown to most Australians. Similarly, as a subject for historical study the Portuguese presence over several centuries in the far-flung islands of what is now known as Indonesia is largely ignored by Australian academic institutions. Indonesian history as taught in Australia tends to concentrate on the twentieth century independence struggle and subsequent events. European influences are viewed largely in terms of the Dutch and British colonial empires. The Portuguese, insofar as they are mentioned at all, are generally dismissed as an historical footnote.

The one exception to this today is East Timor, brought forcibly to Australia's notice by international politics and a series of dramatic events following Portugal's abrupt departure from its former colony. And in truth the drama of civil war, Indonesian invasion and eventual liberation played out in East Timor during the last two decades of the twentieth century represented nothing less than the final act in the near 500-year history of Portugal's presence in the East.

How did a tiny European country of fewer than two million people manage to establish and maintain, albeit precariously, one of the world's great seaborne empires? Why were so many Portuguese adventurers ready to risk their lives on long and arduous voyages, in fever-ridden lands and in battles where they were often vastly outnumbered? Their motives and actions often appear strangely contradictory: on the one hand professing Christian piety and devotion; on the other displaying no compunction about inflicting cruel punishments and even at times reciprocal atrocities on enemies of other religious faiths.

And finally, what of the Portuguese and Australia? Did Portugal's navigators, so skilled in their craft and so close to Australian shores for so long, never venture this way? Can we really believe that they were never tempted to explore Australia's beckoning coastline? Given their enthusiasm for exploration and profit, and in particular the lure of the fabled 'Island of Gold', such an assumption seems entirely unwarranted.

It is the conclusion of this book that Portuguese navigators did indeed explore and map most of Australia's coasts, as well as those of a substantial part of New Zealand. Their voyages were classified as state secrets, any infringement of which was punishable by death—and this is another reason why no satisfactory accounts of them have come to light. In the course of their voyages along Australia's coasts, however, these mariners inadvertently left behind indisputable evidence of their presence, much of which is presented for the first time in this book. As for the early maps, these show conclusively that not just a few, but in fact most of the major topographical features on the east, west and south coasts of Australia were charted and given Portuguese names nearly a century before the coming of the Dutch and some two and a half centuries before the arrival of Captain Cook.

To confess my own background: I am an Anglo-Australian, born not all that far from Captain Cook's home county, Yorkshire, so I can hardly be accused of harbouring anti-British prejudice. James Cook was a great navigator and commander of men, and nothing can ever diminish his reputation. Nevertheless, it serves no purpose to close our eyes to demonstrable facts—the central one of which is that it was Portuguese mariners who charted Australia first.

Ultimately it will be for the reader to judge. This book is not written as an academic treatise. It is aimed at a wider audience and with that in mind the objective has been to make it as readable as possible. This does not mean, though, that truth has been sacrificed to any desire for sensationalism. Controversial evidence concerning Portuguese voyages in the vicinity of Australia has been carefully appraised in order to try and establish what really happened. In this process, many hypotheses which did not stand up to critical evaluation were discarded. That said, there are gaps in the historical records which can only be filled by conjecture. And this is surely a reasonable approach, provided that conjecture is clearly indicated to be such and provided that it stays within the general bounds of known facts.

Author's Note

When circumstances seem to justify it I have opted to use anglicised spellings of Portuguese names. For instance the difficult (for English-speakers) Portuguese sound 'n' pronounced at the back of the throat and written ã, I have where it seems appropriate rendered as an English-style 'n'. Thus (King) João becomes the more familiar 'King John'. Similarly, the first name of the outstanding Portuguese captain Mendonça I have written as Christopher, not 'Cristóvão'. And here, after all, we have ready precedent. For we all refer to the discoverer of America not as 'Cristóvão Colombo' but, of course, Christopher Columbus.

Sketch maps

The sketch maps in this book comparing coasts in the Vallard Atlas with actual Australian coastlines have been compiled with particular attention to accuracy. The methodology followed has been to reproduce a section of the relevant Vallard map and then place alongside it what is taken to be, on the basis of cartographic evidence, the same stretch of coastline shown on a modern map. Given the small scale of the maps, and in order to form a reasonable basis for comparison, some simplification of minor details in the modern map has often been unavoidable. The point to be emphasised, however, is that there has been no distortion of coastlines, either ancient or modern, in order to try to make them fit any preconceived theory.

Nautical distances

Distances at sea are usually expressed in this book in terms of nautical miles, a unit of measurement still in general use. A nautical mile is a little over 1.85 kilometres, longer than the old statute mile for land distances, which measured just over 1.6 kilometres.

In this book, the term 'miles' means 'nautical miles' when used in the context of sea distance. On land, distances have usually been expressed in kilometres, except in references to accounts from earlier centuries, when miles, rather than kilometres, were the standard measurement.

In the sixteenth century, ships' navigators calculated distances in leagues. Today a league is reckoned as being approximately three nautical miles, but it was not always so. In the sixteenth century, Portuguese, Italian, Dutch and English leagues were all different. A league in those days was simply the distance an average ship could sail under average conditions in one hour. Hence it was a variable figure: the faster the 'average' ship could sail, the longer the league.

Portuguese ships in the sixteenth century were among the fastest of their day, and perhaps for that reason a Portuguese league at that time seems to have been considerably longer than its modern equivalent. Evidence from the writings of Portuguese chroniclers suggests that a Portuguese league was the equivalent of approximately five statute miles (about eight kilometres).

A fathom, the old unit for measuring sea depth, was equal to six feet (1.8 metres).

PETER TRICKETT

1

The golden land of Marco Polo

The golden age of discovery was the age when the impossibility of sailing across the vast and until then mysterious oceans of the world suddenly disappeared as if touched by a magic wand...the newly discovered nautical science opened the golden gates of the seaways that led to America, to India and to the Pacific.

Armando Cortesão, Lisbon, 1939

As the sun rose over the Pacific on 20 May 1770, His Majesty's barque *Endeavour*, under the command of Lieutenant James Cook, was edging northward under full sail along the eastern seaboard of the mysterious southern land which we now call Australia. It was late autumn in the Southern Hemisphere but in the latitudes the *Endeavour* had reached, just below the Tropic of Capricorn, the sea breeze was agreeably mild even at this time of year. Sailing past a low, wooded coast which he took to be the outer shore of a long sandy cape, the commander focused his telescope on a headland where a large crowd of naked, black 'Indians' had gathered to stare in wonder as the great sailing canoe passed majestically by.

Cook believed himself to be the first European ever to sight these shores, a view generally accepted to this day. But the English captain was mistaken. Unbeknown to him another European had sailed along this same coast two and a half centuries before—but in the opposite direction. He was a Portuguese captain, Cristóvão (Christopher) de Mendonça. He was in command of a fleet of four ships, and he was on a top-secret mission ordered by Portugal's ruler, King Manuel I, to discover Marco Polo's fabulous 'Island of Gold'.

Disregarded and in some cases even scorned by orthodox academics, the evidence of Mendonça's extraordinary voyage has mounted piece by piece until today it can no longer be ignored or denied. Brought together here in its entirety for the first time, it shows that he explored and charted not just the east coast of Australia but also most of the south coast facing the great Southern Ocean—as well as the North Island of New Zealand. Either he or other Portuguese captains also mapped almost the entire length of Australia's west coast, 100 years before the Dutch first sailed in these waters.

So well-planned and skilfully executed were the original Portuguese voyages of exploration that had the momentum been continued Australians might today be speaking Portuguese instead of English—and with that, instead of a preoccupation with English rugby football, Australia might have blossomed years ago as a world soccer power rivalling Portugal's former colony, Brazil!

The primary evidence of Portugal's maritime exploits in the seas around Australia is contained in a collection of extraordinary yet little-known early sixteenth century maps. Known as the Vallard Atlas, these maps depict all the far-ranging Portuguese voyages of that time—including those to the coasts of Australia—in astonishing detail.

But that is by no means all. In recent times, many other intriguing pieces of evidence have come to light on Australia's coasts, among them a sixteenth century fishing net sinker of Portuguese origin, recorded sightings by Aborigines of strange intruders in 'stone' armour, and the existence on Australia's west coast of a now-vanished clan of black people who lived in stone huts and apparently spoke Portuguese. And last but by no means least, the extraordinary discovery of quantities of broken pottery in the nets of fishing trawlers operating in deep water near an island off Australia's south-east coast. With these fragments have been found two almost intact, marine-encrusted earthenware pots, one of which has been dated to about the year 1500 and with similarities in shape and texture to pots found in shipwrecks of the Spanish Armada. At least one Australian expert in oceanography is convinced that these finds point to the likely existence of an unknown Portuguese shipwreck.

As for the maps, a former cattle ranch in suburban Los Angeles might seem an unlikely place to look for key evidence of the first European voyages in the seas surrounding the unknown southern continent which we call Australia—but this district of San Marino is today the home of the internationally renowned Huntington Library. Endowed by a wealthy American railway magnate, Henry Huntington, early last century, the Huntington houses some of the world's rarest antique books and manuscripts, among them treasures such as an original 1410 edition of Chaucer's *Canterbury Tales* and the 1455 Gutenberg Bible. More important than any of these from an Australian perspective, however, is the Vallard Atlas, a collection of sixteenth century Renaissance maps, each of them a unique chart drawn painstakingly by hand in an age when the world seemed incomparably more vast than it does today and when travel to distant lands was a dangerous undertaking requiring not days but months or even years.

Bound in red morocco leather, this masterpiece contains fifteen maps, all of them drawn on parchment in sumptuous colour. It was completed in France no later than 1547 during the reign of King Francis I, a contemporary of both England's King Henry VIII and Queen Elizabeth I.

The origins of the Vallard Atlas are thoroughly documented, leaving no shadow of a doubt as to its authenticity. It was compiled in Dieppe, an important Channel port on the Normandy coast and takes its name from Nicolas Vallard, a wealthy Dieppe businessman about whom we shall discover more in a later chapter. Vallard is thought to have been not the author of the atlas, but rather its first owner. However the identity of the cartographer has never been established and it is quite likely that Vallard, who is known to have been knowledgeable in geography and nautical affairs, may have at least played a guiding role in its preparation.

The atlas remained in French aristocratic hands until 1816, when its then owner, Charles Maurice de Talleyrand-Périgord, Prince de Bénevent, sold it to the first of several British collectors. In 1849 it was purchased by the English antiquarian Sir Thomas Phillipps. The Huntington Library acquired it in 1924. Such is its value and fragility that it is stored in a special climate-controlled vault, with access strictly limited to researchers who are able to demonstrate a special need to view the originals.

This is one reason why the Vallard Atlas is relatively little-known outside the United States. What little Australian-based research there has been into the possibility of early Portuguese voyages of exploration to the coasts of Australia has concentrated almost exclusively on evidence available in the libraries of European capitals, notably London and to a lesser extent Paris. But while these sources have yielded centuries-old maps also showing an enigmatic southern land usually called 'Jave la Grande', the outlines of these maps are so vague and distorted, and their nomenclature so meagre that critics have been able to dismiss them as nothing more than a figment of the imagination.

The Vallard Atlas is in a different category. It contains fifteen separate maps covering most of the then known world, including the Red Sea and the Persian Gulf, India, South-East Asia, the coasts of Africa and Brazil, the east coast of North America, the West Indies, Central America and the eastern coasts of South America as far as the Strait of Magellan. The main regions of the globe not included are North Asia, the Pacific and the west coast of the Americas—and this for the very good reason that they were at that time either unexplored by Europeans or, in the case of the west coast of South America, a jealously guarded Spanish domain about which it was difficult for foreigners to obtain knowledge.

Each double-page map in the atlas is approximately the size of a modern A3 sheet. For the viewer, the first overwhelming impression is of the brilliance of the maps' colours, which remain vivid even after the passage of four and a half centuries. Coasts and islands are shown in shades of green, red, blue and gold, with place names neatly inscribed in alternating colours of dark blue and red. The oceans are embellished with drawings of elegant compass roses, sea monsters and ships under sail. Sumptuous, multi-coloured illustrations of potentates' courts and other fabulous scenes enliven the interiors of the continents.

A peculiarity of the Vallard Atlas, shared by just a few other atlases of the period, is that all of its maps are drawn with south at the top of the page. The most obvious consequence of this is that modern readers, accustomed since childhood to the concept of north being invariably at the top, will generally find it necessary to turn the Vallard maps upside down in order to begin to make sense of them.

That done, we will see that with two major exceptions—and this is of particular importance—every map in the Vallard Atlas depicts a readily identifiable coastline or continent. The two apparent exceptions are maps of a country or continent bearing the name 'Terra Java'. Some of the coastlines of this strange southern land have a tantalising resemblance to sections of the real coasts of Australia—yet there is also something disquieting about them.

The first Terra Java map depicts a long north-south coastline which could readily be interpreted as the coast of present-day Queensland, were it not that some way further south the coastline abruptly veers almost due east, forming the northern shore of a huge triangular-shaped promontory which does not remotely correspond to any geographical feature on the Australian east coast. Similarly, the second Vallard map depicts a west-facing coastline, parts of which have an undeniable resemblance to sections of the western Australian coast. The problems with this map are twofold: first, the coastline begins directly under the southern tip of Sumatra; and secondly, after heading in the correct direction of the Australian coast—south-west—for nearly half its length, it suddenly and inexplicably veers far to the east.

What we will discover is that all three of these anomalies have strange yet ultimately simple explanations. And once they have been corrected we will be left with coastlines which are clearly and recognisably those of eastern and western Australia. Coastlines, moreover, which are crowded with place names of rivers, islands, bays and headlands: names which are unmistakably Portuguese in origin, which correspond to the actual Australian coastline in great detail and which in many cases can even be shown to provide an accurate geographical description.

To this now can be added documentary evidence drawn from sixteenth century Portuguese chronicles and other historical sources, as well as recent dramatic discoveries at several places on the Australian coasts—one of which, raises the possibility that there could be a Portuguese shipwreck waiting to be discovered in the waters off south-eastern Australia.

Why were the Portuguese here, what were they looking for and why is almost nothing known about their exploits?

The answer can be summed up in a single word—gold. It was the fabled 'Island of Gold' they were seeking and they were in a race to find it before their arch-rival, Spain. Their voyages were therefore cloaked in secrecy—a policy rigidly enforced by the Crown on pain of death. But despite all this, evidence of their discoveries did eventually leak out. It fell into the hands of a hostile European power, France, and it is from this source that much of our evidence about their discoveries will be drawn.

As for the Island of Gold, it was the writings of Marco Polo which set in motion the European quest for this legendary kingdom. Returning to his native Venice in 1295 after twenty-four years of arduous travel which took him through Central Asia and Cathay to the court of the Grand Emperor, Kublai Khan, Polo stirred the imaginations of European adventurers with his description of a fabulous golden kingdom lying in the seas far to the south of China. In the book which he wrote describing his travels, he extolled the riches of the mysterious land of Lochac, a place where 'gold is abundant to a degree scarcely credible'. It was an alluring description. But where was this fabled land to be found? According to Polo:

> Upon leaving the island of Java, and steering a course between south and south-west, seven hundred miles, you fall in with two islands, the

> larger of which is named Sondur and the other Kondur. Both being uninhabited, it is unnecessary to say more respecting them. Having run the distance of fifty miles from these islands, in a south-easterly direction, you reach an extensive and rich province that forms part of the main land, and is named Lochac.[1]

Marco Polo's directions leave much to be desired. Writing of Java, he confides that 'according to the reports of some well-informed navigators' this island is 'the largest in the world, being in circuit above three thousand miles'. Some scholars have decided that 'Lochac' (written in another version of his manuscript as 'Boeach') was Cambodia, even though Cambodia lies north of Java, not south of it, and Polo himself visited Cambodia, which he calls Champa. In actual fact, if we follow the Venetian's instructions as carefully as possible we find ourselves moving in the general direction of Australia. Travelling some 700 miles (over 1000 kilometres) south-west of Java, as he directs, we reach the vicinity of the formerly uninhabited Cocos Islands. The one and only 'extensive' land south-east of these isolated islands is Australia though the Australian coast is not 50 miles distant, but more like 2000 miles (3000 kilometres) away. Of the land of Lochac, Polo wrote:

> Its inhabitants are idolaters. They have a language peculiar to themselves, and are governed by their own king, who pays no tribute to any other, the situation of the country being such as to protect it from any hostile attack. Were it assailable, the grand khan would not have delayed to bring it under his dominion...the country is wild and mountainous and is little frequented by strangers, whose visits the king discourages, in order that his treasures and other secret matters of his realm may be as little-known to the rest of the world as possible.[2]

Much of this could be applicable to the Kimberley region of north-west Australia: inhabited by people who were 'idolators', who spoke a language unrelated to either Malay or Chinese and who were hostile to strangers. Inhabitants, moreover, of a wild, rugged country protected by shoals, reefs and fierce currents from hostile attack. But it would be unwise to jump to conclusions. For Marco Polo goes on to say that this country also produced large quantities of 'brezil wood' as well as gold, that elephants were found there and that its people cultivated a delicious fruit called *berchi*. None of this could be said to apply to Australia. In the introduction to the account which he dictated, *The Travels of Marco Polo*, the Venetian adventurer promises that 'this book will be a truthful one', but he also he admits that it includes both things to which he was an eyewitness and 'things heard from others'. The land of Lochac (also spelt Lochach) seems to fall into the latter category; Polo was writing from hearsay rather than from his own experience, and in this as in some other instances he confused facts from different regions.

Nevertheless, the vision of his mysterious land (under the name of Boeach) persisted in Europe as the 'Golden Country of Beach' and it was almost universally believed to lie somewhere south of Java, not north of it. In many Renaissance maps it is shown as lying at the northern edge of a great southern land whose position approximated that of Australia. The great Flemish cartographer Gerard Mercator in his world map of 1569 depicted 'Beach provincia aurea' (Beach the golden province), as well as a 'Lucach regnum' (the kingdom of Lucach) and 'Maletur regnum' (the kingdom of Maletur). These mythical realms were copied with variations by Dutch cartographers. Even as late as Abel Tasman's day, in the mid-seventeenth century, 'Beach' was still envisaged as the gold-rich province of a continent occupying the southern latitudes somewhere between Cape Horn and the meridian of the Solomon Islands (The Islands of Salamon).

But long before Mercator's famous map appeared, Portuguese voyagers had already secretly begun searching for Marco Polo's Land of Gold. Having pioneered the sea route to India and the Orient at the very beginning of the sixteenth century, they had two overriding commercial aims: control of the East's lucrative spice trade and the acquisition of gold. The spice trade was soon in their hands. Gold was to prove more elusive.

The need to find new sources of the precious metal was becoming urgent because by the latter part of the fifteenth century Europe's own gold deposits were almost exhausted. Gold was vital not just for its own sake, but also for the manufacture of coinage. And without sufficient coinage, the expansion of commerce and finance in Europe was in danger of being seriously jeopardised. The need for gold was particularly acute in Portugal, where the price of gold had risen more than a hundredfold in the space of just a few decades. The country had up to that time been too poor even to mint its own gold coinage, having to use instead the currency of richer neighbours, including the 'infidel' Morocco.

But, the sceptic may well ask, if the Vallard maps really contain such startling evidence that Australia was discovered and charted by Portuguese explorers in the early sixteenth century, why have they apparently remained unknown and unremarked on for so long? Why indeed? It is a reasonable question.

In fact, their resemblances to Australia have *not* gone unnoticed—they were first pointed out some 150 years ago. In 1856 Sir Thomas Phillipps, an English antiquarian who had purchased the Vallard Atlas a few years earlier, saw something in it which seemed to him so obvious that he published a copy of the eastern 'Terra Java' map under the title *The First Map of Australia*. Five years later Richard Henry Major, Keeper of Maps at the British Museum, followed this with a treatise, *The Discovery of Australia by the Portuguese*, in which he concluded that:

> The land described as La Grande Jave [Terra Java] on the French maps...can be no other than Australia; and that it was discovered before 1542 may be almost accepted as demonstrable certainty, but how long before is not clear.[3]

Phillipps's publication of his 'First Map of Australia' seems to have passed with little notice. Major's findings attracted more attention but much of it was hostile. This was particularly the case when he subsequently made the fatal mistake of publicly ascribing the discovery of Australia to a Portuguese voyage in 1601—only to find out later that this planned voyage had never actually taken place. Discredited in the eyes of his critics, Major retired from the academic battlefield. Little more was heard of 'La Grande Jave' for another thirty-five years until George Collingridge, an art teacher and enthusiastic historian living in Sydney, published in 1895 a substantial book in which, inter alia, he claimed that the Portuguese had discovered part of the west coast of Australia and the Spanish part of the east coast—although he produced no real evidence for the latter claim.[4]

Collingridge, who called his studio in the Sydney suburb of Hornsby *Ye Jave-la-Grande*, was dismissed by the history establishment as an eccentric, despite the wealth of material he presented. In the 1970s Kenneth Gordon McIntyre, a retired Melbourne lawyer, took up the cudgels where Collingridge had left off, but this history warrior's published work unfortunately contained some serious errors which allowed it to be debunked in academic circles. A small book on the same theme a few years later by Lawrence FitzGerald, a former Melbourne surveyor, went largely unnoticed.[5]

Why is it that these advocates of a Portuguese discovery of Australia have failed to make any impression on the academic mainstream? The explanation seems to lie in three things: firstly, ingrained opposition to the very suggestion that the European discoverers of Australia could have been anyone other than the Dutch, followed of course by the British; secondly, the stringently enforced Portuguese policy of secrecy about their voyages; and last but not least the fact that those few researchers who canvassed the radical idea of an earlier discovery than that of the Dutch all sought their evidence in documents which were more readily available than the Vallard Atlas—but also, unfortunately, greatly inferior to it.

Henry Major relied primarily on the 'Dauphin' map in the collection of his own museum—a map sometimes referred to as the 'Harleian' because in the eighteenth century it was for a time owned by Edward Harley, Earl of Oxford and one of the principal Lords of the Admiralty. It was purchased by Sir Joseph Banks, who had sailed on Captain Cook's first voyage, and presented by him to the British Museum in 1790. The name 'Dauphin' derives from the fact that it was made about the year 1536, commissioned by the French King Francis I for his son the Dauphin, afterwards to become Henry II. The cartographer is unknown.

Major bemoaned the fact that in his Dauphin map 'the Portuguese names are but few'. He had heard about the Vallard Atlas and was anxious to see it but in the introduction to his book *Early Voyages to Terra Australis, now called Australia*, he complained that he had been 'unsuccessful in efforts to gain a sight of this atlas, or even a fac-simile'. The Vallard Atlas was at that time still in the possession of Sir Thomas Phillipps, but for reasons best known to himself, that gentleman kept it jealously guarded under lock and key in his country house at Middle Hill, Worcestershire. Major vexedly relates that because of Sir Thomas's stubborn refusal to cooperate he was compelled to rely on the memory of a friend, Sir Frederick Madden, who had been given an opportunity to examine the atlas some years previously. Needless to say, Major found this a highly unsatisfactory alternative. Had he been able to see the Vallard Atlas for himself he surely would have felt that it fully vindicated his Portuguese discovery theory.[6]

Given Australia's own predominantly British background, it was only natural that later historians should seek evidence of early European voyages to Australia in the familiar European centres of learning—the British Museum, the British Library, the Rylands Library of Manchester and the Bibliothèque Nationale, in Paris. But while these institutions do have in their collections antique maps of somewhat similar origin to the Vallard and which also depict a continent with a puzzling quasi-resemblance to Australia, unfortunately none of these maps contain the supportive detail necessary to provide conclusive proof of a pre-Dutch discovery.

Not so with the Vallard maps now in California's Huntington Library. Meticulously and beautifully drawn, they present an astonishing wealth of accurate coastal detail. The western coast, the east coast and even large parts of the southern coast of the mysterious continent called Terra Java are all delineated and liberally bestowed with Portuguese-origin placenames.

The two main Vallard Atlas maps of this southern land contain no less than 120 such Portuguese names for geographical features which, as we will see, belong indisputably to the east, west and southern coasts of Australia. In comparison, the same part of the most widely quoted of the other Dieppe maps, the Dauphin, has a mere twenty-six names—and many of these are so incomplete as to be almost meaningless. The two Vallard maps between them contain, in addition, another twenty-five placenames relating either to Indonesia or mysterious islands in the Pacific.

What it all adds up to is eloquent cartographic evidence of epic voyages of discovery—voyages whose chronicles have mostly, though not entirely, been lost, but which on the basis of the charts they produced must have been fully equal in skill and daring to those of Captain Cook some two and a half centuries later.

As for Cook (he was promoted to Captain after his first voyage), the tally of 120 Australian coastal names recorded on the two Vallard maps, sixty-six of which relate to the eastern and adjoining southern coast, can stand comparison with the approximately 100 names bestowed by the great English navigator on his voyage up the same eastern coast in 1770. And most remarkably of all, as we will see in later chapters, some of Cook's names can even be matched with Portuguese names describing identical coastal features on the eastern Vallard map—providing further convincing proof of this map's authenticity.

It would be a marvellous thing indeed to discover a ship's log or other written documents describing, in particular, the astonishing Portuguese voyage made all the way down the east coast of Australia and far beyond. But despite the current lack of any first-hand material of this kind, the wealth of information contained in the Vallard maps is sufficient to enable us to follow the course of this and other voyages, understand the significance of the atlas's place names, and finally to identify many of these names with present-day Australian landmarks—including some that are very well-known indeed.

First, though, in order to appreciate how these voyages were made, who probably made them, and the reason why some errors crept into the final maps it is necessary to understand not only the geopolitical situation at the time but also some of the more important features of early sixteenth century navigation and map-making techniques—including their inherent limitations.

Portugal's emergence as an international superpower at the beginning of the sixteenth century could fairly be described as meteoric. Almost as rapid was its decline into a minor role on the world stage a mere century or so later. In between, in a brief blaze of glory, the ships of this tiny Atlantic nation of less than two million people founded a great trading empire and explored to the furthest corners of the globe—including, as we shall see, the coasts of Australia and even further afield.

A decisive factor in Portugal's history was its 500-year occupation by Arab and Berber invaders from Morocco, beginning in the early part of the eighth century. But Portuguese nationalism was never extinguished and, after holding out in mountain strongholds in the north, Christian armies gradually reconquered their lost domains. The final defeat and expulsion of the Muslim invaders came in the Algarve, on the Mediterranean coast, in 1249.

The centuries of foreign domination led in post-liberation Portugal to an almost pathological hatred of the 'Moors'—as all Muslims would henceforth be known. Yet paradoxically it was precisely the Islamic legacy of learning—in particular the disciplines of science, mathematics and navigation—which would make Portugal's spectacular maritime expansion possible.

As for Portugal's quest for gold, this can be seen to have begun as early as 1415—the year in which the longbow men of England's Henry V routed the French at Agincourt. In that same year Portugal won its first momentous victory on foreign soil by capturing the Moroccan city of Ceuta, on the southern shore of the Strait of Gibraltar. This strategically located walled city had a history going back to Roman times and had been used by the Arabs to control shipping in the western Mediterranean. Ceuta was taken with surprising ease. Its Moroccan governor had made the crucial mistake of sending home a strong Berber force helping to defend the city after being led to believe that the Portuguese invasion fleet of 240 ships had turned back. Many of the remaining Muslim defenders were slaughtered, while incredibly the Portuguese losses totalled a mere eight. After a looting spree, the city's main mosque was turned into a Christian church.

Ceuta (today a Spanish enclave) was a wealthy trading port and an important Mediterranean terminus for caravans bringing gold and other precious commodities from the African lands south of the Sahara. The Pope declared the city's seizure to be a holy crusade, while Portugal's King John I hailed it as a great blow against the 'heathen' Moors. But in reality an equally important aim was access to the wealth of the gold caravans.

However, it was one thing to capture an alien city and quite another to persuade hostile Muslim merchants from the Sahara to come and trade there. The king's third son, Prince Henry, one of the leaders of the Ceuta invasion, urged a wider military campaign to conquer Morocco. When his father rejected this idea as unrealistic, Henry turned his attention to accounts by Moorish captives in Ceuta of lands south of the Sahara where the desert caravans journeyed to a 'River of Gold', said to have its source at the Nile. According to one such account, the banks of this river were populated by a race of ants the size of cats which obligingly dug up the gold and left it for humans to gather. A fantastic tale it might seem today, but to a European world still steeped in medieval fable and superstition such stories were by no means unbelievable.

Prince Henry is said to have had a copy of Marco Polo's manuscript, in which case he would have been well aware of the Venetian traveller's fabled Land of Gold, reputed to lie in the Orient. Determined first to discover the truth about the mysterious African lands, Prince Henry—hailed in English chronicles as Henry the Navigator—from then on devoted himself to the development of Portuguese maritime power. The Prince poured money and resources into improving the arts of navigation, cartography and seamanship. Under his patronage annual expeditions sailed ever further down the west coast of Africa.

Progress at sea was rapid. Cape Bojador, on the western edge of the Sahara, where the 'Land of the Blacks' began, was rounded by one of Henry's captains in 1434, exploding once and for all the superstition of medieval seamen that any ship which dared to pass this cape would disappear forever in the dreaded 'Torrid Zone', where the sea was rumoured to boil. By the time of Henry's death in 1460 the coastline of West Africa had been sailed and charted as far as Cape Palmas, where the coast turns east to form the Gulf of Guinea. The Atlantic islands of Madeira, the Canaries and the Azores had been colonised. Most importantly of all, Portuguese mariners had by now learned to sail confidently out of sight of land, trusting to their compass bearings and accurately calculating their latitude at sea with the aid of two revolutionary navigational instruments, the quadrant and the astrolabe.

His many achievements notwithstanding, Henry the Navigator has tended to be excessively romanticised. He is often portrayed as a heroic scholar, pursuing a disinterested passion for the advancement of maritime science and navigation—but this is only a partial truth. Throughout his life the Prince was obsessed with the idea of a crusade against the 'infidel' Muslims. His religious fervour had its reward when in 1456 Pope Calixtus III confirmed him as Governor of the Order of Christ, at the same time giving the Order spiritual jurisdiction over all regions conquered by Portugal, then or in the future.

Henry's African voyages also aimed to turn a profit. Soon they were being financed by trade in ivory, gold dust—and above all, slaves. In fifteenth century Europe there was nothing in the least reprehensible about trafficking in human beings. Dealing in slaves had been an accepted practice since at least the foundation of the Roman Empire, and even in Renaissance Europe it still had the full sanction of the Roman Catholic Church. The first cargo of slaves from West Africa was shipped back to Portugal in 1441, and before long Lisbon had become Europe's biggest slave market. Prince Henry had no ethical qualms about taking his share of the spoils. A contemporary chronicler records, for instance, that when a cargo of 230 black Africans arrived for auction in 1445 the prince rode into Lisbon to personally collect the forty-six slaves who were his one-fifth entitlement.[7]

At the time of Henry's death in 1460 Portugal was importing 30 000 slaves a year. Henry's captains still had not discovered the elusive 'River of Gold', but they did find something else which they named the *Mina de Ouro* (Mine of Gold). Despite the name, though, it was not a goldmine but simply a place on the African coast where Portuguese goods could be traded for alluvial gold. This region was later called the Gold Coast (now Ghana).

The Portuguese push south down the coast of Africa was given new impetus with the accession to the throne of King John II in 1481. The first major advance was made by Bartolomeu Dias, a navigator second in reputation only to Vasco da Gama. Up to this point, fast and highly manoeuvrable ships known as *caravelas*, or in English caravels, had been employed exclusively for the voyages of exploration. But caravels had only a small carrying capacity and the increasing length of the voyages was leading to supply problems. When Dias sailed from Lisbon in 1487 he took with him a larger supply ship in addition to his two caravels. After sailing down the African coast and reprovisioning the caravels in a harbour of what is now Namibia, Dias left the supply ship in the charge of a small crew and sailed far out to sea, heading south-west to avoid headwinds on a course which kept him well out of sight of land. Finally, at about forty degrees south his ships picked up the prevailing westerlies and turned east to regain the African coast.

This was a major navigational feat, proving that ships could now sail on an accurate course for long distances in the open ocean, relying solely on their newly developed navigational instruments. Unknown to Dias, when his ships turned east they were already some distance south of the furthest point of Africa and they passed the cape without sighting it. They made landfall several hundred kilometres further east and explored the coast as far as the Great Fish River (in present-day South Africa), where warm currents from the north seemed a sure sign that the way to India now lay open. Here, however, wearied and anxious about their dwindling provisions, his crews demanded that they turn back.

Dias sighted Africa's southernmost point and rounded it with difficulty on his homeward journey. After picking up the supply ship his little fleet returned triumphantly to Lisbon in December 1488. Dias had called the great southern headland 'Cape of Storms', but King John strongly disapproved of so forbidding a title. With an eye to encouraging future voyagers, he ordered it renamed 'Cape of Good Hope'—a name which will acquire new significance later in a totally unexpected Australian context.

In 1495 John II died and was succeeded by his brother-in-law, King Manuel I. Two years later a new Portuguese fleet was assembled in Lisbon harbour under the command of Vasco da Gama. It consisted of four ships, three of them new, well-armed and provided with a large cargo of trade goods. The fourth vessel was an old supply ship destined this time to be broken up and its cargo divided among the rest of the fleet. The expedition's aim would be to close the last gap in the sea route to India.

The story of da Gama's epic voyage has been retold many times, so there is no need to dwell on it. Recently, though, evidence has been found suggesting that there were other secret Portuguese expeditions to the Cape—and perhaps even beyond—in the supposedly blank decade between Dias' voyage and that of da Gama. Some historians now believe that secret voyages could explain the assurance with which da Gama ordered his flotilla to set a south-westerly course across the Atlantic far out of sight of land for three months, turning east only when his ships were nearing the latitude of the Cape of Good Hope.[8] When they finally made landfall just 160 kilometres north of the Cape, da Gama's ships had covered 4500 nautical miles nonstop, a voyage at that time unequalled by the ships of any other European nation. Four months later the little fleet (now reduced to three vessels) sailed into the East African port of Malindi, an important centre for trade between Africa and India. A month later they had crossed the Indian Ocean and on 20 May 1498, they triumphantly dropped anchor at Calicut, the major trading port of south-west India.

It was the beginning of an incredibly rapid period of expansion. In 1511 Portuguese naval forces captured the vital port of Malacca from its Malay ruler, and just a year later they dispatched a flotilla of ships from Malacca to one of their ultimate goals, the Spice Islands. By now, immense riches had begun pouring into the royal treasury of 'Manuel the Fortunate'—as he soon became known. Manuel's European rivals might nickname him the 'Grocer King', but it was a derision tinged with envy. Such were the profits from exotic spices that the cargo of a single ship returning fully laden from the East could, it was said, recoup the cost of the voyage 100 times over. Now only the fabulous Island of Gold remained to be discovered. Even greater riches seemed in prospect.

How had it all come about? The three most important elements in the equation were navigation, ship design and gunnery. For the first two, at least, the militantly Christian Portuguese owed much to the scientific and technological achievements they inherited from their hated former rulers, the Moors.

Of all Portugal's innovations in maritime technology in the fifteenth century, arguably the most outstanding was the spectacular advance made in ship design and construction. The crowning achievement was the caravel, an ocean-going ship which was fast, strong, highly manoeuvrable and able to ride out the heaviest seas. Its Islamic origins are revealed by the fact that even the name, caravel, comes from the Arabic word *quaraquir*, meaning 'merchant ship'. One of its nearest relatives still in use today is the graceful Egyptian felucca. A distinctive feature of the caravel, as with the felucca and the Arab dhow, was its rigging. Unlike northern European vessels which were invariably square-rigged, the caravel employed the lateen rig, a triangular sail first adapted by Italian mariners from the Arabic pattern (hence the name 'latin' or 'lateen') which enabled the ship to sail at an angle against the wind.

A second major innovation in the design of caravels was the development of 'carvel' planking (the word is actually derived from 'caravel'). This was a sophisticated method of construction which involved first building the ship's entire frame and then fastening to it edge-to-edge planking. In Portugal by the beginning of the fifteenth century this shipbuilding method had entirely superseded traditional clinker-building using overlapping planks, a method which remained in widespread use throughout northern Europe. Carvel building required far greater technological skill because the ship had to be constructed to a predetermined design rather than, in the case of a clinker vessel, built intuitively from the keel up, plank by plank. Carvel construction produced a light yet strong hull, and it also allowed the building of far larger ships than was possible using the clinker method.

Caravels, however, had their limitations. Their shallow draught made them ideal for inshore exploration, but it also greatly restricted their carrying capacity. For long voyages, such as those from Portugal to India, this was a serious shortcoming. The answer lay in the development of a new type of larger ship with a deeper draught and more capacious holds. These vessels were called naus—meaning 'great ships'. In English they were generally referred to as carracks. They carried mainly square-rigged sails to take advantage of following winds on long ocean voyages, but with a lateen rig on their mizzen (stern) masts for better manoeuvrability. As we shall see, there is evidence that a ship of this kind was involved in a major Portuguese voyage of exploration to the coasts of Australia.

The first outstanding examples of ships of this type were built expressly for Vasco da Gama's expedition to India in 1497. Of the four ships in da Gama's initial fleet, only the little *Berrio* was a caravel. The second vessel, as we have seen, was an old storeship designed to be unloaded and broken up and its cargo shared between the remainder of the fleet once the Cape of Good Hope had been passed. Drawings of the other two ships, both brand-new naus, show them with large, billowing sails emblazoned with the cross of the Order of Christ. They are also characterised by high, multi-decked forecastles and towering poop decks, both derived from the fore-and-after castles of medieval vessels. These multiple decks made the carracks less manoeuvrable than they would otherwise have been, but on the other hand they provided far more comfortable accommodation for officers and crew than did the spartan caravels where there was usually only a single after-cabin for the captain and perhaps his navigator—with everyone else expected to sleep in the hold or out on deck.

Sixteenth century Portuguese sea route from Lisbon to Goa and Malacca.

And the carracks offered another crucially important advantage: their hulls were big and strong enough to carry a formidable array of cannons. The two carracks of da Gama's fleet mounted twenty such guns between them. In comparison with the gunnery of later carracks—described in some accounts as 'floating fortresses'—it would seem a somewhat puny armament, but at the end of the fifteenth century it represented a revolutionary development in maritime warfare.

Carracks quickly became the standard vessel used for the long voyage from Lisbon to Goa, known as the *Carreira da India*. For the first thirty years of the sixteenth century the carracks were usually of about 400 to 500 tons—that is, offering roughly eight to ten times the carrying capacity of a caravel. Ships of this size and power created something of a sensation in South-East Asia. In

January 1514 we find a Portuguese captain Rui Brito de Patalim, at the Malay port of Malacca, writing to King Manuel that the inhabitants of Java and the Moluccas had all been overawed by the size of the ships in the first Portuguese fleet which sailed to the Spice Islands under the command of Antonio de Abreu in 1511. Brito advised:

> We need large ships even though the inhabitants are frightened of them. I have written to the Governor of the Indies that he should send one or two naus of 500 tuns, because apart from asserting our authority they can take a great abundance of spices, which could not be done with small ships.[9]

The measurement of a 'ton' tended at first to be somewhat arbitrary. It originally referred to how many casks, or 'tuns' (perhaps of oil or wine) a ship could carry. However, by the late 1500s, when the measurement had been reasonably standardised, Portugal was building jumbo carracks of 1000 tons, or in a few cases even leviathans of 2000 tons, by far the largest ocean-going ships in the world at that time (in comparison, Cook's *Endeavour* measured a mere 368 English tons).

But big ships, however well-designed, would have been of little use for long voyages without adequate navigational technology—and here again by the beginning of the sixteenth century the Portuguese were paramount. The Arabs had been responsible for the introduction of mathematics (even giving their name to the numerals themselves). The compass had long been known to Mediterranean navigators and by the end of the fifteenth century it had developed far beyond its primitive form of a needle, magnetised by a lodestone, floating on a wooden chip in a bowl of water. Now the needle had become a magnetised iron wire, glued beneath a circular card marked with the eight principal 'winds' and their subdivisions, which could be up to sixty-four points. This compass card was mounted on a central pin on which it could move freely and enclosed in a glass-covered circular bowl mounted on gimbals to offset the pitch and roll of the ship. The columnar wooden box in which the compass bowl was housed, known in English as a binnacle, is credited as being a Portuguese invention.[10]

The Arabs had also been leaders in the arts of astronomy and cosmography. A practical application of these arts in navigation—but this time of European origin—was the instrument known as a seaman's quadrant. This was a simple device consisting of a quarter circle marked with a scale from one to ninety degrees with a plumb-line, or pendulum, hung from the apex. The upper straight edge of the quadrant was sighted on a star—usually the Pole Star—and a reading taken at the point where the plumb-line intersected the scale. The star's altitude could then be applied to calculate the ship's latitude.

But the quadrant with its swinging pendulum was an unsatisfactory device for use on the heaving deck of a ship at sea, and this led in the late fifteenth century to the development of a superior instrument, the marine astrolabe. This invention is generally credited to Jewish astronomers in the employ of Portugal's King John II who adapted an Arab instrument used for land-based astronomy. The astrolabe consisted of a brass disc graduated in degrees and fitted with a rotating sight bar, or *alidade*, for observing the altitudes. There was no troublesome pendulum to worry about; instead, the astrolabe was suspended at eye level from a ring at its top edge, enabling the navigator to obtain a more accurate star sighting with which to determine the ship's latitude.

But even though it might be more practical than the quadrant, the astrolabe still had to be perpendicular for an accurate reading to be taken. This shortcoming led to the invention in the early sixteenth century of a third device for measuring latitude, the cross-staff. Developed from an Arab device called the *kamal*, the cross-staff was a relatively simple instrument consisting of a straight rod about a metre long graduated on one side in degrees and minutes. The navigator held the end of the staff to his eye and slid a cross-piece fitted at a right angle up or down until it corresponded exactly to the distance between the horizon and the star (or other body) being observed. The star's altitude could then be read off from the scale. The cross-staff had its limitations, in particular, it could not be used for sighting heavenly bodies

Calculating latitude using a cross-staff. (*Explorer Archives*)

lower than twenty degrees above the horizon, or higher than sixty degrees. Nevertheless its simplicity made it a favourite instrument with navigators, particularly in middle latitudes.

But while the Portuguese and other European mariners possessed the means to establish their latitude with a high degree of accuracy—as we shall see when we come to retrace the pioneering Portuguese voyages along the coasts of Australia—the same could not be said of longitude. This was a far more intractable problem, not finally solved until the latter half of the eighteenth century with the invention of a reliable ship's chronometer. Until then, attempts to establish a vessel's longitude in unknown waters far from home were subject to large and sometimes disastrous miscalculations.

In the case of sixteenth century Portuguese navigation this is forcefully illustrated by the account of an English Jesuit, Father Thomas Stevens, who travelled from Lisbon to Goa, in India, in a Portuguese ship. Sailing across the Indian Ocean on the last leg of their voyage, Stevens recorded that the 'running seas' drove them off course, with the result that 'we that thought we had bene neere India were in the same latitude neere Zocotoro [Socotra], an Ile in the mouth of the Red Sea.'[11] In other words, although the ship's navigator had got his latitude correct, in terms of longitude they had ended up on the opposite side of the Indian Ocean: the ship's supposed longitude was fully 1000 nautical miles in error! And this was in the latter half of the sixteenth century, in waters with which the Portuguese were by then reasonably familiar.

The basic problem arose from the fact that almost the only way of estimating longitude in distant seas out of sight of land was by dead reckoning—that is, calculating the distance the ship travelled, together with its direction. But calculations of this sort were inherently unreliable. At its most primitive, a ship's speed was reckoned by dropping a small piece of wood off the ship's bow and seeing how long it took to reach the stern. In the sixteenth century this method was refined by tying the piece of wood to a knotted line and counting how many knots slipped through the sailor's fingers in a given time: this gave the ship's speed in 'knots'. But the ship's *true* speed (as distinct from its speed through the water) was difficult to calculate with any accuracy—particularly if, for example, there was a strong following wind and sea. And then there were the unseen effects of leeway (sideways slippage) and ocean currents to reckon with. The problem of getting an accurate dead reckoning was even more difficult if a ship had to tack into the wind. A device known as a traverse board, with strings and pegs, was used to keep track of the ship's zigzagging progress but it was (fittingly) haphazard.

The compass was always the mariner's best friend but it, too, was not invariably reliable. And then there was the imponderable factor of compass deviation—the difference between magnetic and true north—a problem which was not fully understood in the early sixteenth century.

The effect of all these difficulties will become apparent when we come to look in detail at the maps derived from the early Portuguese voyages along the coasts of Australia. Here the ships' navigators were sailing in unknown seas, far from any places that could serve as points of reference and with no known landmarks to check their actual position. Errors, particularly in longitude, were going to be inevitable. This is a factor which will prove to be very significant and which we need to keep firmly in mind.

For Portugal and Spain, the problem of longitude acquired particular importance after their signing of treaties which purported to divide the world between them into equal spheres of influence.

The first of these treaties, that of Tordesillas, was signed in 1494. Based on an earlier ruling by Pope Alexander VI, it specified that the dividing line between the two Iberian powers' spheres of influence should be the meridian of longitude 370 leagues west of the Cape Verde Islands in the Atlantic. Everything beyond that—including the lands newly discovered by Columbus—was designated the exclusive territory of Spain. The other half of the globe east of the 'Pope's line' was allocated to Portugal. It included all of Africa and the Indian Ocean and also, as it subsequently transpired, the secretly discovered land of Brazil.

So far so good. A problem only arose years later when Portuguese ships reached the Spice Islands and the two Catholic kingdoms began disputing to which of them this rich prize rightfully belonged. Establishing the precise position of the vital dividing line of longitude on the far side of the globe proved impossible. Neither the Portuguese nor the Spanish had any means of determining the true longitude of the Spice Islands. Nor did either of them at the outset have any conception of the vast size of the Pacific Ocean.

Things came to a head in 1521 when ships of Ferdinand Magellan's expedition under the Spanish flag suddenly appeared at the Moluccas, precipitating a naval confrontation and a diplomatic crisis. Spain maintained its claim to the Moluccas until 1529 when the dispute was settled by the Treaty of Saragossa. Under this new treaty Spain's King Carlos (Charles) V ceded to Portugal any rights he might have in the Spice Islands for a large cash payment, and the line of demarcation was arbitrarily fixed at seventeen degrees east of the Moluccas. This question of the disputed line of demarcation will be another key factor when we come to consider why the early maps depicting the coasts of Australia show the Australian landmass well out of its true position and displaced far toward the west.

Finally, what do we know about early Portuguese sea charts? An appreciation of how these charts were made is essential for understanding the origins and peculiarities of the sixteenth century maps portraying the enigmatic great southern continent which, as we will discover, in reality represents Australia.

The European navigator's traditional sea chart was known as a *portolan*. As first developed, it was drawn not on paper (which was not at that time in use in Christian Europe) but on specially treated sheep or goat skin—the original meaning of the word 'parchment'. For use as portolans these pieces of parchment were stretched on a wooden frame—rather like the membrane of a drum.

A Mediterranean invention, the portolan made its appearance in the thirteenth century. The earliest surviving example is the 'Carte Pisane' drawn on a sheep skin parchment about 1275 and now housed in the library of France's Bibliothèque National. Remarkably detailed for its time, this portolan depicts all the coastlines of the Mediterranean as well as a crude representation of southern England. As with later portolans of a more sophisticated kind, it is characterised by networks of 'rhumb lines' representing compass bearings, all radiating from a series of principal reference points. To set his course, a navigator simply laid a ruler from his place of departure to his intended destination and then selected the rhumb line most nearly parallel to it.

Portuguese navigators were masters in the art of making hand-drawn portolan charts, and they continued to use animal skin parchment for this purpose long after maps printed on paper began to make their appearance in Europe toward the end of the fifteenth century. The latter point will be an issue of crucial importance when we come to examine the sixteenth century Portuguese charting of the coasts of Australia.

2

'Strangers with skins like crocodiles'

Gold can take the well-defended place;
Gold, of friends, has knaves and traitors made;
Gold oft compels great lords to actions base,
And captains to their foes are thus betrayed.

Camoens, *The Lusiads*, 1572

In the first decade of the sixteenth century which followed their momentous rounding the Cape of Good Hope and breakthrough into the Indian Ocean, the triumphant Portuguese, in a blitzkrieg naval offensive, swept all before them. Proud Indian princes had been forced to submit. Led first by Vasco da Gama and then by their 'Caesar of the East', Afonso de Albuquerque, the Portuguese seemed unstoppable as their naval armada captured first the strategic port of Goa, on India's west coast, and then the great South-East Asian emporium of Malacca, on the Malay peninsula.

It must be said that these and other victories, impressive though they were in military terms, had their darker side. Da Gama, on his second voyage to India in 1503, led his fleet to the port city of Calicut to exact revenge for the massacre of fifty Portuguese merchants and three friars who had been put to death in violation of a treaty of friendship signed earlier with the city's ruler, the Zamorin. One of da Gama's most notorious acts was to order that prisoners taken from some Indian ships should have their hands, ears and noses hacked off. These were piled into a small boat, which was then sent ashore with a message to the Zamorin inviting him to make curry from its cargo.

It was in some respects a violent age, but even so, some Portuguese leaders seemed to excel in the zeal with which they committed atrocities. Francisco de Almeida, the first Portuguese Viceroy of India, had a no less violent disposition than da Gama. In 1509 he engaged and defeated a powerful Muslim fleet at Diu, on India's west coast. Intent on avenging the death in action of his son, he proceeded to execute prisoners taken in battle: their bodies were dismembered and the parts fired from his ships' cannons into the streets of a nearby city—ostensibly as a deterrent.

Almeida's successor, Albuquerque, also showed a fondness for atrocities. Describing his capture of Goa in 1510, he boasted to King Manuel in a dispatch:

> Then I burnt the city and put everyone to the sword, and for four days your men shed blood continuously. No matter where we found them, we did not spare the life of a single Muslim; we filled the mosques with them and set them on fire…

The following year Albuquerque had assembled his formidable fleet before Malacca, the so-called 'Venice of the East'. The Portuguese met with fierce resistance from the city's Muslim ruler, the sultan, but after a month-long siege the walled city succumbed to a blistering bombardment by Albuquerque's shipborne artillery. Malacca's position, dominating the vital narrow sea lane between the Malay peninsula and the island of Sumatra, meant that Portugal could now expect to have a complete monopoly over the rich trade from China and the East Indies.

And that was by no means all. With the capture of Malacca, the Portuguese possessed a secure base and seaport for voyages east to the Moluccas, the fabled Spice Islands, lying at the furthest reaches of the Indies. By a great stroke of luck, also in 1511 a large Javanese chart of the Indian Ocean and the Indonesian archipelago had fallen into Portuguese hands. In a letter to King

Manuel, Albuquerque enthusiastically described it as 'the best thing I have ever seen'. The first Portuguese expedition to the Moluccas was dispatched just a year later, in 1512. The following year, the two most powerful sultans of the Spice Islands gave permission for the foreign interlopers to set up permanent trading posts.

The way was now open for the Portuguese to begin reaping immense profits from the spice trade. But there remained one other alluring prospect. Africa as a source of gold had turned out to be a disappointment. Now, Portuguese hopes were high that they were on the verge of discovering and exploiting the great riches of the Orient's 'Island of Gold'.

But where was this fabulous land? Marco Polo's sailing instructions indicated that it lay to the south of Sumatra, roughly in the direction of an as yet unknown Australia. Can we really believe that Portuguese adventurers, riding a wave of self-confidence, were never tempted to set out from Malacca in search of this storied Eldorado? Did they in fact do just that? If they reached Australia, might we not expect them to have left behind some evidence of their presence?

The answer to these questions is yes. Early Portuguese adventurers did indeed reach Australia's shores. And yes, they did leave reminders of their presence, evidence in the form of material objects and other clues which, as we shall soon see, will speak eloquently of stirring and violent events which accompanied their arrival.

To discover what these events were, we must now fast forward 400 years—to some strange yet almost forgotten finds made on a remote part of the Australian coast early in the twentieth century.

In July 1916, the third year of World War I, the Australian light cruiser *Encounter* was patrolling the rugged north-west coast of Australia for hostile German warships when it put in to the isolated harbour of Napier Broome Bay, at the northern tip of the Kimberley peninsula. The captain, Commander CW Stevens, and the surgeon, Lieutenant W Roberts, were reconnoitring in the ship's cutter when they noticed what appeared to be two marker posts on a small, unnamed island. Their curiosity aroused, the two men ran their boat ashore on the beach. To their surprise they found that the 'posts' were in fact two antique bronze cannon muzzles protruding from the sand twenty-five paces from the water's edge. Lieutenant Roberts later wrote:

> We saw the two carronades protruding through the sand ⅔rds of each being exposed so that they were easily lifted out. They were 6 feet apart and had the appearance of leading marks...a large number of the ship's company landed and next day shifted sand over practically the whole area for a considerable depth. The only other object found was a small portion of a brass-bound chest. You can imagine the disappointment of the matelots who had visions of buried treasure.[1]

Lieutenant Roberts erred in calling the guns 'carronades'. Carronades are short-barrelled, mortar-like guns first made for the Royal Navy in the 1780s and they bear little resemblance to the long, narrow-barrelled cannons found in the sand at Napier Broome Bay. However, the island on which they were discovered was christened Carronade Island—and Carronade Island it remains to this day.

The commander subsequently deposited his trophies with the Royal Australian Naval Dockyard in Sydney. The two cannons are fairly similar in size and appearance. Both are just over a metre long; one has a bore of 46 mm and the other 33 mm. The most obvious difference between the two is that the gun with the smaller bore is plain and unadorned, while the barrel of its companion is decorated with a crown and other emblems.

The find posed some intriguing questions. Where had the guns come from? What could have happened at Napier Broome Bay and why were the cannons left on the island? The year after their discovery, another Australian naval commander, Captain JF Robins, who had some expertise in the field of early ships' artillery, attempted to answer the first question. In a report to the Navy he wrote:

> In my opinion, after a careful survey they appear to be part of the armament of a Spanish or Portuguese caravel wrecked near there. One

> has as a badge a fancy crown, not a Royal one, also a double knot which is the badge of Seville, Spain. A gun factory was established there in the sixteenth century by Rey Carlos de Espagne [King Charles of Spain], and all guns made there would carry this badge in a similar manner to those made at Woolwich. They supplied all vessels fitting out by Royal Monopoly. The other gun has no badge and was probably a trade gun of the same period. These guns in all probability relate to the early periods of Australian history and probably of de Quiros' time.[2]

Not all Australian academics and historians by any means accept Captain Robins' conclusions—and as we shall see, he was wrong in believing the decorated gun's badge to be Spanish. (Pedro Fernandez Quiros, to whom he refers, led a Spanish expedition from Peru which discovered the Pacific islands of Vanuatu in 1606.) Over the years there has been much debate about the cannons' age and origin which is probably one reason why, despite their obvious historical importance, little effort has been made to put them on public view. For some years now the guns have been hidden away in the Royal Australian Navy's historical collection, languishing in storage on an island in Sydney Harbour while it awaits a permanent home some time in the future. One consequence is that the cannons are deteriorating and their very existence is by now in danger of being forgotten.

Captain Robins' conclusion, nevertheless, was not far from the truth. According to another contemporary report, when the decorated gun was cleaned it was found to be 'stamped with the Portuguese Rose gun mark and the Portuguese Crown. Rubbings of these were sent to Lisbon, where it was stated they were believed to have been cast in the late fifteenth or early sixteenth century'.[3]

Several decades later, in 1959, the curator of the Historical Weapons Department in the Museum of Western Australia, a Mr Halls, examined the markings on the decorated cannon and declared that it was his belief, too, that the badge was 'the Rose and Crown of Portugal'.

This finding is undoubtedly correct. The crown on the Carronade Island gun is a highly distinctive design, and many other examples of it can be seen today among numerous sixteenth century Portuguese cannons on display in the Lisbon Military Museum. The Royal emblem, with its five-pointed crown, dates from the period of King Manuel I (1495 to 1521). Minor variations of detail occur from one gun manufacturer to another, but the basic design remains unmistakable. It can be seen also in many other historical locations—not least among them the Tower of Belém, a fortress which King Manuel had built on Lisbon's River Tagus in 1515 as the starting point for ships sailing for the East. Carved in white sandstone on the front of the tower is a huge emblem displaying the King's coat of arms, topped by this same distinctive Royal crown. The rose design on the touch-hole of the Carronade Island cannon can also be seen on other antique cannons on display in Lisbon.

Other evidence confirms that it was apparently normal practice for Portuguese bronze guns of the sixteenth century, including those used in the East, to be marked with the royal insignia. Evidence for this comes serendipitously from the contemporary Portuguese adventurer Fernando Mendes Pinto who, describing how his ship had captured two Chinese pirate junks, wrote that 'seventeen bronze artillery pieces were found, *practically all of them stamped with the royal emblem*, which that dog [of a pirate] had stolen from three ships on which he had killed forty-six Portuguese' (author's italics).[4] As for the motif engraved on the Carronade Island cannon below the crown, this is clearly a stylised 'Portuguese Rose'.

Conversely, any suggestion that the Carronade Island guns resemble the highly ornate Malay cannons of this period known as *lantakas*, as one Australian archaeologist has claimed, can be ruled out. On any reasonable comparison there is no similarity whatever. The only real question about the Carronade Island cannons is whether they were actually made in Portugal or manufactured for the Portuguese navy elsewhere in Europe—or perhaps even somewhere in South Asia.[5]

The mystery surrounding the two Carronade Island guns deepens when we find that these are not the only antique cannons of European type to have been discovered on Australia's Kimberley coast in the region of Napier

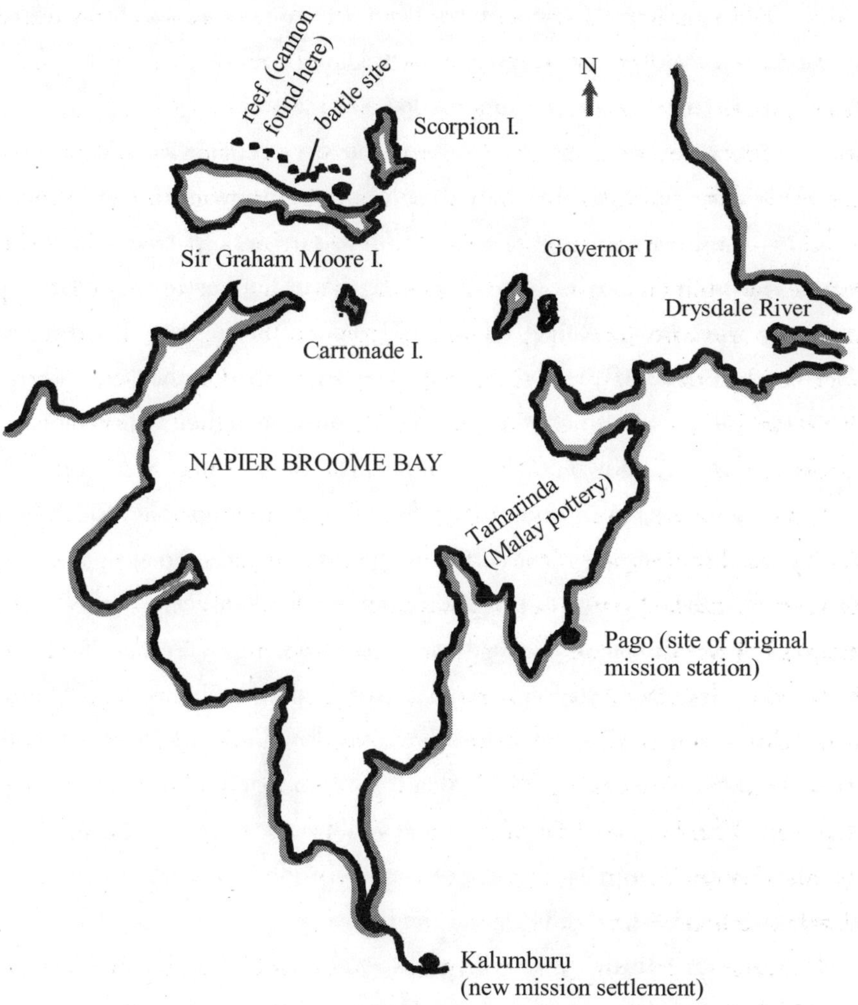

Napier Broome Bay.

Broome Bay. In 1918, two years after the Carronade Island guns' discovery by the Navy, three Australian fishermen on a pearling expedition found a small cannon submerged on a reef in the same bay in about half a metre of water. After cleaning the gun they found it to be made of bronze and decorated with markings. The leader of the fishing party, Cyril Davey, was convinced that the cannon had historical value. He attempted to interest museums in

Sydney, Melbourne and London in buying it, but to no avail: all three institutions rejected his overtures. Disappointed, Davey sold the cannon to a brass foundry in Perth where it was melted down as scrap.

This gun was also referred to in contemporary reports as 'a Portuguese carronade', but once again the 'carronade' part of the description is incorrect. Fortunately, Davey took a photograph of the gun before it was melted down. This shows it to be a narrow, long-barrelled cannon, similar to a European type classed collectively as *culebrinas*, or in English culverins. The chief characteristic of these guns, whatever their size, is that their barrels were generally at least thirty times longer than the diameter of the bore. The smallest culverins, known as falcons and firing a shot weighing between 0.45 and 1.4 kilograms, were commonly carried as swivel-mounted deck guns on ships in the sixteenth century including, of course, the Spanish Armada.[6] Interestingly, the eighty-year-old photograph of the Kimberley gun shows it actually mounted on a swivel, suggesting that this component may have been found with the gun. And for what it is worth, people in Perth with some knowledge of antique guns who examined this cannon before its destruction are said to have given their opinion that it was Portuguese.

Yet another antique cannon was found well over a century ago at Vansittart Bay, the harbour immediately west of Napier Broome Bay. This little-known discovery was made in 1884, some thirty-two years before the Carronade Island guns came to light. As with the swivel gun, the finder was again an Australian pearl fisherman. According to an account prepared in the 1930s by the man's son, Mr KM Wood, of Perth, this bronze cannon was discovered in a deserted Aboriginal camp on an island in Vansittart Bay. The pearl fisherman was astonished to find that it still contained a cannonball and an ancient charge of gunpowder. The gun was said to have been about one metre long with a distinguishing badge, described as a partly obliterated crown, near the touch-hole. Whether this cannon still exists is not known. Mr Wood's father sold it to the German captain of a ship collecting guano fertiliser and that was the last that was heard of it. Today the gun may perhaps be an anonymous exhibit in some European museum. [7]

What conclusions can we come to about these last two all-but-forgotten discoveries? The circumstances in which the swivel gun was found—in shallow water on the reef off Napier Broome Bay—suggest that it was perhaps jettisoned in haste by a ship's crew attempting to lighten their vessel. As for the Vansittart Bay gun, it sounds very much like a companion piece to the larger of the two Carronade Island cannons, which, it will be recalled, was also about three feet long and similarly decorated with a crown near its touch-hole.

The finding in the same general area of no fewer than four abandoned European-type cannons—one of them even primed for firing—speaks of some dramatic event at Napier Broome Bay long ago. What could it have been?

The possibility of a Spanish ship having visited this bay can be discounted. There is no record of any Spanish vessel coming to the coast of Western Australia in either the sixteenth or seventeenth centuries. But what of the Portuguese?

Crucial evidence which will reveal part of the answer to this mystery is to be found in a strange and unexpected place, a centuries-old Aboriginal corroboree dance. This tribal ritual, the details of which by great good luck have been preserved, tells vividly through Aboriginal eyes not only how and when the cannons actually came to Napier Broome Bay—but also, by inference, who brought them there.

For the eyewitness report which we have of this dance drama we can thank yet another Australian pearl fisherman, a Mr HV Howe. His account, written some years after the event and following the Navy's Carronade Island discovery, makes fascinating reading. In 1909, Howe related, he was one of five white men on a pearling fleet working off Sir Graham Moore Island, at the head of Napier Broome Bay. Referring to the Carronade Island cannons, he continues:

> We were fortunate enough to witness a re-enactment of the battle in which the guns were captured. The island Aborigines had met white

men on not more than a dozen occasions in the previous hundred years, but they had been in regular contact with trepang fishing schooners from Timor which visited the coast each year. From their crews many of the tribesmen had learned to speak Malay—which my companion and I also spoke as it was the *lingua franca* of the Asiatic crews on the pearling luggers.

Ability to converse freely with the tribesmen facilitated establishment of friendly relations, particularly with the tribal elders who thoroughly enjoyed giving strangers detailed explanations of the many corroborees we witnessed.

Passed from one generation to another with meticulous attention to preservation of the traditional form, these performances portrayed tribal history and mythology. The one depicting capture of the carronades is here described and explained.

At the foot of the sandhills above the beach on Graham Moore Island the women and children mustered, providing the chorus and orchestra—singing in a continuous undertone, clicking yam sticks together and slapping thighs. In front of them lined up warriors armed with spears and shields, bodies decorated in conventional patterns with feathers stuck to the skin and painted in pipe clay. They stamped and shouted according to established ritual and occasionally a star performer gave a solo item.

From across the offshore reef two large canoes approached the shore, each with a tube of bark projecting over the bow. In them were about thirty men, who chanted as they paddled and who jumped out and pulled the canoes high on the beach immediately they grounded.

They all wore helmets, breastplates and thigh pieces fashioned from bark. Some carried long wooden swords, others had pieces of wood fastened together to resemble crossbows.

Their landing was followed by a chanted parley between their leaders and those of the warriors on the beach—interrupted suddenly by two loud blasts on conch shells. The cannon had opened fire and the battle was joined. The swordsmen charged forward; the crossbowmen hurled short sticks at their opponents and the defenders replied with showers of spears. At intervals the conch shells boomed, and at each blast several warriors fell dead. Occasionally all stood fast while individual champions fought in single combat—the vanquished expiring to an appropriate dirge while the victor boasted of his prowess.

> The songmen chanted a continuous commentary on the progress of the battle. At the end of an hour the last invader fell. The beach was littered with bodies, and the survivors roared a paean of triumph.
>
> Thereupon the dead came to life, victory celebrations commenced and the whole tribe crowded around the campfires on which were roasting about a dozen turtles and a huge quantity of fish and game.
>
> All having had long experience of Aborigines, we white spectators had no doubt whatever [that] we had witnessed the re-enactment of an important event in tribal history. Our belief was confirmed by the full explanation subsequently given us by the tribal elders.
>
> They of course had no idea of the nationality of the invaders, whom they described as white men with skins like turtles and crocodiles—an obvious reference to their armour.
>
> Speaking of the bark tubes in the bows of the canoes, they used the Malay word for cannon, but they would not describe the guns or make any further reference to them. From this we assumed that, impressed by their destructive power, the original captors had adopted them as sacred totems and carried them off to the tribal ritual ground—now known as Carronade Island—on which we were not permitted to land. As totems the guns could not under any circumstances be discussed with strangers to the tribe. We never saw them, but their subsequent discovery by HMAS *Encounter* confirmed our assumptions about them.
>
> The island Aborigines were able to count only to 'two fives' and consequently had no conception of the number of years which had elapsed since the battle. By setting out pebbles on the beach in accordance with the intricate tribal marriage system the elders could however demonstrate the number of generations between living tribesmen and their ancestor heroes whose names and marriage sects had been passed on from one generation to another. The pebbles showed seventeen generations had intervened, and indicated the battle had taken place about 350 years previously—about 1560 AD.[8]

It was fortunate indeed that Howe recorded this remarkable story. But notwithstanding its vivid detail, his account has been either ignored or belittled by mainstream Australian academics. Because his calculation of the date of the battle was based entirely on an Aboriginal tradition, Howe was under no illusion that there was any chance of it being accepted as valid by the historians of his day. Now, however, with traditional indigenous knowledge belatedly being accorded the respect it deserves, the time has surely come for a reappraisal.[9] Howe's own conclusion from the events he had witnessed was that the Carronade Island cannons were, as he wrote, 'undoubtedly the first guns captured by Australians from enemy invaders—and possibly relics of the first European landing on Australian soil'.

Again thanks to his knowledge of Aboriginal lore, Howe knew that important tribal traditions such as this one were passed down from generation to generation 'with meticulous attention to preservation of the traditional form'. That being the case, what are we to make of the extraordinary fact that the Aboriginal 'invaders' are said to have been wearing helmets, breastplates and thigh armour made of bark and that the tribal elders described the real invaders as 'white men with skins like turtles and crocodiles'? [10]

The Aborigines were familiar with Malay fishermen and some even spoke their language, so there could be no possibility of their describing any Malays in this outlandish way. That leaves only one tenable and obvious explanation: the intruders were Europeans, wearing armour of the type in use in Europe in the sixteenth century.

Of particular importance in the historical context is Howe's observation that some of the Aboriginal 'invaders' carried imitation crossbows. The crossbow of medieval Europe reached the peak of its technological development in the fifteenth century. Despite the introduction of firearms it was still in general use in the early sixteenth century because it was more reliable and accurate than its crude firearm contemporary, the matchlock arquebus. Da Gama's crew, for example, carried no firearms—relying on crossbows, swords, spears and boarding pikes. The soldiers of Magellan's fleet in 1519 carried some arquebuses, but they still relied on crossbows as their main distance anti-personnel weapon. By the middle of the sixteenth century, however, when the arquebus was superseded by the bigger and more effective musket, the crossbow quickly became obsolete. This change in weaponry is graphically illustrated in the pages of the Vallard Atlas, published in 1547. Here, all the European soldiers without exception are pictured carrying firearms.

This fact allows us to fix an approximate date for the Graham Moore Island battle. The fifteenth century can be ruled out because the first Europeans to reach South-East Asia, the Portuguese, did not arrive at the East Indies until the beginning of the sixteenth century. On the other hand the battle is unlikely to have occurred after the middle of the sixteenth century because, as we have seen, the European intruders by then would have been using firearms, not crossbows. From this evidence the battle can be assumed to have occurred at some time in the first half of the sixteenth century.

Significantly, this corresponds closely with the date arrived at by Howe's figure based on the Aboriginal genealogical record. His suggested date for the battle—approximately 1560—assumed that the seventeen Aboriginal generations since the event each occupied twenty years (remembering that his own reference date was 1909). This estimate of the time between generations—and of course it was only an estimate—would, if anything, be on the low side. In contemporary European terms a 'generation' is normally reckoned to be about twenty-five or thirty years. For short-lived Aborigines, however, these figures would almost certainly be too high. If each of the seventeen generations was assumed to occupy twenty-five years the date of the battle would be approximately 1484. The mid-point between this and Howe's date of 1560, derived from generations each occupying twenty years, would be 1522.

And who exactly were these foreign intruders? It has been suggested by some academics that the people who left the cannons behind at Napier Broome Bay must have been fishermen from the Indonesian region of Macassar, who came in search of the Chinese delicacy trepang or sea cucumber, probably in the nineteenth century. But this theory is at odds with the facts. Howe's account stresses that the Aborigines who performed the corroboree in which the origin of the cannons was re-enacted had long been on familiar terms with Macassarese fishermen, and some could even speak their language, Malay. Some Aborigines are even known to have been living at Macassar in 1865.[11] This surely rules out any possibility that the indigenous Australians would have confused these well-known Indonesian fishermen with the strange, armoured 'white men with skins like turtles and crocodiles'.

As for the boats of the Macassarese fishermen who visited the Kimberley coast, the first reliable account we have of them indicates that their weaponry and equipment was technologically unsophisticated. Early in 1803 the French captain Nicolas Baudin was surprised to encounter two dozen small Malay vessels, called praus, searching for trepang at Cassini Island, west of Napier Broome Bay. Far from carrying long-range cannons, their armament consisted of nothing more formidable than Malay-type swords or daggers, known as kris, and a few bows and arrows. Baudin also noted that there was only one small compass for the entire fleet—and even that was 'extremely poorly made'.[12]

All this points to just one tenable explanation: the intruders were Europeans, wearing armour of the type in use in Europe in the sixteenth century.

Narrowing the possibilities down further, as we have seen, a Spanish foray can be ruled out because there is no evidence of any Spanish ship having sailed anywhere near Australia's Kimberley coast in the sixteenth century. That leaves only the Portuguese.

But are there any records of a Portuguese expedition which could possibly be identified with the events recorded in the dance drama of the Aboriginal people at Napier Broome Bay? The answer is yes. To discover how this expedition came about and who led it, we must turn now to accounts describing the tumultuous events which accompanied the founding of Portugal's seaborne empire in the east.

By the second decade of the sixteenth century, the east, with its promise of fame and wealth, had become an irresistible magnet for the ambitious younger sons of Portuguese noble families. Contemporary accounts speak of a ferment of excitement among these *fidalgos*, as they were known, arriving at Goa and Malacca from their homeland in search of fortune. Some believed they were on the verge of discovering Marco Polo's legendary kingdom of the Land of Gold, and that fabulous riches lay almost within their grasp.

One such adventurer was Diogo Pacheco, a young *fidalgo* who arrived at Malacca in 1518 in a ship commanded by his older brother Antonio, who was to take charge of the Portuguese fleet at Malacca. Also on board was a third brother, Manuel. The trio are thought to have been relatives of Duarte Pacheco, considered by some historians to have been the truc discoverer of Brazil in 1498—two years before the official 'discovery' by Pedro Alvares Cabral but at a time when for political reasons Portugal needed to keep the find secret from its rival, Spain.

Duarte Pacheco had been a captain in Vasco Da Gama's second fleet of 1502. He played a vital role in establishing the Portuguese Empire of the Far East by volunteering to stay behind on the Malabar coast of India in command of a tiny squadron of three ships to maintain Portuguese naval authority while Da Gama and the rest of his twenty strong fleet returned in triumph to Lisbon (Da Gama's more senior officers had all declined the offer of command). Left to his own devices for two years from 1503 until October 1505 when a new fleet arrived carrying the first Portuguese Viceroy of India, Pacheco had repulsed repeated attacks by hostile Indian fleets of overwhelming numerical superiority. A grateful sovereign rewarded him with high honours when he eventually returned to Lisbon.

The royal esteem in which Duarte Pacheco was held could help explain the Pacheco brothers' rapid advancement in the east. There are some reasons for suspecting that Diogo Pacheco may even have been entrusted by King Manuel with an important secret commission to search for the Island of Gold.

This fabled land was the cause of much speculation among the Portuguese in India. It was believed to lie somewhere south of Sumatra, as Marco Polo had indicated—but just where, no one was sure. That Diogo may have had a specific mission to discover it is given credence by the fact that according to the Portuguese chroniclers his arrival at Malacca was accompanied by rumours that he had already managed to obtain some very important information as to its whereabouts.

But for the time being thoughts of a search for golden lands had to be put aside, for other more pressing tasks faced the Malacca garrison. After capturing this key Malay trading port dominating the sea route to the Orient in 1511 the Portuguese had quickly strengthened its fortifications. Six years later, however, their hold on it was under serious threat. The danger came from the city's powerful former ruler, Sultan Mahmud Shah, who prior to his defeat had held sway over the entire Malay peninsula as well as several states in eastern Sumatra.

After his expulsion from Malacca the sultan had retreated overland with the remnants of his forces to the east coast of the Malay peninsula, and from there sailed south to regroup on the island of Bintan, south-east of present-day Singapore. Here he began developing a rival centre for trade, while the strategic position of his new base enabled him to attack Portuguese ships passing through the Singapore Strait.

Diogo Pacheco's first task was to sail for the Singapore Strait in command of a Portuguese flotilla. His task there was to rendezvous with and escort through the enemy waters a large Portuguese merchant ship, the *Santo André*, returning to Malacca from China with a valuable cargo and carrying the first Portuguese emissary to China, Fernando Peres. The rendezvous was successfully achieved and Pacheco escorted the merchant ship to Malacca without serious incident.

The Portuguese could perhaps have tolerated Sultan Mahmud Shah's attempts to interfere with their shipping, but emboldened by his successes Mahmud advanced up the coast to build a strong fortress at the mouth of the River Muar, a mere twelve nautical miles south of Malacca. His next move, he proclaimed, would be to kill the foreigners and retake Malacca.

For the Portuguese this was an intolerable situation. The sultan's fortress at Muar was defended by strong earthworks topped with wooden palisades and protected by artillery, but for the security of Malacca it was essential it should be attacked and destroyed. The Portuguese launched their assault from the sea early in 1519. It succeeded after fierce fighting in which the fortress was heavily bombarded by the Portuguese ships' formidable artillery. Diogo Pacheco is named in the Portuguese chronicles as having played a leading role in this battle, which is said to have left 800 of the sultan's army dead.

Of particular significance as far as the possible origin of the Carronade Island guns is concerned is the fact that at Muar the Portuguese captured over 300 Malay cannons, among which, according to the chronicles, were many of bronze. Most of the sultan's ships were also seized.

As for Diogo Pacheco, the way was now clear for him to prepare for his voyage to discover the Island of Gold. The Portuguese chronicles provide two separate accounts of this voyage, one by Pacheco's near-contemporary, João de Barros, and the other by Manuel de Faria e Sousa. They differ in some respects and in the case of Barros there are some puzzling contradictions. Since Barros' account is the first and by far the more detailed (he even visited Malacca himself in 1540) it makes sense to start with him.[13]

First appearances to the contrary, it is clear that Diogo Pacheco was not just acting as a freelance adventurer. Barros records that the order for him to be provided with ships, armaments and provisions for his expedition came from no less an authority than the newly arrived Portuguese Governor of India, Diogo Lopes de Sequeira. The governor, in his dispatch to the commander at Malacca, also praises Pacheco as a 'very skilled mariner' and a gentleman with a 'great talent for discovery'.

For his expedition Pacheco is said to have been allocated two ships, one of them a large carrack. This vessel was specially fitted out with cannons; whether they were Portuguese cannons, or whether—as seems possible—some of them were guns captured at Muar or elsewhere is not clear. The other vessel is described as a bargantim and is said to have been commanded by another Portuguese officer, Francisco de Sequeira. The Portuguese term bargantim has a parallel in the English word 'brigantine' (usually abbreviated to 'brig'). In the sixteenth century it referred to a smallish ocean-going vessel, usually two-masted, with a square-rigged sail on the foremast and a triangular fore-and-aft (lateen) sail on the mainmast. With its relatively shallow draught and superior manoeuvrability, Pacheco's bargantim would have been intended for close inshore exploration once the Island of Gold had been reached. As trade items for the anticipated gold, the Portuguese captain loaded a cargo of many sorts of 'cloth of Cambaia', from southern India.

Departing from Malacca, in the latter part of 1519, the ships sailed north-west to the northern tip of Sumatra, then some twenty leagues (about sixty nautical miles) down the south coast of that island as far as the sultanate of Daia. Here (according to Barros) they were hit by a storm and the brigantine was wrecked The sole survivor, a slave, managed to trek back overland to the north Sumatran sultanate of Achem (today's Aceh) where there were some Portuguese to whom he was able to pass on news of the shipwreck.

In the meantime, Diogo Pacheco, in the carrack, was continuing his voyage south-east down the coast of Sumatra—a course which, had he pursued it for long enough, must surely have taken him to the north-west coast of Australia. However, some distance further down the Sumatran coast he turned west to enter the port of Baros. (Barros the chronicler spells the port with two 'r's, the same as his own name, but in other accounts it is normally spelt 'Baros', and to avoid confusion this is the spelling that will be used here). Baros, in Pacheco's time, was noted for trade in gold, jewels and other precious merchandise. Today it can be identified with the small town of Barus, a fairly insignificant place. In the sixteenth century, however, it was a prosperous port visited annually by merchant ships from India, the Malay coast and the northern realms of Sumatra. There were three such ships in port when Pacheco arrived. News of the Europeans' ferocity in warfare had evidently gone before them, for on sighting the Portuguese vessel all three crews fled ashore, taking shelter among the local inhabitants.

Pacheco, on seeing the panic his arrival had caused, hastened to make signs of peace. A parley ensued, with the emissaries of the local ruler demanding to know who the foreigners were and what they wanted. Pacheco's reply is revealing. In a lengthy speech he identified himself as a captain of King Dom Manuel of Portugal, ordered by the king's Governor of India to circumnavigate 'this island' (Sumatra) toward the south spreading the news at all ports at which he called that Portugal was now the ruling power in the great merchant city of Malacca and that King Manuel would henceforth guarantee their security.

Pacheco prudently made no mention of his quest for the Island of Gold. According to the chronicles, the 'King of Baros' and his ministers professed themselves very contented with this explanation and ordered that the Portuguese ship be reprovisioned. Pacheco in response offered to trade his cargo of cloth. The chronicles record that he managed to obtain benzoin (an aromatic resin) and some gold—but only a little.

Of much greater value to Pacheco, however, was the information he managed to extract during his stay from these 'Moors of Cambaia' (Muslim traders from India) concerning the nature and whereabouts of the Land of Gold. Sumatra, they assured him, did not produce much gold itself but a great amount was obtained through barter with the people of the mysterious land to the south, to which merchants of the King of Baros from time to time sailed. Now the chronicler sets out in detail the sailing directions which Pacheco obtained:

> You must set a course approximately south-east from this port, and after you have sailed more than 100 leagues you will come to a cordon of shoals and reefs. Through a gap in the middle of these you will come to a hilly island at the foot of which are palm trees among which live many black people. One can trade with these people for gold at the water's edge—but only at the water's edge, for they do not allow others to seek out where they live. Because of this, nothing is known about the land in the interior, nor what they have there, nor the way of life of these people. Nevertheless, these people give good quantities of gold in exchange for cloth of Cambaia.

Pacheco even obtained advice on what types of cloth to take for barter. Most popular with the black people, he was told, were *mantazes* (blankets), *bretangis* (cotton cloth from India), and *vespicias*—an obsolete word which could indicate some kind of clothing. The black people's favourite colours were said to be blue and red.

All this would have been music to Pacheco's ears but there was also a warning. Even though a lucrative profit could be made by exchanging cheap cloth for good quantities of gold, the dangers of the voyage were so great that many of the merchants who sailed once to this place refused to go again for fear of losing their lives. So formidable indeed were the hazards that, according to his informants, out of every twenty ships that had sailed for the Island of Gold, only one in four had ever returned.

Pacheco was also advised that the three months of the monsoon were the best time to sail from Sumatra to this island. It was also preferable to use small vessels because of the many shoals and reefs there. Between these reefs, he was told, were some navigable channels and straits which changed their appearance every year because of a 'revolution of waters' which the region experienced each rainy season.

This is an apt description of the Kimberley coast of north-west Australia. It fits in almost every detail. The Kimberley coast is indeed reached by sailing 'more then 100 leagues' south-east from Sumatra—quite a lot more than 100 leagues, to be sure, but it is the *only* coastline in that direction. The Kimberley coast is protected by a dangerous cordon of reefs and shoals, just as the Malay captains warned. Any sensible navigator would sail in that direction from Sumatra during the months of the monsoon, for at other times the winds would be contrary. Most telling of all is the Muslim seafarers' extraordinary description of how this coast was affected by a 'revolution of waters' (*revolução das águas*) in the wet season. This, as anyone familiar with the region will know, is exactly what occurs on the Kimberley coast during the monsoon. In that season, dry river beds become raging torrents, the huge tidal range creates fierce sea currents and rapids, and the flow of water is so great that many dry season landmarks are submerged. It seems out of the question that any foreigner who had not actually visited the Kimberley coast during the monsoon could come up with a description so extraordinarily accurate.

The Kimberley coast also fits the Malay description of a hilly country inhabited by 'many black people' who were wary of strangers. But what about the palm trees said by Barros to line the shore? The word used in the Portuguese is *palmares* and in this context it might possibly (though not necessarily) signify 'coconut palms'. But would there have been coconut palms on the coast of Australia? Does the chronicler mean some other sort of palm trees, or are we looking in the wrong place? In fact there is an unexpected twist to the story—and one which might even seem to confirm the truth of what Pacheco had been told. For one of the unusual features of Napier Broome Bay are groves of coconut palms. And they have flourished in this locality for as long as anyone there can remember.[14]

At Baros, Pacheco had got what he most wanted: directions to the Island of Gold. After a cordial leave-taking from his hosts he set a course south-east along the coast of Sumatra as far as the Sunda Strait, between Sumatra and Java. Here he turned north through the strait and sailed back up the northern coast of Sumatra without incident, thus completing his circumnavigation of the island. Although he had returned to Malacca with only a few tangible rewards from his voyage, his resolve to find the Island of Gold was undiminished and for a very good reason, for now he believed he knew the way!

On his return to Malacca, Pacheco received some startling news. During his absence his compatriots had fought a second pitched battle with Sultan Mahmud Shah. The victory at Muar in 1519 had been greeted by the Portuguese at Malacca with great festivities, but the rejoicing had been premature. Mahmud Shah had managed to retreat with much of his army to the city of Pago, some fifty kilometres up the Muar River. Aware now of the lethal power of the Portuguese naval guns, he had strong wooden palisades constructed across the river at fortified strategic points, a precaution which he believed would make it impossible for the enemy's ships to sail upstream. At Pago he built a new fortress defended by stone walls and massive palisades protected by artillery. Next he re-imposed a land blockade of the Portuguese in Malacca and declared that he would either retake Malacca or die in the attempt.

The sultan had believed himself safely beyond the reach of the formidable Portuguese ships' guns, but he had reckoned without the ingenuity of the enemy's naval commanders and the exceptional manoeuvrability of their ships, especially the caravels. In July 1520 an armada of thirty ships fought their way upstream and then subjected the Pago fortress to a massive, cannon bombardment.

Pago is said to have been defended by 2000 well-armed men and artillery; but even the sultan's war elephants were to no avail as the citadel's fortifications were reduced to rubble. As at their victory at Muar, the Portuguese were able to capture the sultan's remaining artillery pieces—more than twenty of them, mostly bronze cannons. Records of the battle tell how returning to Malacca in triumph, the victors drove before them a long column of prisoners of war followed by the sultan's elephants laden with the captured cannons. Before leaving Pago the Portuguese burned it to the ground. We should keep the name of this city in mind because we shall come across it again, in a very unexpected place.

Pacheco confided to the Portuguese governor of Malacca his dramatic news that he now had precise sailing instructions to reach the Island of Gold. With a great booty of cannons and captured ships at his disposal, the governor could well afford to fit him out for a second expedition.

Barros' account of Diogo Pacheco's first voyage takes up some three pages of his history; in contrast the second voyage is covered in a scant two paragraphs. Pacheco, we are told, was this time given another brigantine and a 'ship of the land' (*navio da terra*). This description suggests that it was a locally built, as opposed to a European ship, presumably constructed either at Malacca or the Portuguese base at Goa. With these two ships, Pacheco set out with the monsoon, probably toward the end of the year 1520. According to the chronicler he again reached the port of Baros. Here, according to the chronicler Barros, he was met by four or five ships 'from Cambaia [India] and other places'. These ships refused him entry to the port and began to bombard his vessels.

Now comes the chronicler's final, crucial paragraph. The scene has abruptly changed. The two Portuguese vessels are in shallow water just off an unnamed beach; there is a contrary wind and a crowd of hostile native people are attacking them, some in canoes. Pacheco, in command of the larger ship, hails the crew of the more manoeuvrable brigantine and urges them to throw a tow-rope so they can pull his ship off the beach and prevent its being seized. The response from the smaller vessel is not recorded. Barros simply writes that the brigantine got out to sea, leaving the other ship stranded on the shore. The only people who managed to escape, he adds, were some Malay seamen from Malacca, who made their way across the interior of the land, emerging on the opposite coast where they found a ship to take them back to Malacca. 'And thus narrated,' concludes Barros, 'is the misfortune of Diogo Pacheco, who was the first of us [Portuguese] to lose his life *por descobrir* this Island of Gold'.

The exact meaning of Barros's final sentence is of crucial importance. The few English-language historians who have looked at this passage have interpreted *por descobrir* as meaning that Pacheco died 'attempting to discover' the sought-after island—but two native Portuguese speakers consulted by the author because of their expertise in both the Portuguese and English languages say that this translation is definitely wrong. The correct meaning, they say, is that Pacheco died *'because he discovered'*, or perhaps even more accurately, *'for having discovered'*, the Island of Gold.

And indeed if we look closely at Barros's original text we find that in his final paragraph describing the battle on the beach he does not mention Sumatra or the port of Baros at all. Instead, he refers first to 'the island' and finally 'this Island of Gold'.

If this were not proof enough, unequivocal confirmation that Pacheco did indeed meet his fate on the far-off coast of the Island of Gold comes from a second Portuguese chronicler, Manuel de Faria e Sousa. He gives a description of Pacheco's exploits, but it is far briefer than that of Barros and very different in tone. Writing a century later, but with the advantage of access to official documents in Lisbon, Sousa confirms that Diogo and his brother Manuel were two of the six Portuguese captains who played a leading role in the successful attack on the Malay fortress at Muar. Turning to Diogo's voyage to what he terms the 'Isle of Gold', however, he refers to this expedition with undisguised scorn:

> All of what was done in those days is more appropriate for a cash-book than a chronicle. Diogo Pacheco lost his life with many others in a shipwreck while searching with two ships at the Isle of Gold. He planned to list as revenue the gold of that island—but Death listed as expenditure his own life and the lives of almost all of those who accompanied him.[15]

Its ironic tone notwithstanding, Sousa's account confirms unambiguously that Diogo Pacheco lost his life while searching *at* the Island of Gold—not 'for' it. Sousa is a meticulous recorder of events, so there is absolutely no reason to think that this could be a careless slip of the pen. Pacheco, he confirms for us, was in the process of searching what can only be the coast of Australia with his two ships when he met with disaster. Sousa's description of how Pacheco and nearly all his companions lost their lives in 'a shipwreck' at this 'Isle of Gold' is fully consistent with Barros's story of how one of Pacheco's ships was stranded on a far-off beach, and how the commander and his companions died trying to repel a fierce attack by hostile natives.

That scene recounted by the Portuguese chroniclers surely sounds familiar. Where have we heard something like it before? The parallel scenario is, of course, the battle recorded in song and dance by the Aboriginal people of Napier Broome Bay.

Most of the known facts about the Portuguese expedition match the Aboriginal account in detail. Pacheco had two vessels: the Kimberley tradition also featured two hostile ships. The Aboriginal re-creation of events portrayed the foreign vessels firing cannons, as the Portuguese doubtless would have done. We know from the records of Vasco da Gama's voyage that Pacheco and his officers would have worn armour and their men would have been clad in breastplates and leather jerkins—thus appearing to the Aborigines as though they had the skins of turtles or crocodiles. In combat the Portuguese would have used swords and crossbows, just as the corroboree dancers portrayed. And finally, despite their technologically superior weaponry the foreigners were overwhelmed and killed, as both the Aboriginal tradition and the Portuguese accounts agree.

Finally, the Barros chronicle indicates that Pacheco's ship (including, obviously, its armament) was captured by 'the people of the land'. This would explain the presence on Carronade Island of the two cannons, which had evidently been seized as war trophies and set up as totems on the Aborigines' sacred island. It would also account for the discovery of the third (loaded) cannon in a deserted Aboriginal camp and why a fourth cannon was found submerged on the shallow reef at Napier Broome Bay. In the tribal re-enactment witnessed by HV Howe, it will be recalled, the two canoes of the Aboriginal 'invaders' came into sight paddling over this very reef. The most obvious explanation for the discovery of the submerged cannon here is that it might have been thrown overboard by the crew of the brigantine to lighten the vessel in their bid to escape.

Several other questions remain unanswered. It seems unlikely that the battle re-enacted by the Aborigines of Sir Graham Moore Island would have begun spontaneously, without any cause. What could have led to the hostilities? Had the Portuguese adventurers already made contact with these or other Kimberley coast Aborigines in their search for gold? If so, was that contact less than peaceful?

And then there is the key question which Barros fails to address: what became of the brigantine? The chronicler records that this ship did manage to escape to the open sea—but nothing further. Having got away safely, it should have been a relatively straightforward matter to sail back to Malacca once the winds were favourable. It sounds suspiciously as though the brigantine was the real means by which the surviving Malay seamen managed to return. Was the survivors' account of an overland trek and the chance finding of a homeward-bound ship simply a cover story concocted to avoid accusations of desertion and treachery?

Why, we might also wonder, does Barros's account contain ambiguities of this sort? At least part of the answer may well be found in the fact that his writings had to be submitted for official approval prior to publication. The Portuguese censors would have been sure to expunge the disclosure of a state secret such as an obvious reference to Pacheco having discovered the actual whereabouts of the 'Island of Gold'. A cunningly worded, oblique statement would have been necessary in order to slip the truth past the censors and into print.[16]

It is an extraordinary thing, nevertheless, that the sailing instructions we are told Pacheco was given by the traders at Baros seem to have been intended to lead him precisely to Sir Graham Moore Island. This long, hilly island does indeed lie in a gap between a cordon of shoals and reefs at the entrance to a bay—Napier Broome Bay—just as Pacheco's instructions stated. And in pre-European times this island was indeed the home of a tribe of black people—the very people whose descendants staged the battle re-enactment witnessed by the pearl fisherman HV Howe.

Other enigmatic signs have been found in this Kimberley region pointing to the arrival of mysterious foreigners. In 1838 a party led by the English explorer George (later Sir George) Grey were exploring the region of the Glenelg River in the Kimberley south-west of Napier Broome Bay when they came across a strange Aboriginal painting on the roof of a cave. It was a human-like form known to Aborigines as a 'Wandjina'—a powerful spirit being. There are many such paintings in the Kimberleys, executed usually in red, yellow and black pigments on a dense white background. Traditionally Wandjinas are depicted only by a head or a torso, without mouths and with a prominent ring or halo around their heads. The unusual thing about the Wandjina found by Grey is not just that it was a full-length figure—it was a figure clad in what

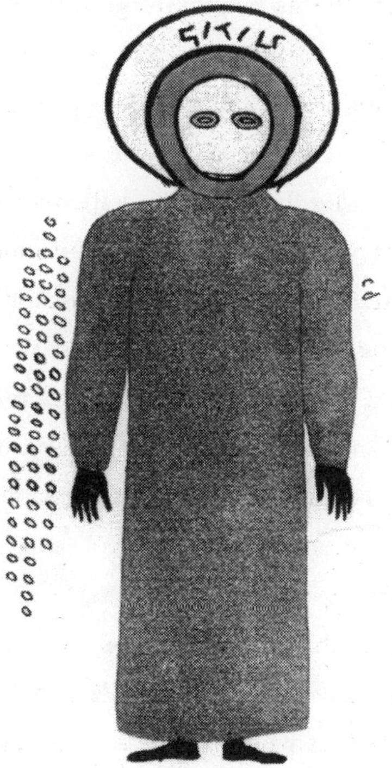

Sir George Grey's drawing of the enigmatic figure which he discovered, painted on the roof of a cave, during his expedition to the Kimberley region south of Napier Broome Bay in 1838. The garment was coloured red, and it reminded Grey of a European priest's cassock.
(Sir George Grey, *Journals of Two Expeditions*)

Grey took to be a European priest's cassock and cowl. Grey, a man of considerable intellect, was later to receive a knighthood and become successively Governor of the British colony of South Australia, and then Governor of New Zealand. Of the strange cave painting he wrote that it was:

> ...the figure of a man, ten feet six inches in length, clothed from the chin downwards in a red garment which reached to the wrists and ancles; beyond this red dress the feet and hands protruded ... [17]

To Grey, the figure brought to mind 'the [Biblical] description of the Prophet Ezekiel'. His sketch of the painting has from time to time been reproduced. The rock art figure itself has deteriorated markedly over the 160 years since Grey saw it, but in his original sketch the likeness to a priest's cassock seems unmistakable.

Grey thought that the painting was already 'ancient' when he saw it. What could have inspired it? The explanation several commentators have supported is that it depicts a priest from a European ship, either Spanish or Portuguese. A Spanish priest can be ruled out because there is no record of any Spanish ship having visited this coast in earlier centuries, but a Portuguese identity for the figure would not be surprising. Although it is perhaps unlikely that Pacheco's small ship would have carried a priest, men of the church often sailed with major expeditions, both to serve the religious needs of the crew and to convert the heathen to the True Faith if the opportunity arose. This could well have been the case on later Portuguese voyages along the western coast of Australia.

Grey's 'priest' figure is relatively well-known. Not so two other Wandjina paintings on rock faces near Napier Broome Bay, though they are equally puzzling. Copies of these figures were drawn by an Aboriginal man early last century for the archives of the Benedictine mission settlement at Kalumburu, on the shore of Napier Broome Bay. Both figures have a headdress resembling a halo and menacing, slit eyes. Both are also wearing what might be described as a large bib or breast plate, under which there seems to be a short tunic ending at the waist under a belt. Beneath their belts the figures have knee-length trousers and what appear to be leggings.

What are we to make of these drawings? Why are the Wandjinas depicted wearing these strange garments when the Aborigines themselves for the most part went naked?

At least one leading Australian authority on the Wandjina art form believes that Wandjina figures portray contacts with foreigners. He suggests that the tradition associating Wandjinas with power over thunder and lightning may well be an Aboriginal explanation for the power of the firearms used by foreign intruders. This in turn would help explain why the foreigners were seen as supernatural beings. [18]

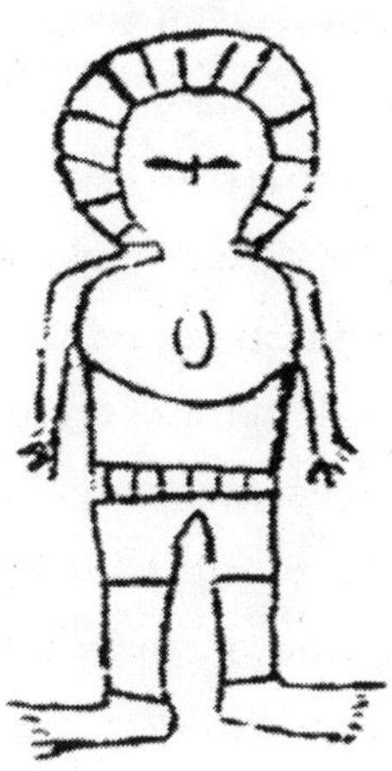

This Wandjina figure is a copy of part of a rock painting, drawn by an Aboriginal man for monks at the Kalumburu mission station at Napier Broome Bay. The figure is wearing what seems to resemble a breastplate, and also what look like belted pantaloons and leggings in the style of sixteenth century European soldiers. (*New Norcia mission*)

What sort of foreigners might the Napier Broome Bay Wandjinas represent? A possible answer is to be found when we look at a typical Iberian sailor's outfit in the early sixteenth century. This consisted of a loose shirt, often with a hood, over which was worn a rough woollen pullover held in by a belt at the waist. The sailors usually also wore floppy, pyjama-like pants made of coarse linen which reached just below the knees. This style of dress has distinct similarities to the garments depicted in the Aboriginal paintings. The halo-fringed headdress of the Wandjinas is not unlike a hooded shirt, or possibly an iron helmet, with the wearer peering through the helmet's eye slits and the halo representing the helmet's raised crest. The 'bibs' worn by the two figures could similarly be interpreted as armoured breast plates, with their upper garments perhaps an attempt to depict coarsely knitted wool, leather or perhaps even chain-mail tunics—combinations all fitting the Aboriginal elders' description of the intruders as 'white men with skins like turtles and crocodiles'. Belted knee-length trousers of the type apparently being worn by these Wandjinas were also part of the dress of some sixteenth century European soldiers.[19]

All this might perhaps seem fanciful were it not that recently some of the oldest Wandjina rock paintings have been scientifically dated. This was possible because the Aboriginal artists drew their Wandjinas' eyes with charcoal, and enough charcoal still remained at the beginning of the present century for radiocarbon dating tests to be carried out. The results put the date for the creation of these Wandjinas at around 1530 plus or minus thirty years—a remarkable finding, given that it fits almost exactly the arrival of Diogo Pacheco's Portuguese expedition on the Kimberley coast.[20]

Furthermore there is compelling pictorial evidence of another early European contact further along the Kimberley coast at Bigge Island, the largest island of the Bonaparte Archipelago. Here, a set of important rock paintings record the landing of what is unquestionably a party of foreign seafarers. The figures in these old paintings are known as 'kaiara'—beings who were similar to the Wandjina but more benign. Aboriginal tradition records that the kaiara also arrived from the sea but took no part in a great legendary battle between the Wandjinas and the native people.

One suggestion has been that these kaiara figures represent pearl fishermen, but another more likely explanation is that they are Dutch seamen. The Dutch captain Abel Tasman, sailed along this north-western coast of Australia with three ships on a reconnaissance voyage in 1644. One of the kaiara paintings depicts a row of figures dressed in clothes not unlike those of the Napier Broome Bay Wandjinas, but with what seem to be pipes in their mouths. Aboriginal tradition refers to these prominent objects as 'lily roots', but the resemblance to tobacco pipes is unmistakable: so much so, indeed, that one can even see in this naive painting the same elegant, long-stemmed tobacco pipes recovered from early Dutch shipwrecks on the west Australian coast. The log of Tasman's voyage has been lost, so we have no knowledge of where on the coast he might have called to replenish his fresh water, but Bigge Island would be a likely place.[21]

If the Bigge Island Aborigines were moved to portray pearlers or seventeenth century Dutch visitors in their rock paintings, then why would the people of Napier Broome Bay not have had a similar reaction to foreigners? There is nothing to associate the Napier Broome Bay Wandjinas with the Dutch voyage, so if the two strange Napier Broom Bay Wandjinas are indeed paintings of foreign intruders the foreigners would have to be Portuguese. And not just any Portuguese. It is food for thought that these Wandjina figures might actually represent soldiers of Diogo Pacheco's expedition who were killed on the beach at Sir Graham Moore Island during that fatal encounter in 1520.

There is one other puzzling, unexplained fact about Napier Broome Bay. When the first Christian missionaries arrived at the bay in the early 1900s they set up their mission at a place which the Aborigines called Pago. Where have we come across this name before? It was, of course, the name of Sultan Mahmud Shah's citadel on the Malay peninsula which the Portuguese captured and burned in 1520—the very year in which Diogo Pacheco, just a few months later, sailed on his fateful expedition to the Island of Gold.

Can this be just coincidence or is there a connection somehow? The Malay city of Pago rose to prominence only after the Portuguese captured the sultan's former capital, Malacca, in 1511. One of the Portuguese chroniclers describes it as having 'sumptuous houses'.[22] But Pago's fame was short-lived. After conquering the city the Portuguese reduced it to ashes and it sank into oblivion, never to be rebuilt. Even its name disappeared. So if there was some link between the Malay city of Pago and the Pago of Napier Broome Bay, that connection could only have occurred at about the time of Diogo Pacheco's voyage.

But what could the link have been? The Pago of Napier Broome Bay had a sheltered harbour, and importantly it offered a reliable supply of fresh water. Could some Malay seamen from Pacheco's ship have been spared after the battle and called this place after their lost city? Or was it perhaps given its name by other Malay visitors? To complicate the question still further, there is also a Portuguese word 'pago' meaning 'pay' or 'wages' but it is difficult to see any connection between this and the Napier Broome Bay locality.

We may recall that according to Barros's account, Pacheco was not only given instructions on how to sail to the Island of Gold—he was also told that Malay traders themselves sometimes risked the dangerous voyage to this mysterious land because of the lure of gold. Was this just a fairy tale? Not long ago it probably would have been dismissed as a fiction, if only because it would imply that Australian Aborigines were capable of engaging in trade—and this would have been considered a preposterous idea.

But times have changed, and such notions are today seen as often having been coloured by prejudice. It is now proven that over more recent centuries Aboriginal tribes in northern Australia traded on a regular basis with Malay voyagers from Macassar in southern Sulawesi—a place which also happened to have close connections with the Portuguese dating from the mid-sixteenth century. Other Malay praus came from as far away as Madura island, off the north coast of Java. These two-masted boats sailed direct from the Lombok Straits to the Kimberley coast, which they called Kai Jawa, an open sea voyage of some 1300 kilometres taking a full month.[23] A journey of this length and difficulty is comparable with the voyage the Sumatran gold traders of Pacheco's day would have had to make to reach Napier Broome Bay.

In comparatively recent times these Malay visitors (the term 'Malay' including people from what is now called Indonesia) came primarily to gather trepang, a prized delicacy on the Chinese market, but they also bartered with the local Aborigines for such items as turtle shell and timber. And there was also a trade in women. Last century, a respected Aboriginal elder from the Cape Voltaire region, south-west of Napier Broome Bay, told how when the 'Malays' came, Aboriginal men would send them women—one each.[24] In return the Aborigines were given trade goods such as iron axes and knives, glass, rice, tobacco, alcohol and cloth. Some intrepid Aborigines even sailed back to Macassar on board the foreigners' ships and there is evidence of intermarriage. Place names of Macassan origin are numerous on the northern Australian coasts, as are Macassan words in Aboriginal vocabularies.[25]

If Aborigines were sophisticated enough to trade with Malay visitors seeking turtle shell, timber and women, then why not with earlier Malay merchants seeking gold? No independent evidence has yet come to light proving that some Kimberley coast tribes actually traded gold for cloth, as the Sumatran account given to Diogo Pacheco suggests—but equally there is no evidence to disprove it. And the fact is that at least fifteen Malay contact sites of varying antiquity have now been positively identified on the north-west coast of the Kimberley. At Tamarinda, on the opposite shore of the bay on which Pago is located, archaeological finds have included pottery, cooking hearths, coins, musket balls and fish hooks.[26] According to the respected Australian archaeologist Ian Crawford, the number and size of Malay/Indonesian sites on the Kimberley coast suggest 'very extensive contacts over the past several hundred years'.

Sir Graham Moore Island, strategically situated at the entrance to the Napier Broome Bay, is today recognised by archaeologists as having been a prime Malay contact point. Unlike most Aboriginal tribes, whose seagoing craft consisted only of rafts or frail bark canoes, the Juarinanda people who formerly lived on Sir Graham Moore Island possessed large dugout canoes. One of these canoes has been preserved in the Western Australian Museum. Its seaworthiness had been enhanced by a splashboard sewn to the hull with rope—a technique suggestive of Malay and Arab shipbuilding.

And it has also been established that the Juarinanda spoke their own unique 'trade' language.[27] Curiously enough, in the local Aboriginal language of Napier Broome Bay the word 'pago' nowadays can mean either the place, Pago, or a women's fighting stick. Which came first—whether the fighting stick was named after the place, or vice versa—is not known today. There is, however, one possible clue which could be a link to the sixteenth century battle on the beach: a European missionary who has lived in the area for many years sees a resemblance between the shape of the women's fighting stick and that of the early European firearm, the arquebus.[28]

As for Diogo Pacheco, ironically we can now say with certainty that he was on the right track, for in 1883 the rugged Kimberley region became the site of Western Australia's first gold rush. Most of this gold was in alluvial deposits which could be scooped up from river beds, and small quantities are still panned even today. From this we may perhaps surmise that in Pacheco's time the local Aborigines, with their intimate knowledge of the land and its waters, may not have found it all that difficult to obtain alluvial gold for trade with Malay seafarers, just as the Portuguese adventurer had been told.

Diogo Pacheco has a rightful claim to the distinction of being the first European ever to set foot on Australian soil. But although his expedition adds an intriguing new chapter to Australian history, it was of little value in furthering Portugal's search for the Island of Gold—for the simple reason that he was fated never to return. And from that fact it also follows that he could have made no contribution to the Portuguese mapping of Australia's coasts.

That task, however, would soon fall to others. Pacheco's voyage was merely a prelude.

3

Braving the Coast of Danger

Between us and destruction was only a dismal Valley, the breadth of one wave. Yet in this truly Terrible Situation not one man ceased to do his utmost and that with as much calm as if no danger had been here. All the dangers we had escaped were little in comparison to being thrown upon this Reef, where the Ship must be dashed to pieces in a Moment. A Reef such as is here spoke of is scarcely known in Europe. It is a wall of Coral Rock rising almost perpendicular out of the unfathomable Ocean.

Captain James Cook, describing the *Endeavour's* hair's-breadth escape from disaster off Australia's Great Barrier Reef, 16 August 1770

Even as Diogo Pacheco was making the first Portuguese circumnavigation of Sumatra in 1519, King Manuel I in Lisbon had decided that it was high time to mount a major expedition in search of the Island of Gold. The king's decision was prompted by some alarming news: one of his own captains had defected to his rival, King Carlos (Charles I) of Spain, and was preparing to lead a Spanish fleet to the Spice Islands by sailing westward by way of the Americas. The renegade was Ferdinand Magellan and by the beginning of 1519 Magellan's plans were well advanced. A few months earlier King Carlos had drawn up a formal agreement appointing him captain-general of an expedition of five ships and undertaking to make him governor of all the lands he might discover.

For King Manuel this development posed a dilemma. He himself was in large measure responsible for Magellan's defection, having scornfully refused the captain's request for an increase in his modest stipend in recognition of his years of loyal service during Portugal's military campaigns in the East—during which he had suffered several wounds. Magellan had gained King Carlos's backing for his bold and risky expedition by assuring the King that it would enable him (Magellan) to prove that the Spice Islands lay within Spanish territory as defined in the Treaty of Tordesillas. For Portugal, this would have been a devastating blow. All the immense profits then being amassed from the spice trade would have belonged instead to its fiercest competitor.

And that was not the only concern. Not only had Magellan already sailed under the Portuguese flag as far as the Moluccas; he was also believed to be well aware of the stories concerning the fabled Island of Gold. If the opportunity arose, no doubt he would claim this territory for Spain as well. Portugal already had a firm foothold in the Spice Islands, but King Manuel was uncomfortably aware that his captains still had not even discovered the whereabouts of the Island (or Islands) of Gold, let alone established a presence there. Magellan must be forestalled if the possibility of riches perhaps equalling or greater than those of the Spice Islands were not to be allowed to slip through his fingers.

Thus it was that when in 1519 a great armada of seventeen large ships sailed from Lisbon for the East, the king's most trusted captain, Pedro Eanes, carried with him important sealed orders for the Governor of India, Diogo Lopes de Sequeira. Eanes' ship reached Goa in June 1520. When the orders were opened, one of them informed the governor that one of the captains in this fleet 'in whom he [the King] has trust' had been appointed by royal command to lead a voyage of discovery to the 'Islands of Gold'. These islands, the royal order continued, lay 'beyond the island of Sumatra...many persons who have travelled in those parts of India give great hopes that they can be discovered'.[1]

The captain commissioned by the king for this most important task was Christopher de Mendonça. The Portuguese records tell us very little about Mendonça as a person. We have not a single portrait of him. All we know of his background is that he was a young nobleman, son of Pedro de Mendonça (pronounced 'Mendonsa') who was the *Alcaide-mor*, or Governor, of Mourão—one of a chain of fortified towns guarding the often turbulent Portuguese frontier with Spain. (The name '*Alcaide*' betrays its Moorish origins: although the Portuguese word is written in the Roman alphabet, it is pronounced exactly the same as the Arabic *al-Qa'ida*, made infamous in the twenty-first century by Osama bin Laden. The word is difficult to translate exactly into English. The best equivalent is perhaps 'the Organisation'—in this case, the Organisation of the King's Nobles entrusted with the defence of the Realm.)

Pedro de Mendonça was himself a captain who had served the King with distinction. In 1504 he commanded a carrack in a fleet of twelve such ships which sailed to India to secure the Indian Ocean trade routes and bring to heel the rebellious ruler of Calicut.

Mourão, the family's ancestral home, is today a modest little town of narrow streets lined with orange trees and tiny white houses notable for their odd-looking chimneys resembling giant pepper pots. The fourteenth century castle where Mendonça would have grown up is still there on the hilltop, looking out across the River Guadiana which for part of its winding course forms the frontier with Spain. King Manuel had the castle's fortifications strengthened in 1510, fearing a possible attack. The towering walls, flanked by ancient cannons, still look as strong as ever, but the buildings inside have fallen into ruin. A reminder of the Mendonça era is a plaque on the wall of a square just outside the castle walls which proclaims it to be the 'Square of Governor Furtado de Mendonça'.

As for the expedition Christopher de Mendonça had been chosen to lead, this was clearly regarded by King Manuel as being of the highest importance, for he was allocated a fleet of no less than four ships for the voyage—the same number that Vasco da Gama had set out with on his epoch-making expedition to India in 1497. Most commentators seem to have assumed that Mendonça's fleet consisted solely of caravels, but this is a misconception. Not only is there no evidence to support such an assumption, the fact is that while the nimble caravel had proved its worth in coastal exploration, it was far from ideal on its own for long voyages. As we have already seen, its shallow draught meant that its cargo-carrying capacity was severely limited, as was the number and size of guns it could carry.

The larger nau, or carrack, on the other hand had shown itself well suited for major expeditions of this kind. Da Gama had used a nau as his flagship. Its capacious hold could accommodate an ample supply of provisions and trade goods, and it could carry a formidable array of cannons. And on this point we have to remember that Mendonça would have been fully expecting to find himself at some stage in a naval confrontation with Magellan's fleet, where heavy guns would play a decisive role.

The Portuguese chronicles indicate that on his outward voyage from Lisbon, Mendonça was in command of a carrack—and research for this book has confirmed this fact graphically for the first time. The confirmation is to be found in a little-known painting of the 1519 armada which appeared a few years later in a Portuguese manuscript called *The Book of Lisuarte de Abreu*. This painting, executed at Goa in ink and watercolour, depicts all seventeen ships of the great armada under full sail in three lines abreast, together with the name of each ship's commander. (See illustration between pages 196 and 197). A strip of parchment has evidently been cut off the left-hand edge of the picture (now in the collection of New York's Morgan Library) but fortunately enough has survived to make out the name…*stovão de Medosa* above the large ship shown at the left-hand end of the middle line. Given the Old Portuguese form of spelling used (for instance, the letter 's' instead of 'ç'), this is undoubtedly Christopher de Mendonça.

The painting is roughly done, and unfortunately the name of Mendonça's ship is not given. It is nevertheless shown unmistakably as a three-masted carrack, similar in size and appearance to most of the other vessels in the fleet. It has a high, beaked forecastle and towering poop deck, with rows of gun ports on its sides. The carrack's great billowing sails are emblazoned with the red cross of the Order of Christ. The apparently almost identical carrack of Pedro Eanes is depicted in the bottom right-hand corner of the same painting.

How these two ships would have looked in real life is better shown in a mid-sixteenth century Flemish oil painting in the collection of Britain's National Maritime Museum. This large artwork, 'Portuguese Carracks off a Rocky Coast', depicts the carrack *Santa Catarina de Monte Sinai* majestically under full sail in the midst of other vessels in 1521 while bearing King Manuel's daughter, Princess Beatriz, to France for her marriage to Charles III, Duke of Savoy. (See colour plate following page 180.)

A state-of-the art vessel like this—perhaps the very one pictured in the cruder de Abreu painting—would have been the obvious flagship to lead Mendonça's Island of Gold expedition. The huge sail area carried by vessels of this class allowed them to achieve surprisingly fast speeds when running before a fair wind. Vasco da Gama's ships are said to have been able to cruise at between six and eight knots in these conditions. Given that Mendonça's ship would have been the product of twenty years' technological development since the prototype carracks of da Gama, we might reasonably expect that Mendonça's flagship would have been capable of at least eight knots with a favourable wind. This, by way of comparison, would have been about twice the speed of Cook's *Endeavour*—a converted North Sea collier described as 'slow and sturdy'.

The great Portuguese armada of 1519, including the carrack commanded by Mendonça, had sailed from Lisbon in April—well ahead of Magellan, who was not ready to put to sea until nearly the end of September. Magellan's late departure was a major reason why he failed to reach the southern tip of the Americas before the onset of autumn storms and was forced to winter over on the Patagonian coast. It was not until the end of November 1520 that he was able finally to break out into the Pacific after discovering the strait between Patagonia and Tierra del Fuego which today bears his name.

But in the Portuguese Far East the authorities would have had no knowledge of this delay. When the news of the impending Spanish expedition reached Goa and Malacca from Lisbon something close to panic ensued. Preparations were begun immediately for the construction of forts at strategic points in the Portuguese sphere of influence. Of particular importance was the protection of Ternate, the chief Portuguese trading post for the lucrative Spice Islands trade. Preparations for building a fort at this far-off spot were given the highest priority but the fort was not actually completed until 1523, two years after the dramatic arrival at the nearby island of Tidore of the two battered, surviving ships of Magellan's fleet.

Assuming that all of the ships of Mendonça's armada reached Goa at about the same time as Eanes's carrack, the dates for the voyage indicate that they must have rounded the Cape of Good Hope too late to catch the favourable south-west monsoon winds. These blow from the vicinity of Madagascar toward India between April and September. After September the wind direction reverses.

One of the fastest passages to Goa recorded in the early part of the sixteenth century was in 1524, when a fleet sailed from Lisbon on 9 April. Favoured by particularly good following winds, the ships reached Mozambique four months later on 14 August 1524 and pressed on immediately to Goa, where they arrived at the end of September, just before the monsoon winds were due to change. But that was exceptional. The fleet with which Mendonça was sailing spent several months at Mozambique awaiting the next south-west monsoon. Mozambique (in Portuguese Moçambique) was in those days just a port, though eventually it would give its name to the country. It was a place where much-needed provisions could be obtained—and here, too, slaves could be bought cheaply. No women were permitted to embark at Lisbon on the India-bound ships, so it would hardly seem surprising that after being cooped up for months in crowded quarters with no female company, many of those on board who had money to spend, eagerly seized the opportunity of buying one or more slave girls for their personal use. This, though, is a matter about which the official chronicles maintain a discreet silence.

Mendonça's fleet must have sailed again with the first winds of the summer monsoon in order to have reached Goa in June 1520. 'Golden Goa', as it soon came to be called, was rapidly becoming a major naval base as well as a prosperous trading centre. Slaves were bought and sold like animals. Accounts relate how both male and female slaves were stripped naked on the auction block in the Rua Direita to exhibit their good points. To prevent trickery, girls offered as virgins were examined by women before they were purchased by their Portuguese masters. New European arrivals at Goa soon discovered that in this cosmopolitan city, sexual licence prevailed. Visitors reported that many men kept entourages of female slaves, known as 'mosses' (a term referring to their Mozambique origin), who were kept busy all day 'caressing and massaging' the bodies of their masters. As a scandalised Italian Jesuit missionary was to write to his superior, St Ignatius Loyola in Rome some years later:

> The sin of licentiousness is so widespread in these regions that no check is placed upon it...the Portuguese have adopted the vices and customs of the land without reserve, including this evil custom of buying droves of slaves, male and female, just as if they were sheep, large and small. There are countless men who buy droves of girls and sleep with all of them, and subsequently sell them. There are innumerable married settlers who have four, eight or ten female slaves and sleep with all of them, as is common knowledge...

Mendonça, however, had important business to attend to in Goa. He would doubtless have been given a confidential briefing by Governor Lopes de Sequeira concerning the secret, sealed orders handed over by Pedro Eanes. It was probably also at Goa that Mendonça would have been provided with two of the additional ships required for his Island of Gold expedition.

As for these other vessels, an exact description of them has only in the past year or two come to light in Portuguese archives. Mendonça's second ship was a caravel and the third a bargantim (sometimes written bergantim: in English 'brigantine'). This is the same type of vessel which, as we saw in the previous chapter, was used by Diogo Pacheco on his fateful voyage to the Island of Gold.

The source for this information is a Lisbon historian, José Alberto Leitão Barata, who discovered references to Mendonça's 'Ilha do Ouro' (Island of Gold) expedition in official Portuguese archives originating from Mozambique. And interestingly, these records reveal that the following year, at the south Indian port of Cochin, a fourth ship was added to Mendonça's fleet. This vessel is referred to in the archives as a 'parau'—a widely used South Asian and Malay sailing vessel better-known in English as a prau, but also going under the names of 'proa' and 'parahu'. This particular prau is described as being the private property of another Portuguese captain already in the East, Gonçalo Homem. In requisitioning it for the Island of Gold expedition, the Portuguese Governor of India, Diogo Lopes de Sequeira, ordered that it be refitted at the King's expense to take part in what was described as this 'important voyage of discovery'. The records indicate that Homem (about whom little else is known) captained his ship in person when it sailed on the Island of Gold expedition.[2]

Homem's swift and manoeuvrable prau would have been a valuable addition to the Portuguese fleet for close inshore exploration, in company with the brigantine. But is seems not to have been part of the fleet initially.

The brigantine, unlike the prau, was a vessel of European design, adapted for use as a small trading vessel in Asian waters. Usually built in the East to Portuguese specifications, brigantines normally carried two masts—the foremast rigged with a square sail and the mainmast with a triangular lateen sail. Brigantines varied in size. The smaller versions were fitted with banks of oars so that they could be rowed when there was no wind, whereas the larger ones were big enough to carry anything from twelve to twenty guns. And very importantly, as we have seen in the case of the Pacheco expedition, they had a shallow draught and a flat or nearly flat bottom—a feature which will acquire unexpected significance later in the story of Mendonça's voyage.

From Goa, or Cochin, a friendly port further south, it would have been a relatively straightforward voyage across the Indian Ocean to the strategic port of Malacca, on the Malay peninsula, which had been in Portuguese hands since 1511. When he reached Malacca, though, Mendonça found that it was not only the possible danger posed by Magellan's expedition that was causing concern among his compatriots. Hostile Asian rulers were proving to be a constant threat and none more so than the fiercely independent Muslim sultanate of Achem (today's Aceh), in northern Sumatra, against whom the Portuguese were in a state of war.

The commander at Malacca, António de Brito, had decided that his most urgent task was to complete the building of a fort at Pedir, the harbour of a friendly state in Sumatra which bordered Achem. If the official Portuguese chronicles are to be believed, in spite of King Manuel's express command that Mendonça's should search for the Island (or Islands) of Gold, de Brito ordered the captain and his fleet to make a detour to Pedir (the chronicler Barros says that at this stage Mendonça had three ships—which also indicates that the prau was added to the fleet later). Mendonça's voyage to Pedir could have been a useful first stage in the quest for the land of gold, but if the chroniclers are to be believed, after engaging and defeating some Achinese galleys his ships' crews were detailed to help with construction of the fort. This work is said to have taken so long that by the time it was completed, in the early part of 1521, Mendonça had missed the favourable monsoon winds which would have carried his flotilla toward the south-east, so that he had no alternative but to take his ships back to Malacca. Here he found de Brito about to sail for the Moluccas to build the Ternate fortress.

Where did Mendonça go next? The chronicler Barros indicates that Mendonça's voyage of discovery to the Island of Gold is something to which he will refer again later but puzzlingly he fails to do so. Of Mendonça's activities over the next six years there is not a word, either here or in the other Portuguese chronicles. He re-emerges from the shadows only in 1527–28, when the chroniclers name him as the captain of a carrack which has just reached Goa from Portugal. His ship, we are told, was one of an original fleet of five, two of which were wrecked on the coast of Madagascar. His mission this time was to assume command of the important Portuguese fortress of Ormuz (Hormuz), at the entrance to the Persian Gulf.[3]

But what had Mendonça been up to in all that time since his return to Malacca from Pedir in Sumatra in early 1521? It is most unusual for the name of a captain of importance to totally disappear from the annals for such a lengthy period. Why does the official chronicler, Barros, suggest that he will relate what happened on the voyage to the Island of Gold—and then tell us nothing?

There is absolutely no reason to believe that Mendonça would have failed to carry on with his voyage. He had, after all, been commissioned for this vitally important task and provided with a fleet of four ships to carry it out by no less an authority than King Manuel, Portugal's supreme ruler. To have wilfully disobeyed the Sovereign's command would hardly have been a sensible course for any captain who wished to retain his head—let alone his reputation. And on the subject of motivation, we might note that before sailing from Lisbon on their historic voyage, da Gama and his men all swore an oath that death was the only alternative to success. In Mendonça's case, an act of wilful insubordination most certainly would not have brought the high honour indicated by his promotion to the post of commander at Ormuz. And besides, if the stories about the Land of Gold were even half true, it was a voyage that might well bring him fame and fortune.

So why was its outcome not recorded? The most obvious explanation for the blank in the official chronicle is that Mendonça's voyage produced discoveries which were classified as a state secret—and knowing that, Barros may have felt it necessary to maintain a discreet silence. The alternative possibility is that he did write something, but his account was censored.[4]

Some sceptics bent on dismissing at all costs the notion of a Portuguese discovery of Australia have disputed whether there really was an official policy of secrecy—claiming that this is merely a self-serving device to justify the fact that no first-hand record of a voyage to the Island of Gold has come to light. In fact, nothing could be further from the truth. Portugal's obsession with secrecy on such things as charts and voyages of discovery is attested to by all authoritative European historians. The mandatory punishment for revealing such state secrets was death—and it was a policy that was rigorously enforced.

One example will illustrate the fate of those bold or foolish enough to defy the king's command. Toward the end of the fifteenth century a Portuguese ship's pilot and two sailors defected to Spain after an exploratory voyage to West Africa. Fearful that they would 'disserve the king' by revealing navigational secrets, the Portuguese secret service had them followed. The two sailors were summarily beheaded, while the pilot was brought back to Lisbon 'with hooks in his mouth' to be executed. His body was then hacked into quarters, one of which was displayed at each of the city's four gates to discourage any other intending turncoats.[5]

One outcome of this rigorously enforced cult of secrecy was that not a single book revealing major Portuguese discoveries of any kind was published in Portugal during the whole of the first half of the sixteenth century. This was in stark contrast to the relative openness about such matters in Spain. As one historian has observed, had Magellan sailed in the service of his homeland rather than that of Spain, even his great voyage around the world might have been lost to history.[6]

And that brings us to another cogent reason why the details of Mendonça's voyage had to be kept secret. In recent times, evidence has been found in the Portuguese archives that when de Brito sailed for the Moluccas in 1521 he took with him secret royal orders declaring that 'the King will defend what has been discovered in his name'. If he should encounter Spaniards in the Moluccas, de Brito was ordered, he should 'force them to submit, annihilate them and burn their ships so they would never return to Spain to tell the tale'.

In the words of the discoverer of this secret document, the respected twentieth century Portuguese historian Armando Cortesão, this draconian order leaves no doubt that the King would have been equally determined not to let word get abroad about any discovery of the famous Island of Gold, of whose reality he had finally been convinced.[7] Mendonça in all probability would have carried secret instructions similar to those of de Brito, ordering that if he fell in with the traitor Magellan and his Spaniards on the coast of the Island of Gold he was to take ruthless measures to eliminate them.

Finally, the most powerful reason of all for secrecy in the case of Mendonça's expedition would have been the fact that once he had returned and reported its remarkable outcome to Governor Lopes de Sequeira it would have been perfectly clear to everyone that at least the eastern part of his voyage had intruded far into territory which under the Pope's ruling officially belonged to Spain. This would have been a most serious infringement of the treaty of demarcation—one which, had it become generally known, could even have led to war.

If the official chroniclers are to be believed, Mendonça's expedition to the Island of Gold, planned to depart in late 1520, had been delayed until it was too late in the season to set out. As we will see later, there are reasons for doubting this claim. But in any case the evidence points to Mendonça having sailed back to India in 1521, at least as far as Cochin where we know that Gonçalo Homem's refurbished prau was added to his fleet. There must have been a good reason for this late addition. Could it perhaps be that contrary to the story the official chroniclers tell, Mendonça had already made a secret voyage to the west coast of the Island of Gold? And that he had come back convinced of the need for a small, highly manoeuvrable vessel for close inshore exploration on these dangerous, unknown coasts?

At all events, waiting in the port of Cochin for favourable winds as the year 1521 progressed, we can imagine that the commander would have been carefully readying his fleet for the long and hazardous new voyage that lay ahead.

If other major Portuguese voyages of discovery in this period are anything to go by, the expedition would have been well equipped to cope with whatever might eventuate. The voyage of da Gama provides an instructive precedent. As with Mendonça, da Gama's expedition was instigated personally by King Manuel and financed by the royal treasury—and nothing was skimped on or left to chance. Da Gama's ships carried food to last for three years. Daily rations for each man were as generous as they were varied. They included (in contemporary English terms) a pound of salt beef or half a pound of pork, one and a half pounds of biscuit, one and a half pints of wine and an allocation of flour, salt, oil and vinegar—the latter consumed with dried fish. Other stores included dried sardines, prunes, raisins, almonds, cheese, garlic, beans, lentils, mustard, sugar and honey. This reads like the menu for a royal banquet when compared with the standard rations for British men o'war of a later century, described by one prominent naval surgeon as comprising 'putrid beef, rancid pork, mouldy biscuits and flour'.

The more exotic items on the Portuguese ships' menus, it must be added, were intended not just for the officers' table but for the ordinary sailors as well. Portuguese crewmen were as a rule also expert fishermen, with the result—as we shall see when we come to retrace Mendonça's voyage—that they were ready to supplement their provisions with whatever fresh bounty the sea might offer. So if Da Gama's expedition can be considered a model, Mendonça's ships would also have carried a surgeon, drugs for the sick and a priest for confession and the conversion of heathens.

There is no record of the weapons Mendonça's fleet carried, but his flagship at the very least would have been armed with heavy cannons, probably twenty or more, bearing in mind the possibility of a naval confrontation with the westward-bound Spanish ships of Magellan's *Armada de Molucca*. As for smaller weapons, we can get some idea of what would most probably have been provided from a list of the arms Magellan had when his fleet of five ships set sail from Seville in September 1519. Hand-held weapons included 1000 lances, 125 swords, 60 crossbows, 4320 crossbow bolts and 120 javelins.

However the greatest enemy on long voyages in the sixteenth century (and much later) was very often not hostile natives but scurvy. The Italian adventurer Antonio Pigafetta, who sailed with Magellan and kept a journal of the voyage, wrote that during their three months and twenty days' crossing of the Pacific the crew were reduced to eating:

> ...biscuit which was no longer biscuit, but powder of biscuits swarming with worms...It stank strongly of the urine of rats. We drank yellow water that had been putrid for many days. We also ate some ox hides that covered the top of the mainyard...and which had become exceedingly hard because of the sun, rain and wind. We left them in the sea for four or five days and then placed them for a few moments on top of the embers, and so ate them; and often we ate sawdust from boards. Rats were sold for one half-ducat apiece, and even then we could not get them.
>
> But above all the other misfortunes the following was the worst. The gums of both the lower and upper teeth of some of our men swelled, so that they could not eat under any circumstances and therefore died.[8]

Nineteen of Magellan's men died of scurvy during this time, and another twenty-five or thirty fell sick. Nearly three centuries later François Péron, naturalist with Nicolas Baudin's French expedition to Australia, on board *Le Géographe* when it reached Port Jackson in June 1802, would echo Pigafetta's account. After describing tremendous storms which the ship encountered in Bass Strait, he wrote:

> These dangers were not to be compared with the shocking ravages which the scurvy spread among our people. Several of our men had already been committed to the deep: already more than half of our seamen were incapable of service: only two of our helmsmen were able to get on deck. The daily increase of the epidemic was alarming to an extreme degree. And in fact, how should it be otherwise? Three-quarters of a bottle of stinking water was our daily allowance: for more than a year we had not tasted wine...the biscuit served out to us was full of insects; all our salt provisions were putrid and rotten.[9]

Scurvy we now know is caused by a deficiency of vitamin C, but of course early sea captains had no knowledge of this medical fact. Practical experience nevertheless indicated to the more perceptive among them, even as early as the sixteenth century, that fresh vegetables and citrus fruit seemed to help ward off the dreaded disease. In 1593 the Elizabethan seafarer Sir Richard Hawkins wrote of 'scurvey' that:

> ...it possesseth all those of which it taketh hold with a loathsome sloathfulness, *even to eate*...it bringeth with it a great desire to drink, and causeth a generall swelling of all parts of the body, especially of the legs and gums, and many times the teeth fall out of the jawes without paine...in twentie years since I have used the Sea, I dare take upon me to give accompt of ten thousand men consumed with this disease.

To which Hawkins thoughtfully added: 'That which I have seene most fruitfull for this sicknesse is sower Oranges and Lemmons'.[10]

Scurvy could have been expected to be a problem on the long voyage into unknown seas planned by Mendonça. As to how he may have coped with it, the eminent British historian CR Boxer has noted that 'the importance of fresh provisions in combating scurvy had been vaguely realised from the time of the early Portuguese voyages'. Some other reports go even further, suggesting that in the field of scurvy prevention Portugal was far ahead of other seafaring nations. When in 1601, for instance, the noted English captain Sir James Lancaster set out with four ships on a pioneering voyage to the Spice Islands he warded off scurvy by taking with him 'bottles of the juice of lemons, which he gave to each one, as long as it would last, three spoonfulls every morning'. And where did Sir James get his knowledge of this remedy? In Portugal, where he spent his youth.

In the case of Mendonça's voyage, a further consideration is that as we have seen from the example of da Gama, Portuguese ships sailing under royal patronage tended to be exceptionally well provisioned. If scurvy did affect Mendonça's crew toward the end of their voyage, we can at least say with certainty that the problem was not so serious as to prevent his ship making a safe return.

As for the navigation equipment he would have carried, again we can get some idea from the inventory for Magellan's fleet. Magellan was provided with six compasses (one for each ship and one spare), thirty-six spare compass needles, twenty-one wooden quadrants, seven astrolabes and eighteen hour-glasses—the latter the only instruments then available for measuring the time.

The forerunner of the modern barometer was not invented until 1643. Instead, Portuguese pilots in the sixteenth century employed a curious form of 'live' barometer consisting of a leech swimming in a bottle of spring water. The position of the leech in the bottle on any given day was held to predict the weather. The system was more sophisticated than it might sound. There were no fewer than nine recognised positions for the leech, and navigators memorised a set of nine rhyming verses to remind them of what each position foretold. As a rule of thumb, the finer the impending weather, the lower in the water the leech would swim. Conversely, if the leech climbed high in the bottle it would be a warning of stormy weather on the way.

Most significant of all, though, in the context of Mendonça's voyage and its outcome is the fact that Magellan carried with him twenty-four specially prepared parchment sheets made of preserved animal skin. These were the blank sheets on which all newly discovered coastlines were to be drawn. Mendonça would undoubtedly have carried a store of similar parchments—probably goat skin—for the same purpose, for this was the standard Portuguese method of shipboard cartography.

At first sight it might seem strange that these animal skin portolan charts remained the norm despite the growing use of paper in Europe (the first printed maps had appeared in 1477) but there were good reasons for this apparent anomaly. Portugal and Spain at first had only limited knowledge of the new printing processes, but in the case of Portugal a far more important reason was the official veil of secrecy which was rigorously applied to all discoveries in the East. Portugal's rulers were particularly averse to the idea of printed maps, which they feared would inevitably fall into the hands of political and commercial rivals. And there was also a practical consideration: animal-hide portolans had an important advantage over the new printed paper charts because of their durability under adverse conditions at sea. For all these reasons, portolans continued to be used as the best and most accurate navigational aids in Portuguese ships throughout the sixteenth and even well into the seventeenth century.[11]

The system of command on Portuguese ships of Mendonça's era was different to that of English vessels. As a general rule, the vital qualification for appointment to the command of large ships such as the India-bound carracks was an individual's rank as a nobleman or gentleman. But these members of the upper class might have little practical knowledge of the sea when they set out on their first voyage, so it was the pilot who had total responsibility for the ship's course and navigation.

If we were to look at a comparable situation on board Cook's *Endeavour*, the aristocrat Sir Joseph Banks would have been designated captain, with the commoner, James Cook, having the title of pilot or navigator. But of course in the Elizabethan era the English, too, had their gentleman captains, Sir Walter Raleigh and Sir John Hawkins, to name just two. On the whole, the Portuguese system of command seems to have worked well, for the skills of their best pilots were universally admired. And by no means were all high-born Portuguese captains nautical ignoramuses. Christopher de Mendonça most certainly was not in that category, as the extraordinary achievements of his voyage will show.

We can probably assume Mendonça and his pilot would have had access to the same information as Diogo Pacheco: that is, that the Island of Gold could be reached by sailing south-east from the southern coast of Sumatra. Mendonça, on his voyage to Pedir, had already sailed at least as far as Sumatra's north-western coast—perhaps a good deal further. Rather than go this way again, on his second voyage it seems highly likely he would have taken a more direct route from Malacca, sailing southward through the Malacca Strait past Singapore and then coasting eastward through the Java Sea to about the longitude of Flores or Timor.

This would have made sound practical sense, since his main aim now was to explore the eastern regions of the Island of Gold. And here we might also recall that one of his chief objectives was to forestall Magellan and his Spanish fleet, which he would have expected to be approaching from the east.

The Java Sea route was by now well-travelled by Portuguese ships heading for the Spice Islands. For Mendonça it would have offered two important advantages over Pacheco's route. Firstly, it would have avoided the weeks of extra sailing needed to double back from Malacca around the northern tip of Sumatra in order to reach that island's south-western coast before striking out for the Island of Gold. And secondly, Portuguese captains had already found that the monsoon winds from India enabled them to sail from Malacca as early as February or March to load cargoes of spices in Java.

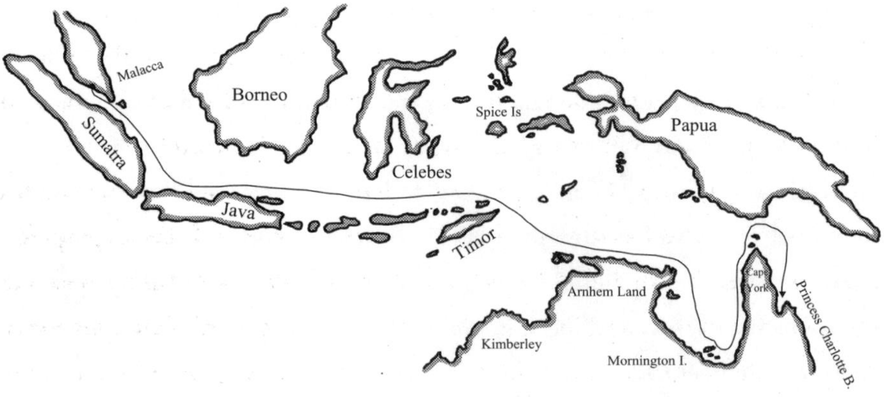

Mendonça's probable route from Malacca to Torres Strait and Princess Charlotte Bay.

We know for a fact that Mendonça and his fleet—by now enlarged to four ships—were back at Malacca on 10 January 1522. This key piece of information has come to light only recently, another chance outcome of the archival research in Lisbon by historian José Barata. After reprovisioning in Malacca, Mendonça would have been ideally positioned for an early departure in the direction of Java, taking advantage of the favourable monsoon winds. Assuming this route was the one his ships followed, they almost certainly would have called in at one or another Javanese port in order to trade, top up their provisions and take on fresh water. Sailing still further east, Mendonça would have needed only to steer south when he reached Timor to be sure of making landfall on the coast of the Island of Gold, either at the Kimberley region or near the large islands of Bathurst and Melville.

Where might he have gone next? Here our best course might be to turn to the Vallard map of eastern Terra Java to look for clues.

To make sense of this map, the first thing we have to do is invert it because, as we have already seen, this cartographer always placed south at the top of his page. Once we turn the Vallard map of Terra Java upside down things become a lot clearer. One point we can make about this enigmatic land is that in outline the north-eastern section of its coast now has an unmistakable resemblance to a large part of the east coast of Australia—or more specifically, to the entire coast of Queensland and northern section of the coast of New South Wales. The south-eastern section of the Vallard chart's coast is however a different matter and we will examine this in detail later.

In the main map of Terra Java we cannot see this land's northernmost coastline because it has been cut off by the map's border. Of the geographical features in the far north that are depicted, a peninsula whose tip has been cut off at the top of the page, is a tolerable likeness of Cape York Peninsula, were it not for the fact that immediately east of it are two large islands neither of which exists in this location. The outermost of these is a wedge-shaped island with the name *Tmor*. This is surely supposed to be Timor—which actually lies off Australia's *west* coast, not the east coast! What can it possibly be doing in this out-of-the-way place?

The answer to the puzzle becomes clearer when we turn to an adjoining map in the Vallard Atlas, that of South-East Asia. The northernmost portions ofTerra Java excised from the map we have just been looking at are all depicted in full just above the southern edge of the South-East Asia map, and we can now see plainly that this entire north coast is not the coastline of Australia at all, but rather the north coast of Java and two neighbouring islands to the east of it, Sumbawa and Flores. There can be no doubt at all about this: not only is the main coastline on the map recognisable as the actual coast of Java; many of the place names written on it are readily identifiable—for example *Japarra* (modern-day Japara), *Tubany* (Tuban), *Curibaia* (Surabaya), *Canal Do Cunda* (Sunda Strait) and *C.* [Cape] *da Frolles* (Flores), to name just a few.

For reasons which we will explore in more detail later, all of the Dieppe cartographers, including the author of the Vallard maps, were convinced that the great landmass of Terra Java was a southern extension of these Indonesian islands. And there are in fact some superficial similarities between the coastlines of northern Java and Australia's Arnhem Land. If the Dieppe cartographers had come into possession of Mendonça's charts they might well have decided that his portolan depicting the region of today's Arnhem Land was merely an inferior version of the already well-mapped Java coast—and therefore discarded it.

The Vallard map of South-East Asia shows both Java and Sumbawa absurdly elongated toward the south, extending right to the margin of the page. We can be sure that the second protruding landmass is supposed to be Sumbawa because one of the place names on it is *Simbana* and another is *c: apee (Cap Apee)*—clearly a reference to the prominent volcano, Mount Api. If we now turn to the Terra Java map we see that the extended 'islands' of Java and Sumbawa join together at the southern end of a deep gulf, on the western shore of which is a realistic drawing of a turtle (interestingly, the turtle is not shown on the corresponding part of the South-East Asian map). A palm tree and a group of seminaked black men, some with spears and one riding what might be a camel, can be seen trekking across the interior of the map's Java. Flores, the third large Indonesian island depicted on both charts, is also shown extending far to the south, but unlike Java and Sumbawa it is not joined to the Terra Java continent—evidently because there was no vacant part of the Terra Java coast remaining to which it could be fitted. Instead, a narrow strait has been left at its southern tip. Depicted further east again is the island of Timor, hopelessly adrift in the Coral Sea.

The Vallard cartographer had apparently concluded that in order to reach the east coast of Terra Java from the Malacca Strait a ship would have to sail along the north coast of the Indonesian island chain and then turn south through the strait between Sumbawa and Flores. As it happens, this may have been more or less the route that Mendonça actually *did* take—but of course he would then have found himself not at the coast of eastern Australia, but either near the Kimberley coast or, more likely, off the coast of Arnhem Land.

There is intriguing evidence to support the idea that the cartographer did in fact have another chart showing the north coast of Australia which he discarded. Near the top left-hand corner of the Terra Java map, Bali and Lombok, the two smaller islands between Java and Sumbawa, are correctly placed (though on a reduced scale), but to the south of them is a wide gulf ending in solid land. We know there is no land or gulf south of Bali, so what can this waterway represent?

A clue is provided by the large island shown near the gulf's lower end. Unlike Bali and Lombok, which have the names *C: Arianj* (probably 'Cape Orange') and *Illa Bona* (Good Island), the island near the bottom of the gulf has no name. Comparison with a present-day map shows that both in size and shape it has a very clear resemblance to Mornington Island, the large island lying in the lower reaches of northern Australia's Gulf of Carpentaria. The gulf depicted in the Vallard map is much narrower than the Gulf of Carpentaria. This discrepancy could partly be due to the problem of measuring longitude, but it could also be viewed as a deliberate attempt by the Vallard cartographer to make the gulf's entrance fit the much smaller sea gap between Java and Sumbawa.

The Terra Java map also depicts two small islands lying a little further south of Mornington Island. The nearest of these would correspond to

Bentinck Island, with the other perhaps Sweers Island—where in 1802 the British navigator Matthew Flinders on his exploration of the Gulf of Carpentaria found part of the quarterdeck of a wrecked ship and near it gruesome evidence of a cannibal feast, in the form of seven human skulls and piles of bones lying in the remains of old fireplaces.

Could this grisly scene have anything to do with Mendonça's expedition? It seems unlikely, if only because there would have been little if anything left of any wooden ship's superstructure after nearly three centuries' exposure to tropical sun and monsoon rain. Flinders concluded that the quarterdeck was the remains of an East Indiaman. Nevertheless, we can reasonably assume that Mendonça's fleet would have come this way. For all they knew, this wide gulf might lead a ship all the way to unexplored southern seas. It had to be investigated.

On the mainland coast below its three unnamed islands, the sixteenth century map shows a huge river emptying into the lower reaches of the gulf. Several large rivers do indeed flow into the bottom of the Gulf of Carpentaria, but none of them end in an estuary of anything like this size. The positioning of the river in the Vallard map suggests that it could be an attempt to depict the Leichhardt River, which flows into the Gulf of Carpentaria at approximately the right location, near a headland called Kangaroo Point. Alternatively, the fact that the waters at the head of the gulf are shallow and dangerous for sailing vessels might perhaps more realistically lead to a conclusion that the estuary in the Terra Java map is largely a product of guesswork, perhaps a composite of three or four actual river mouths, sketched from the deck of a ship keeping a safe distance well out at sea.

From here the Portuguese fleet would have had to swing north to follow the coast of Cape York Peninsula. The inshore waters here are also shallow, and the coast mainly low-lying and somewhat featureless—facts fully consistent with the coastline depicted on the sixteenth century map. The map also indicates that as they turned north the Portuguese lookouts sighted a large cape near the south-eastern end of the Gulf of Carpentaria. This is evidently Point

Burrows the most prominent headland in this part of the gulf. In the Vallard map it is depicted as a large unnamed cape, and it appears in exactly the right position—almost due east from the southern coast of Mornington Island.

Proceeding further north along the coast of the peninsula Mendonça's fleet would have been travelling in the opposite direction to that taken nearly a century later by the first Dutch ship to explore the Cape York coast, the *Duyfken*. The most southerly point reached by the Dutch expedition was another headland some 400 kilometres north of Point Burrows which the Dutch captain, Willem Jansz, named Cape Keerweer (Cape Turnabout). This may well be a small cape which appears on the Vallard map at about the right place—but it is difficult to be sure, because it is at about this point that the northern end of the Cape York peninsula has been excised and replaced by the east and north coasts of Sumbawa.

Jansz's chart of the Cape York coast and nearby islands, made in 1606, is sketchy and incomplete, testifying to the fact that the Dutch skipper was sailing cautiously and taking no risks—hardly surprising in view of the fact that he had only one small ship and that eight of his crew had earlier been ambushed and killed by cannibals on the south coast of New Guinea. Later, while using a ship's boat to explore the Batavia (now Wenlock) River, near the northern end of Cape York, the Dutch had suffered another setback when one of their number was fatally speared by a band of Aborigines.

Jansz (referred to in some accounts as Janszoon) had been hoping to find a sea passage between New Guinea and the shore of what was to become known as New Holland. From documents which came to light years later he seems to have thought he might actually have seen an 'opening' between the islands which lie between New Guinea and Australia, but for some reason he failed to investigate. At Cape Keerweer he gave up the quest, heading back to the nearest Dutch fort at Banda in the Spice Islands.

Only a few months later another European mariner, Luis Vaez de Torres, would sail through what may well have been this very same 'opening' but in the opposite direction to the one Jansz was hoping to take. Torres was in command of the second ship of a Spanish expedition led by Pedro Fernandes de Quiros, whose mission was to find the Island of Gold and claim it for Spain. As with Mendonça, Quiros's sought-after golden land was the legendary site of King Solomon's Mines, and he believed that on a previous voyage under Alvaro de Mendaña he had been on the verge of locating it in the South Pacific at what Mendaña had already named the 'Islands of Solomon' (today's Solomon Islands). Sailing from Peru, Quiros and Torres eventually reached what is now Vanuatu, but the expedition ended chaotically with Quiros surreptitiously setting sail back to South America. Torres, finding himself deserted by his commander, decided to head for the Spanish fortress at Manila.

Significantly, both Quiros and Torres were of Portuguese origin, and there is some reason to suspect that thanks to his Portuguese connections Torres may already have known about the existence of the strait that now bears his name—for on nearing the easternmost point of New Guinea and finding that headwinds from the east were preventing him making for Manila by the usual route around New Guinea's north coast he confidently turned his ship around and headed west along the island's southern shores. This would have been a rash and inexplicable action if he had believed that solid land lay ahead.

Interestingly, Torres himself apparently never made any claim to have been the actual discoverer of the strait. It was not until a century and a half later, in 1762, that he was credited with the discovery, by an Englishman, Alexander Dalrymple, a Fellow of the Royal Society, who came across an archival account of the Torres voyage during the British occupation of Spanish Manila.

If Torres had heard of the existence of a strait between New Guinea and Australia, where could that knowledge have come from? The answer is fairly obvious. By Torres's day Portugal had fallen under Spanish rule, so the jealously guarded Portuguese maritime archives would have been open to Spanish investigators. Or alternatively, knowledge of a strait might simply have been a valuable piece of information that was passed confidentially among Portuguese captains by word of mouth.

Yet more compelling evidence that the existence of a strait was known well before Torres sailed through it is to be found in four world maps produced by Flemish cartographers in the latter part of the sixteenth century. The first of these was published by Abraham Ortelius in Antwerp in 1570, the second by Rumold Mercator in 1587, and the last two by Cornelius Wytfliet in 1597 and Cornelius Doetszoon a year later. All four maps clearly depict a sea passage lying between New Guinea and their equivalent of the northern tip of Australia.

From an Australian viewpoint Cornelius Doetszoon's map is of particular interest. Doetszoon is the least well-known of the Flemish cartographic quartet. His original map is preserved in the National Museum of Japan, in Tokyo. A rare copy (none of his works appear in Australian catalogues) can be seen on display in Manila, in the National Museum of the Filipino People, where it accompanies an impressive exhibition of artefacts recovered from the wreck of a Spanish galleon, the *San Diego*, sunk during a battle with two Dutch ships on 14 December 1600.

Doetszoon's map is thought to have been used by the Dutch captains as a sea chart for their attack on Manila. It was intended primarily as a chart of the East Indies, but it also depicts unmistakably the Kimberley coast of Australia, Arnhem Land and the Gulf of Carpentaria—the latter in considerable detail. As with the other Flemish maps, Arnhem Land has been given the name *Beach* (echoing Marco Polo), while the Kimberley coast south-west of it is called *Lucach*. Here, a prominent feature in the Doetszoon map which does not appear in the other three is the distinctive shape of the wide Joseph Bonaparte Gulf.

In the middle of all four maps' equivalent of the Gulf of Carpentaria there appears a very large, roundish island with the name Java Minor. This is clearly fictitious (the real Java is shown in approximately its correct place as Java Miaior), but there is nothing necessarily fictional about the two smaller islands depicted off the western coast of Doetszoon's gulf. The more northerly of these equates with Groote Eylandt, while the island drawn near the southern end of the gulf fits the location of Mornington Island. Several large rivers are also shown emptying into the head of Doetszoon's gulf—a depiction consistent with the original chart having been drawn toward the end of the monsoon season, when the rivers in the Gulf of Carpentaria would have been in full flood.

Most interesting of all, though, is the map's eastern section. Doetszoon correctly depicts Nova Guinea as an island, with a clearly defined strait separating it from the coast of the land which we now call Australia. To this mainland he has given a name which is at once both familiar and totally unexpected—*Terra Nullius* (Land of No one). And this was a full two centuries before the British were supposed to have invented this term for Australia!

Ortelius and Mercator both seem to have entertained some doubts about the veracity of the information on which they based their depiction of a strait between Nova Guinea and Terra Australis, Mercator musing in a commentary note: 'Can it [New Guinea] be an island, or is it part of Australis?' Wytfliet, however, had no such reservations. In the text accompanying his map he states categorically:

> Terra Australis is the most southern of all lands. *It is separated from New Guinea by a narrow strait*. (Author's italics)

Even Wytfliet's map, the third of the quartet showing a strait, was printed in Europe nine years *before* Torres's alleged discovery! Wytfliet appears to have been privy to some important and surprisingly accurate new information about Terra Australis. He writes:

> It [Terra Australis] is maintained by some to be of so great an extent, that if it were thoroughly explored, it would be regarded as a fifth part of the world... Its shores are hitherto but little known, since after one voyage and another that route has been deserted, and seldom is the country visited unless when sailors are driven there by storms.[12]

'After one voyage and another...' What Wytfliet is clearly telling us is that earlier in the sixteenth century there were two deliberate, planned voyages to *Terra Australis*—as opposed to the accidental, storm-driven visits he also mentions.

All four Flemish maps appeared long before ships from that northern part of Europe first began exploring the seaway between Papua and the north-east shores of Australia. As we have seen, the first Dutch voyage to that region, that of the *Duyfken*, did not take place until 1606. Yet Ortelius's map showing a strait between New Guinea and the northernmost tip of Australia had already been published in Antwerp some thirty-six years previously.

So, what are we to make of Wytfliet's reference to earlier voyages? Its significance seems to have escaped the notice of mainstream academics in Australia and elsewhere, even up to the present time—but not so George Collingridge, the maverick nineteenth century Sydney historian, who was quick to recognise one of its implications. 'It is a known fact,' he wrote, 'that the Dutch appropriated to themselves Portuguese and Spanish documents and charts, which, when altered to serve their purpose, made them appear to be the actual discoverers, whereas, in reality, the countries described in such documents and charts had at the time never been visited by them.' [13]

On all the known evidence, at least one of the voyages of discovery which Wytfliet reports and whose results all four Flemish cartographers record in their maps can scarcely be other than that of the Portuguese expedition of Christopher de Mendonça. And that, of course, would have to mean that Mendonça, not Torres or any Dutch captain, was the true discoverer of Torres Strait.

And the plain fact of the matter is that in order to reach Australia's east coast from the Gulf of Carpentaria, Mendonça would have to have come through the strait depicted by Wytfliet because, quite simply, there was no other feasible way. In theory he could have doubled back and sailed in a huge loop right around the west and north coasts of New Guinea—but only in theory, because in reality such a route would have been so time-consuming and impracticable as to be out of the question.

There is no absolute certainty about the exact route Torres took through the 100-mile-wide strait which bears his name. Some historians believe he exited through the Endeavour Strait, the passage along which Cook sailed in 1770, but this seems questionable, given that his aim was to head west, not south. Similarly there is no way of telling which route Mendonça followed, for the simple reason that the chart his navigator would doubtless have drawn of the Cape York area has been substituted in the Vallard map by the north coast of the Indonesian island of Sumbawa.

Mendonça's fleet would probably have taken something like three months to cover the distance from Malacca to Cape York. By now, with any luck, the north-west monsoon with its dangerous storms and cyclones would have faded away. Winds would have been mainly light and variable, the seas relatively calm, and visibility far better than in the monsoon season.

Approaching Cape York from the south-west Mendonça might perhaps have been tempted to enter the Endeavour Strait, but it seems more likely he continued his course north, skirting the main Torres Strait islands until he was able to turn east into the relatively open waters of Torres Strait proper. From there he would have had to sail well over 100 nautical miles further east before he would be clear of the perilous maze of reefs and islands lying between Cape York and the outer edge of the Great Barrier Reef.

This is an area of shallow seas in which the two smallest ships of his fleet, the brigantine and the prau, would have proved invaluable for their ability to scout ahead and find safe channels. The Portuguese fleet may well have sailed out through what is now known as the Bligh channel—the route taken by William Bligh, captain of the British ship *Bounty*, during his epic 3600-mile open-boat voyage to Timor in 1789 after he and loyal crewmen had been cast adrift by mutineers near the islands of Tonga. Only after leaving the Bligh Entrance behind them would Mendonça's ships have been able to turn south in relative (though by no means total) safety, following the outer line of the barrier reef as it edged gradually closer to the Australian coast.

There is one puzzling feature in the Vallard's Terra Java map which on investigation supports the theory that this is the course the ships did indeed take. That feature is the shoreline of the fictitious southern extension of the island of Flores, shown running more or less parallel to the Cape York peninsula, but at some distance from the peninsula's eastern coast. Intriguingly, detailed examination of the western coastline of this fictitious 'Flores' shows that it closely resembles the *actual* line of the outer Great Barrier Reef all the way from the latitude of Cape York to the major promontory of Cape Melville, some 300 nautical miles further south. Off Cape Melville the gap between the reef edge and the mainland reaches its narrowest point—only about twelve miles, or some twenty kilometres. Just the same thing can be seen at the corresponding place in the Vallard map. The explanation would seem to be that the Vallard cartographer has mistaken a charted outline of the barrier reef for the coast of an island, which he has decided must be Flores, and then sketched in an imaginary east coast to complete the picture.

Assuming that the fictitious 'Flores' west coast does indeed represent the outer edge of the barrier reef—where the reef drops precipitously to the sea floor—this means Mendonça's ships must have been sailing south through the open expanse of the Coral Sea outside the reef, in order to be able to chart its shape and direction accurately. And this would also explain why no recognisable features such as large capes, bays and rivers—of which there are several—are depicted on the east coast of the Vallard Atlas's equivalent of Cape York Peninsula (the northern half of this peninsula has been excised on the Terra Java map, but it appears in full in the accompanying map of South-East Asia). As we have already seen, staying at a safe distance beyond the Great Barrier Reef, Mendonça's ships could initially have been anything up to 160 kilometres from land, but they would have been approaching steadily closer as they sailed south. By the time they neared Cape Melville the mainland coast would in all probability have been only about thirty or forty kilometres away.

Significantly, at this very point the Great Barrier Reef and the western coast of the Vallard map's 'Flores' both bend abruptly toward the east for a short distance. The fictitious 'Flores' west coast ends here. As for the barrier reef, on this bend are two gaps each about three kilometres wide. This is where we might expect Mendonça's ships to have turned toward the mainland. Approaching as they would have done from the north, the beckoning entrance to these passages would have been straight ahead. With its two smaller ships again scouting ahead, the fleet would have been able to sail without any great difficulty through one or other of these openings into the deep, sheltered waters beyond.

As they neared the gap in the reef, Mendonça's men would have seen ahead of them a range of steep, thickly forested hills stretching away southward to the far horizon. This is the northern end of the Great Dividing Range, Australia's mountain backbone which runs for almost the entire length of the continent, ending at the faraway Snowy Mountains, almost within sight of the great Southern Ocean.

The map derived from Mendonça's voyage shows us that the Portuguese expeditioners now found themselves in a wide, north-facing semicircular bay. Depicted flowing into this bay is a large river, which the map calls *Rio Grand*.[14] At its outer eastern end, the semicircular bay is shown ending in twin headlands separated by a smaller bay which is also north-facing. The headland furthest east is the more prominent of these two and has been given the name *Cap Mata*—in Portuguese *Cabo Mato*, or 'Forest Cape', a name which we shall encounter again later.

These are all very distinctive geographical features, and in virtually every detail they fit the area of Cape Melville and Princess Charlotte Bay. The semicircular bay in the Vallard map is almost identical in shape to Princess Charlotte Bay, at the south-east end of Cape York peninsula, while *Rio Grand* (Big River) is equally clearly the North Kennedy River, which rises on the distant foothills of the Great Dividing Range and flows into the sea through a broad, mangrove-fringed estuary over a kilometre wide at the centre of Princess Charlotte Bay—exactly as depicted in the sixteenth century map.

A smaller bay shown immediately further east in the Vallard map corresponds to Bathurst Bay, which terminates at its eastern end in the

The Vallard map of north-eastern Australia compared with a modern map of Australia's Queensland coast.

prominent headland of Cape Melville—Mendonça's 'Forest Cape'. The north-projecting Cape Melville rises to peaks more than 400 metres high. Protected today as a national park, it is clad in rugged native bush, just as the Portuguese name suggests. It probably still looks much as Mendonça and his men would have seen it nearly 500 years ago.

A particularly interesting detail is the depiction in the Terra Java map of twin islands lying just off *Cap Mata* and coloured red. These are in exactly the right position to be the two main islands of the Flinders Group, Flinders Island and Stanley Island, with scattered reefs correctly indicated to the north of them. In the earliest British charts these islands were shown as an extension of the mainland, and it was not until 1821 that Captain (later Admiral) Phillip Parker King discovered the error and named them the Flinders Group.[15]

Aborigines living here could hardly have failed to notice the strange and wondrous Portuguese vessels passing by. It is possible there could in fact even be graphic evidence of them in ancient Aboriginal rock art still surviving in what is known as the 'Ship Shelter' on Stanley Island, which nowadays looks out across the commercial shipping channel to and from Torres Strait. Here are numerous paintings of sailing ships, some with two, three and even four masts, mostly depicted in symbolic form. Archaeologists who have examined the rock art have been unable to date the older paintings, or determine exactly what types of ship they might represent.

We can probably assume that Mendonça's flotilla would have anchored for some days in the agreeable surroundings of Princess Charlotte Bay while the crew replenished the ships' fresh water casks in the 'Rio Grand' and tried their luck at fishing. Whether or not they sighted any Aborigines, they would doubtless have seen the smoke of fires, confirming that this remote, unknown land was indeed inhabited.

Descriptive place names like 'Rio Grand' and 'Cap Mata' are something we will encounter throughout Mendonça's voyage. Rather than calling coastal features after wealthy aristocrats and navy bigwigs, as Cook and Flinders were often wont to do, the Portuguese for the most part fortunately chose

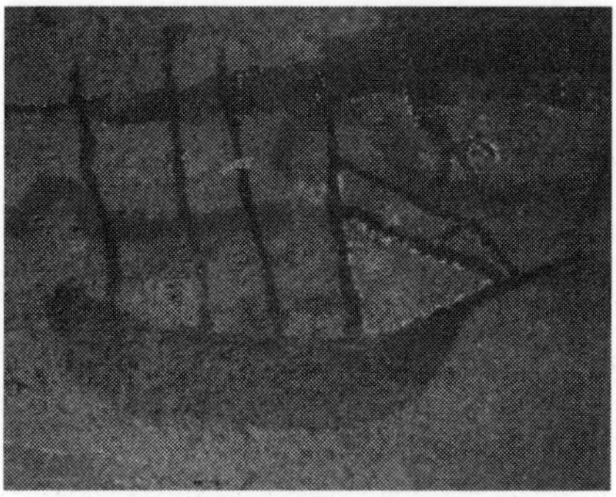

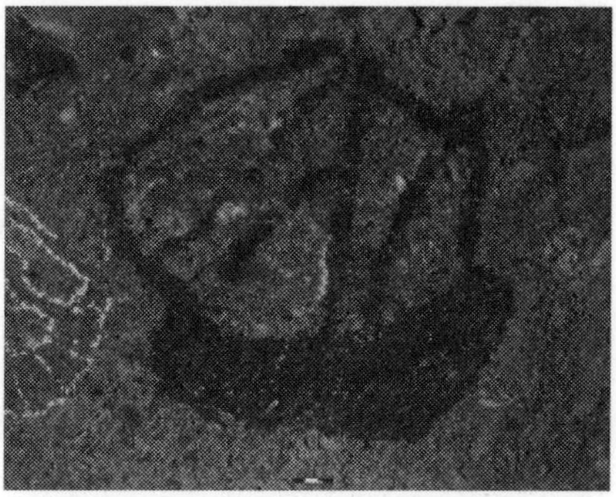

Two of the many Aboriginal ship paintings preserved in the 'Ship Shelter' on Stanley Island, near the entrance to Princess Charlotte Bay, north Queensland. The island looks out over the main shipping channel inside the Great Barrier Reef, along which the Portuguese ships would have had to pass. The age of the older ship paintings in the shelter is unknown. (*Grahame Walsh*)

names which described the things they were actually looking at and this is what makes it possible to identify these places 500 years later.

As for navigation, by now that reassuring celestial beacon, the Pole Star, would have long been out of sight for Mendonça's ships. Night-time navigation would have to rely on other stars and in particular the Southern Cross. For Portuguese captains, though, this would have been a familiar situation, thanks to their years of voyaging to the southern tip of Africa. The first known drawing of the Southern Cross appears in a document written in 1516 by an Italian adventurer, Andrea Corsali, who sailed in a Portuguese ship to Goa, but by then the constellation was already well-known to Portuguese pilots.

Heading south from Cape Melville and evidently now staying inside the Great Barrier Reef, we can trace the progress of Mendonça's flotilla past a coast which he named *Terra Alta*, or 'High Land'. This again is an accurate description. Not only is Cape Melville 400 metres high; the hills beyond it rise to 600 metres, and beyond them again are the still higher peaks of the Great Dividing Range. A little further south is Cape Flattery, rising to 274 metres and named by Captain Cook while the *Endeavour* was inching its way northward through the reef-bound channels of this part of the coast. Cook gave it the name 'Flattery' because he had at first flattered himself that it indicated a passage to the open sea. From its shape, Cape Flattery is readily identifiable in the Vallard map. Mendonça would also have sighted nearby Lookout Point, the high promontory which Cook was to climb with his telescope on 11 August 1770 in the renewed hope of locating a navigable passage.

Beginning on the southern side of Cape Flattery the Vallard chart depicts another highly distinctive coastal feature: a scalloped bay with a hook-shaped headland at its southern end. In the context of a present-day map, this is unmistakably the large bay terminating at Cape Bedford, which has a similar distinctive shape. This bay is so little frequented that even in the latest maps and Admiralty charts it has no recognised name. The Vallard map, however, does give it a name: *G Ferro*—'G' being the standard Vallard abbreviation for 'Gulf'. In Portuguese this would be 'Golfo Ferroso', meaning 'Iron Gulf'. Does this make any sense? In fact the name fits the locality perfectly. Not only do iron ore deposits occur on this coast: further north is a range of hills which is actually called the Iron Range. The name *Golfo Ferro* probably indicates that here Mendonça experienced a magnetic compass deviation.[16]

Just past *Golfo Ferro* the Portuguese lookouts sighted the mouth of a river, which the Vallard map calls *Rio Darena*. This name in modern Portuguese would be *Rio de Areia*, or 'River of Sandbanks' though in sixteenth century Portuguese it could have been written just as it appears in the map. It is a spot Cook knew well, for it is none other than the Endeavour River at Cooktown, where the *Endeavour* was beached on a sandbank after its near-disastrous grounding on a coral reef.

A particular feature of the Endeavour River's entrance is its extensive white sandbanks. The fact that the Portuguese name should provide so accurate a description suggests that Mendonça may have sent a small boat through the reef passages for a closer look. His flagship, though, probably gave this area a wide berth, as the approaches to the river mouth are marked by a dense cloud of dots and crosses from every direction—an indication of dangerous reefs and shoals. It is a warning which Cook, stranded there for nearly two months while his ship's hull was repaired, would have wholeheartedly endorsed.

The Portuguese fleet next sighted a river which in the Vallard map is called *Rio Primero*—in modern Portuguese 'Primeiro', which would usually give the meaning of 'First River'. The name is something of a puzzle. Clearly this was not the first river they had found. 'Primeiro', however, can also have the secondary meaning of 'foremost', or 'principal'—so that may well provide the answer. *Rio Primero* is located just south of a promontory shaped like a broad arrowhead. A test of the map's accuracy here is to see whether any promontory matching this shape exists in this section of the Queensland coast. The answer is that one does indeed—the distinctively shaped Cape Kimberley. *Rio Primero* can on this basis be identified as the Daintree River, flowing into the sea from what is today a World Heritage listed rainforest, the Daintree Wilderness, lying immediately south of Cape Kimberley.

Continuing no doubt with extreme caution through the narrow channel between the Barrier Reef and the mainland, the Portuguese ships came next to another prominent cape. This feature is unnamed in the Vallard chart, but close to it is an island called *Illa Grossa* (Big Island). The headland resembles Cook's Cape Grafton, near the port city of Cairns. As for *Illa Grossa*, this corresponds to Fitzroy Island which lies a little south of Cape Grafton, just as depicted in the Vallard map.

Cook, who named Fitzroy Island, described it as 'tolerable high and about 4 Miles in circuit'. The size of the Vallard island has been exaggerated, but if we have interpreted the Portuguese fleet's position correctly there should be another island of roughly the similar size a little north of Cape Grafton corresponding to an unnamed island shown at this location in Mendonça's chart. And sure enough there is—Green Island, also sighted and named by Cook, who wrote that it was 'a low green woody Island...distant 3 or 4 Leg^{S} from Cape Grafton'.[17]

A final confirmation that we are in the right spot in relation to a modern map is that the distances travelled so far check out with impressive accuracy:

Rio Darena on the sixteenth century map is approximately halfway between *Cap Mata* and *Illa Grossa*: Cooktown similarly lies midway between the two modern equivalents: Cape Melville and Cairns.

South of *Illa Grossa*, alias Cairns, Mendonça's lookouts sighted a large, dangerous-looking shoal. The Vallard map gives it the name *Basse Grand*. The present-day Portuguese equivalent of 'basse' is *baixo* (pronounced as 'baishu') generally meaning 'low' but also with the nautical meaning of 'shoal'. The shoal the Portuguese sighted was most probably the Sudbury Reef, an extensive reef lying close to the main shipping channel south-east of Fitzroy Island. Giving the reef a wide berth, a little further south they sighted a river mouth which they named *Rio Bassa* (Shoal River) identifiable as the Johnstone River at Innisfail, today an important sugar-growing region.

Unknown to Mendonça, his ships were now entering a particularly hazardous stretch of the Great Barrier Reef coast: a region of myriad islands, reefs, dangerous currents and hidden shoals. This is reflected in the fact that the next section of the Vallard map is drawn with less precision. The Portuguese were far better equipped than Cook for finding a way through the coral labyrinth, assuming they still had four ships instead of Cook's one. This being the case, Mendonça would once again have sent his smaller, shallow-draught vessels ahead to scout for safe channels. Even so, the dangers and difficulties were formidable.

The winds on this coast would have been somewhat less favourable for Mendonça than for Cook, blowing frequently from the east and south-east. Nevertheless, in summer these winds are regularly interspersed with periods of favourable northerlies and north-easterlies. And even easterlies would have been no great problem, since Portuguese ships of this period were able to sail reasonably well on a broad reach—that is, with the wind sideways-on. On occasions when the winds came from dead ahead the normal practice would have been to anchor and await a favourable wind change, a manoeuvre which should not have presented too many difficulties in the relatively shallow waters inside the main Barrier Reef.

All the same, it must have been of great concern to Mendonça and his men to find that day after day they were having to painstakingly thread their way through a maze of hidden reefs where one incautious move or wrong decision could mean disaster. It is easy for us to forget that in the first half of the sixteenth century there was no such thing as a telescope, and ships did not even have steering wheels—these were not invented until the beginning of the eighteenth century. In Mendonça's day all ships, even the largest, were steered by crewmen heaving on the tiller. In bigger vessels this tiller was a massive baulk of wood protruding through the lower poop deck. Sailors working down here could not see where the ship was heading, so orders had to be shouted down to them by the steersman, standing on the quarterdeck.

The effort needed to steer a large carrack in bad weather was prodigious. A graphic account of the scene below decks is given in an account by Henrique Dias, an apothecary who sailed from Lisbon for Goa in the year 1560 in the ill-fated carrack *São Paulo* :

> Because the vessel was heavy and built in India it was hard to control and reacted sluggishly. In every heavy weather at least forty to fifty men had to be at the helm. Some hung onto the tiller, others worked with block and tackle. On each side of the helm twenty men and their foremen pulled...to swing the rudder and avoid gybing. Because this is the greatest and most eminent of the five dangers, besides fire, water, shoals and enemies, which are dreaded by seamen.

The *São Paulo* ended up wrecked on the coast of Sumatra. The danger of gybing occurs if a following wind suddenly catches the wrong side of a ship's sails; it can result in dismasting, or in extreme cases even a capsize.

Like Cook, Mendonça and his companions had no way of knowing how far the great coral reef would continue. They must surely even have wondered at times whether there would be any way out to the south or whether they would find themselves trapped in a dead end. Their feelings are evident in the next place name which appears in the Vallard map south of present-day Cairns—*Costa Dangeroza* (Coast of Danger). Significantly, this name appears in more or less the same place in other Dieppe maps as the 'Coste Dangereuse' or 'Perilleuse' (in modern Portuguese the equivalent word is 'Perigosa'). In all the other Dieppe maps this is one of the very few placenames—never more than ten in number—appearing on the entire eastern coast; in one of the earliest maps it is the one and only name. This suggests that it was an important warning relating to the dangers of the Great Barrier Reef coast as a whole, rather than to any specific part of it.

The next significant coastal feature sighted by the Portuguese seems to have been the puzzlingly named *Rio Timor*. This is apparently the Tully River, one of the largest rivers flowing into the Barrier Reef waters. Dunk Island and what appear to be two large reefs are shown offshore in this vicinity. South again Mendonça came to a prominent twin headland which the Vallard cartographer calls *Cap Double*. It is tempting to speculate that this 'Double Cape' might be the same as the 'Double Point' sighted and named by Cook just south of Innisfail, but Cook's headland seems too far north and also too small to correspond comfortably with the cape drawn in the Vallard map. A better match would be the twin headlands of Hinchinbrook Island—Cape Richards and Cape Sandwich—with the forty-kilometre-long Hinchinbrook Island not unreasonably taken to be an outlying part of the mainland. Bush-clad Hinchinbrook Island is separated from the mainland only by a narrow channel, and from out at sea this would have been hidden behind the island's 1000-metre-high mountains.

Next sighted and recorded by Mendonça were evidently the Palm Islands. In the sixteenth century chart they have been given the name *Tres Illes* (Three Islands). The main part of this group in fact consists of five islands, but an explanation for this apparent discrepancy lies in the fact that three of these islands overlap, so that from the deck of a ship some distance away they could well have appeared to be a group of three. The largest of these islands, Great Palm Island, was to achieve notoriety in the early twentieth century when it was proclaimed a native reserve and Aboriginal people from the mainland were forcibly resettled on it.

A little further on the Portuguese sighted the mouth of a river which has the name *Rio Ferondo*. In the original version it would have been *Rio Ferroso*, meaning 'ferrous' or 'iron' river. Its true identity is not hard to determine. The river sighted and named here by Mendonça is clearly the Ross River, flowing into crescent-shaped Cleveland Bay, site of the present-day port of Townsville.

Any possible doubt about this identification is dispelled as soon as we look at Cook's original name for Cape Cleveland, lying at Cleveland Bay's eastern end. Cook called it 'Iron Head'. Just north of the Ross River, moreover, lies Magnetic Island, named by Cook because he blamed it for causing his magnetic compass to swing wildly, a problem which he attributed to the presence of iron ore. Magnetic Island is clearly shown in Mendonça's chart, together with what is evidently the 550 metre-high 'Iron Head', south of Ross River, mistakenly drawn as a second island close to the mainland. It takes no great feat of imagination to conclude that Mendonça called Ross River the 'Iron River' because he experienced the same magnetic phenomenon as Cook.

Proceeding south-east again, this time through the wide and relatively safe channel lying close to the mainland coast, the Portuguese could hardly have failed to sight the large island of Whitsunday, lying across the far end of this 120-mile-long passage. And sure enough, it appears in the sixteenth century map, with the appropriate, if uninspiring name of *Illa Grossa* or 'Big Island' (the same name Mendonça had bestowed on the large Fitzroy Island further north). Whitsunday Island (so named by Cook because he sighted it on Whitsunday) is in fact by far the biggest island on that part of the coast and it is today a popular tourism and yachting resort. In Mendonça's chart Whitsunday and the smaller Hook Island, separated only by a very narrow channel, would seem to have been drawn as one. The fact that Illa Grossa is shown as an island rather than a headland suggests that the Portuguese ships sailed through the strait which Cook was to name Whitsunday Passage, between the islands and the mainland.

What happened next on Mendonça's voyage presents a tantalising mystery. South of *Illa Grossa* on the Vallard map we can see a very large bay depicted with the puzzling name *Baie Perdeue*. 'Perdeue' seems to be an old version of the French adjective 'perdu', meaning 'lost'. Comparison with modern Portuguese suggests that the original name might have been 'Baia Perdida', meaning 'lost bay', or 'the bay [which was] difficult to locate'. This same feature is prominent in other Dieppe maps, with the similar name *Baye Perdue*. In those maps it is one of the other very few coastal features actually named, so presumably it, like 'Costa Dangeroza', must have been considered important. But what exactly does the name imply? What was 'lost' or 'difficult to locate' here?

If the original Portuguese name was 'Perdida' it suggests two possibilities. The first is that Mendonça's ships sighted this large bay, then later lost their bearings as they tried to get back to it, perhaps in bad weather. The other more intriguing possibility is that the name was intended to indicate that in this bay *something* was lost. If so, what? Could it be that some men were lost—or perhaps even a ship?

A shipwreck actually seems the more likely when we look at 'Baie Perdeue' in the context of old rather than modern Portuguese. In the sixteenth century, Portuguese chroniclers make frequent use of two words similar to 'perdeue' which are now obsolete—'perdeo' and 'perderã'. Both are used to indicate shipwreck. And even until quite recent times 'perdida' also had the secondary meaning of 'wrecked'. Could the name 'Baie Perdeue' indicate that here one of Mendonça's ships came to grief?

And what of the bay to which the name refers? Where was it? It is not easy to identify on a modern map, perhaps because of the very problems hinted at by the name. The northern shore of the bay drawn in the Vallard map would seem to indicate Repulse Bay, since this is the only substantial bay on this part of the coast which has a large, south-trending peninsula (Cape Conway) resembling the headland drawn in the Vallard chart. Assuming this headland to be in fact Cape Conway, the smaller island outside the cape would be

Lindeman Island—or perhaps the Lindeman Group represented as a single island. Repulse Bay was given its name by Cook because shoals 'repulsed' his efforts to enter it. Could this also have been how the bay shown on the Dieppe charts was 'lost'—if that, rather than 'shipwreck', is what the name means?

Repulse Bay, however, is much wider and more open to the sea than the 'Baie Perdue' shown in the Dieppe charts. Cape Hillsborough, at the southern entrance to Repulse Bay, might appear to have a slight resemblance to the southern headland of the Vallard bay, but the comparison is not very convincing because Cape Hillsborough seems too small and too distant from Repulse Bay's northern arm to be the cape shown in the Vallard map. More importantly, Repulse Bay has no broad estuary winding far inland which would match the very prominent estuary of 'Baie Purdeue'. Apart from its northern headland, in fact, the 'lost' or 'wreck' bay more closely resembles in size and shape either Broad Sound or Shoalwater Bay much further south, both of which do have long, sinuous estuaries.

It almost seems as if either Mendonça's chart maker or the French cartographer have somehow confused or combined Repulse Bay with one of these other bays—neither of which otherwise appears on the map. This would also explain why the coastline appears to have been truncated at this point. Identification with Shoalwater Bay is perhaps the more likely, if only because the prominent southern headland of the 'lost' or 'wreck' bay somewhat resembles the shape of Shoalwater's Cape Townshend.

Another clue is that the southern headland of 'Baie Perdeue' has the name *Cap Veloza* (Swift Cape). The 'swiftness' presumably related to the currents here and 'Cap Veloza' is surrounded by a semicircle of dots which might perhaps indicate a dangerous tidal flow. Identification of 'Cap Veloza' with Cape Townshend is supported by a mishap which befell Matthew Flinders here nearly two centuries later. In August 1802, charting a strait at Cape Townshend, Flinders was moved to give it the name 'Strong Tide Passage' after he lost the cutter he was using for inshore surveys when it was swept away by the fierce current. Flinders reckoned the rise and fall of the tides here to be more than ten metres.

The main arm of the Vallard's bay is shown leading east rather than south as it should do if this is Shoalwater Bay, but leaving aside this cartographic error it seems that one of Mendonça's smaller ships must have ventured far into the estuary to have charted it in such detail, including what seems to be a depiction of the Cannibal Islands, lying inside the entrance to the bay.

What about the possibility of a shipwreck? Many old wrecks have been recorded in this part of the coast, but none have ever been positively identified as dating from before the eighteenth century. Perhaps the most interesting of them was reported by Captain Owen Stanley, commander of the British naval ship HMS *Rattlesnake*, in 1848. While surveying at the head of Port Molle, on Long Island in the Whitsunday Passage, he noted the remains of a wooden hull lying in muddy sand. Years later, six 30-pound (thirteen kilogram) iron cannonballs were found on a hill not far away. Some stories about the 'Port Molle galleon', as it came to be known, also claimed that Spanish silver and gold coins were found in the vicinity. An archaeological investigation in the 1980s, however, reached a more prosaic conclusion. It showed that the wreck was that of a British merchant ship, the *Valetta*, which was deliberately run aground on the beach in 1825 after being holed on a coral reef. Spanish coinage was the currency in use during the early years of the British colony at Port Jackson, while the cannonballs were identified as the product of gunnery practice by a Dutch naval vessel in 1863. And Port Molle in any case lies outside the area indicated in the Vallard map as 'Baie Perdeue'.

Were Mendonça's ships overtaken by a storm at this part of their voyage? One thing that could possibly suggest a link with the Portuguese expedition is an early Aboriginal tradition of an unknown ship having been hurled onto a beach by a storm somewhere in the vicinity of Cape Conway many generations previously. There are also legends of the crew of a beached ship having been attacked and killed by Aborigines.[18] Given that the stretch of coast from Repulse Bay to Shoalwater Bay covers 140 nautical miles and contains a vast number of islands and reefs, the possibility that the remains of an ancient, undiscovered wreck might lie somewhere in these remote waters cannot be entirely discounted.

What we can be quite sure of, though, is that if there was indeed a shipwreck the vessel involved was not the fleet's flagship, for as we shall see later, there can be no doubt whatever that Mendonça continued his voyage. However, the absence in this section of the map of some coastal features, including apparently the large inlet of Broad Sound, suggests that south from Cape Conway the Portuguese fleet sailed much further offshore than did Cook, thus avoiding the maze of the Cumberland and Northumberland Islands through which the *Endeavour* painstakingly threaded its way.

The next large group of islands Mendonça drew on his chart do not have a name, but their identity is clear enough. The landmark in question is a large island surrounded by five smaller ones, lying south of 'Cap Veloza'. This group is clearly identifiable as Great Keppel and the lesser Keppel Islands.

Today a popular holiday resort, the Keppels are located some fifty nautical miles south of Shoalwater Bay. They vary greatly in size, and six would not be a bad estimate of their number. Despite the tiny scale on which it is drawn, there is even an undeniable resemblance between the largest of the islands in the sixteenth century map and the shape of Great Keppel Island. Mendonça's largest island has five main headlands, just as does Great Keppel. Drawn particularly prominently is what is surely the distinctive crescent-shaped bay on Keppel's ocean side between the headlands of Big Peninsula and Coconut Point. Also recognisable are the four islets of Humpy (south of Coconut Point), Middle Island, Miall and probably Sloping Island. The larger North Keppel Island seems to be depicted behind Sloping Island, although a little out of position.

As the first Europeans ever to have traversed one of the world's most treacherous coasts, Mendonça and his men must by now have experienced much danger and uncertainty. With the enigmatic 'lost' or 'wreck' bay now behind them, where exactly were they and how far south had they sailed?

We can answer that question with confidence because they were nearing a point which stands out as a vital reference line in the Vallard map—a marker which will enable us to confirm the fleet's exact position in terms of latitude. That marker is the Tropic of Capricorn.

Like all other sixteenth century mariners, Portuguese navigators in distant seas lacked any way of checking their longitude, but two newly developed instruments, the astrolabe and the cross-staff, gave them a ready means of calculating latitude. On the deck of a heaving ship these instruments might lack the precision of the sextant—a much later invention—but in favourable conditions such as those of the sheltered waters inside the Great Barrier Reef they could be expected to give a reasonably accurate fix of a ship's latitude. Capricorn should therefore provide another important test of the accuracy of the Vallard map. If it is reliable we should be able to identify from it actual coastal features of the Queensland coast on or near the line of the tropic.

So what do we find? The Tropic of Capricorn is drawn boldly and unambiguously on the Vallard map. A short distance north of it are the islands which we have identified as the Keppel group—and the Keppel Islands do in fact lie some twenty nautical miles north of Capricorn, a distance fully consistent with what we can observe on the sixteenth century maps.

Another prominent feature in any modern map is Keppel Bay, on the mainland south of the Keppel Islands. This wide bay lies exactly on the Tropic of Capricorn. It also forms the entrance to the Fitzroy River, on which today stands the city of Rockhampton.

What do we see in the map derived from Mendonça's chart? Here, too, a river is depicted south of its version of the Keppel Islands, shown in the same location as the Fitzroy River, flowing into the sea just fractionally above the line of the tropic. This river has the name *S Fransois* ('S' being the standard Vallard abbreviation for 'Saint'). In the original Portuguese this would have been *Rio Sã Francisco* (River of St Francis).

Out to sea, the line of Capricorn in the Vallard map passes through the middle of a reef-fringed island which is depicted in some detail. It must have been somewhere near here that Mendonça's navigator, using an astrolabe or cross-staff, made his crucial observation of the latitude of Capricorn. From its distinctive shape it is not difficult to determine that this island is Heron Island—or more accurately, Heron Island together with its extensive reef and probably the adjacent Wistari Reef. Although just a small coral cay, in fine weather Heron Island is visible from far off, thanks to its tall groves of native trees. North of it in the sixteenth century map we can also see what is either North West Island or nearby Wreck Island, both of which are also surrounded by large reefs.

Heron Island in reality lies just under the meridian of Capricorn, while Wreck Island and North West Island are some six to nine nautical miles north of it. In other words, the Tropic of Capricorn as it appears in the Vallard map is in almost exactly right—placed perhaps two or three nautical miles south of its correct position. This minor discrepancy would also help to explain the slight northward displacement of the mouth of Mendonça's 'St Francis' (Fitzroy) River.

Taking into account the technological limitations of early sixteenth century navigational instruments, though, this must count as an astonishing feat. The Vallard map's Capricorn demonstrates that in terms of latitude, at least, Mendonça and his officers were able to fix their position on the east coast of Australia with impressive accuracy.

Continuing southward from Capricorn, the Portuguese next sighted an estuary to which they gave the name *S Nicollas*—probably meaning 'Bay of St Nicolas'. This would seem to be Port Curtis, the entrance channel between Curtis Island, lying hard up against the mainland, and its smaller southern neighbour, Facing Island. This entrance is today a main shipping route to the aluminium-exporting port of Gladstone.

The feast day of St Nicolas is 6 December. It might be tempting to think that this was the day Mendonça's ships sighted Port Curtis, but other evidence indicates this could not have been the case. As we have just seen, the Fitzroy River, only a little further north, had been named after St Francis—and his feast day is 4 October. It should have taken hardly more than a day to sail from the 'River of St Francis' to the 'Bay of St Nicolas', not two months. One of the dates must surely be wrong. As we will soon see, evidence further down the coast points to the correct date being 4 October. The 'Bay of St Nicolas' must in that case have been given its name for some other reason—and that reason is most likely to be found in the fact that St Nicolas was venerated as the patron saint of sailors. What would have been more natural than to give thanks to the Saint for deliverance from the perils of the Great Barrier Reef?

Sailing further south again with Mendonça, a second *Rio Grand* now appears on the map. This is probably the southern entrance to a wide channel which runs for some fifty kilometres behind Facing and Curtis islands and which was probably mistaken for the lower reaches of a large river. The fact that this waterway is drawn with a certain amount of detail suggests that one or more of the Portuguese ships explored it for some distance. The next river depicted, whose name seems to be *Rio Patmo*, would correspond with Bustard Bay. Further south again is a small cluster of islands, the largest of which equates reasonably well with Lady Musgrave Island and its extensive surrounding reef. The islet off its northern side would be Fairfax Island, with the second small island to the southward probably Lady Elliot Island.

Bustard Bay (whose next-door neighbour is Turkey Beach) was given its name by Captain Cook because of its abundance of turkey-like bustards and other wildfowl. After anchoring in this 'very secure' bay on 23 May 1770, the British captain ventured in one of the ship's boats up a channel leading to a large lagoon. Here he noted seeing:

> ...black & white Ducks, and Bustards such as we have in England one of which we killd that weigh'd 17 and a half pounds which occasioned my giving this place the name of Bustard Bay.[19]

As for the unfortunate bustard, the naturalist Joseph Banks, writing in his journal after a satisfying dinner, remarked that 'it turnd out an excellent bird, far the best we all agreed that we have eat since we left England...'

Rio Patmo, the name given to what appears to be Bustard Bay in the Vallard map, has no immediately obvious meaning. It is perhaps faintly possible that 'Patmo' could be a garbled version of 'Pedro' (Peter), making this the 'River of St Peter'. A second more likely and fascinating possibility is that the name might originally have been 'pato'. In Portuguese, 'pato' is the word for a duck. Could this be Rio Pato—Mendonça's 'Duck River'?

The idea is not as far-fetched as it might seem. At about the same time, on the other side of the world, Mendonça's adversary Magellan was bestowing a very similar name on a bay he had discovered, and for a similar reason. Probing southward along the coast of Patagonia in search of the elusive passage to the Pacific, Magellan's Spanish ships came to a wide inlet with islands sheltering what appeared to be large flocks of ducks. Magellan named it *Bahia de los Patos* (Bay of the Ducks). Magellan's Venetian chronicler Pigafetta speculated that the plump, easily caught birds must be a type of flightless goose, but his description has left historians in no doubt that they were in fact penguins!

As for Mendonça on his voyage down the eastern coast of Australia, if the Portuguese captain had entered Bustard Bay what would be more natural than he, like Cook in a later century might have thought to record in his chart a name indicating that this was a place where a ship's company could obtain fresh provisions in the form of ducks and other waterfowl? It would be something of a coincidence, to be sure—yet on occasion truth can indeed turn out to be stranger than fiction. And we will shortly be coming to other extraordinary discoveries which will more than bear this out.

4

Mendonça's calling card: a sinker from the bay of prawns

On this point of steeridge, the Spaniards and Portuguese do exceed all that I have seen, I mean for their care, which is chiefest in navigation. And I wish in this...we should follow their examples. In every ship of moment, upon the half-deck or quarter-deck, they have a chair or seat out of which whilst they navigate, the Pilot, or his Adjutants (which are the same officers which in our ships we term the Master and his Mates) never depart, day nor night, from the sight of the compass; and have another before them; whereby they see what they do, and are ever witnesses of the good or bad steeridge of all men that do take the helm.

SIR RICHARD HAWKINS, *OBSERVATIONS*, 1622

Fraser Island, lying off the Queensland coast south of the Tropic of Capricorn, is the world's largest sand island. One hundred and sixty kilometres long, it is the product of sand carried by coastal currents which has built up over millions of years around what were once isolated rocky islets. Today much of the island is heavily forested; it contains numerous freshwater lakes and supports a rich variety of native wildlife, such as squirrel gliders, dingoes, tortoises, bandicoots and marsupial mice.

Captain Cook is credited with being the first European navigator to see Fraser Island; the *Endeavour* sailed past it heading north on 20 May 1770, with Cook naming a prominent headland near the island's northern end 'Indian Head' after observing that a large crowd of 'Indians' had gathered there to watch as the ship went by. But now it can be shown beyond doubt that it was Mendonça's expedition which sighted Fraser Island first, fully two-and-a-half centuries earlier.

How can we be sure? For a start we need only glance at the outline of the long island shown on the Vallard map which has been given the name *Illa Plata* (probably meaning 'Silver Island') to notice something familiar. This is the distinctive V-shaped angle between the island and the mainland—the same angle which Fraser Island makes to the coast 250 kilometres north of Brisbane.

Taking the comparison further, the southern end of Fraser Island, at the bottom of the 'V', lies some twenty kilometres north of a twin headland on the mainland coast which Cook mistook for two islands and named Double Island Point. This same feature is shown in the Vallard map—but correctly, as a headland. And here it has to be said that the wealth of detail recorded by Mendonça is far superior to Cook's chart of the same coast. The British captain wrongly concluded that Fraser Island was attached to the mainland and drew it as a large promontory which he named 'Sandy Cape'. Mendonça made no such mistake. Not only does his map depict Fraser Island correctly as an island, it also shows many less obvious but significant features: for example the prominent headland now called Moon Point, on the island's south-western coast facing the mainland.

If the shape of the northern end of *Illa Plata* looks slightly wrong for Fraser Island, this is evidently because sailing down the seaward side of the island Mendonça's lookout—bearing in mind that he did not have the advantage of a telescope—was unable to see the narrow, low-lying tongue of sand which joins Sandy Cape to the rest of Fraser Island north of the prominent headland Waddy Point ('waddy' being the name of an Aboriginal war club). Hence Sandy Cape appears in the Vallard map as a separate small island.

Once this minor error has been corrected, the true shape of the northern end of Fraser Island becomes apparent. Waddy Point is clearly identifiable on the north-western corner of *Illa Plata*, with Cook's 'Indian Head' correctly shown just south of it. The sea coast south from here is drawn slightly indented instead of more or less in a straight line as in modern maps but this is evidently because the unbroken line of the Seventy-Five Mile Beach which forms the outer edge of the island at sea level would also not have been visible from the ship's deck. Mendonça would therefore have based the shape of his coastline on the contours of the higher sandhills further inland, though a long row of crosses marked on the map all the way down the island's coast indicates he was aware of shallows. No doubt for similar reasons Cook's chart also shows this as an indented coast.

As the *Endeavour* passed Sandy Cape, Cook sighted a low, partly submerged reef or shoal 'stretching out to the northward as far as we could see', its direction marked by long line of breakers. The English captain estimated its length from Sandy Cape at about thirty-two kilometres. Eventually, when soundings from a ship's boat on the line of the shoal showed five fathoms, he deemed it safe to take the *Endeavour* across into the sheltered waters beyond. Cook named the shoal Break Sea Spit.

What is quite clearly this very same shoal, extending a considerable distance northward, is depicted in the Vallard map in impressive detail, delineated by neat rows of tiny crosses. A larger point at the end of the spit, coloured red, in all probability marks an area where the shoal was visible above the breakers.

We can imagine the relief of Mendonça and his men on finding that they had at last left the dangers of the labyrinthine Great Barrier Reef behind them and emerged into calm, clear waters teeming with wildlife. This sixty-four kilometre-wide bay in the lee of Break Sea Spit made a strong impression on Joseph Banks, naturalist on board the *Endeavour*. In his journal entry for 20 May 1770 he paints an idyllic picture:

> This shoal extended a long way out from land...we ran along it...then passed over the tail of it in seven fathom water: the Sea was so clear that we could distinctly see the bottom and indeed when it was 12 and 14 fathom deep the colour of the sand might be seen from the mast head

> at a large distance. While we were upon the shoal innumerable large fish, Sharks, Dolphins &c. and one large Turtle were seen. A grampus of middle size Leapd with his whole body out of the water several times making a Splash and foam in the sea as if a mountain had fallen into it. At sunset a few Bobies [boobies] flew past towards the NW.[1]

From the size of its 'splash and foam', Banks's 'grampus' was probably a whale. We can see a curious depiction of what might also be called a grampus in the Vallard map. Off the Queensland coast there rises from the ocean a huge sea monster that looks like a cross between a swordfish, a whale and a crocodile. It is clearly a mythical beast, but the drawing may well have been based on seamen's tales.

However it was not something as spectacular as a 'grampus' that particularly caught the attention of the southbound Mendonça expedition. Approaching the northern tip of Break Sea Spit three cryptic words appear on the Vallard map: *pomezita, taragontray* and *camonron*. Two of these words obviously defied the French translators, because they have come down to us in pure Portuguese. *Pomezita*, the first of the three, means 'small pumice'. It is an extremely uncommon word. What can it be doing here?

The third name, *camonron*, is less of a puzzle, but no less intriguing. It is an old Portuguese word for a prawn or shrimp—in modern Portuguese, *camarão*, pronounced 'camaranoo'. That same word has become the name of an entire West African country—Cameroon. Exploring down the African coast in the fifteenth century, Portuguese ships ran into shoals of prawns off a land in the Gulf of Guinea and named the coast 'Camonron'. Did a similar thing happen off the coast of Queensland in the year 1522? The answer is yes—and we can be quite sure of it. We can be certain thanks to a million-to-one coincidence involving a modern environmental research program and the fact that Mendonça inadvertently left behind an exquisite calling card.

The modern part of the story begins in September 1976, when three Australian environmental scientists were excavating core samples from compacted sand and pumice layers at Hook Point, on the narrow southern

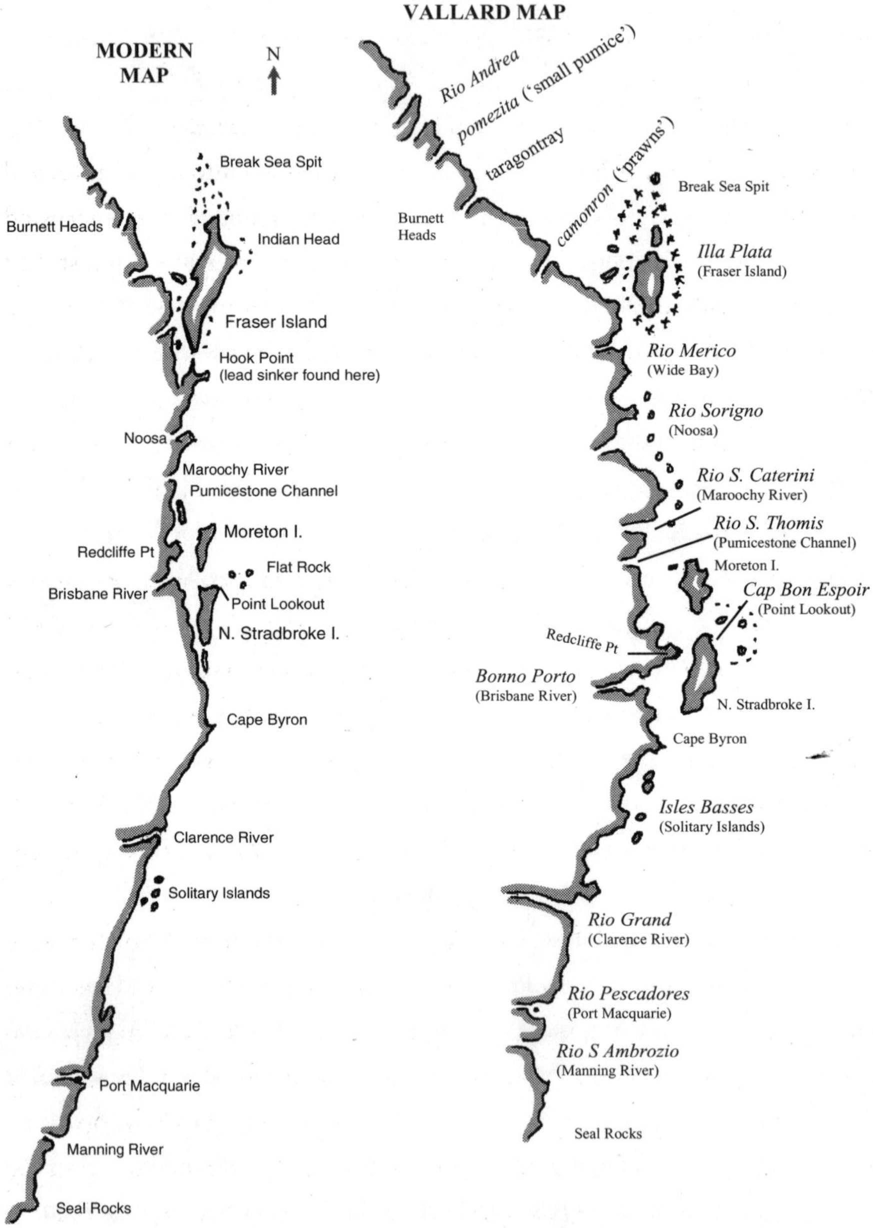

The Vallard map section from Fraser Island to Seal Rocks compared with with a modern map of the same stretch of coast. Other major geographical features depicted in the Vallard map include Moreton and North Stradbroke Islands, Port Macquarie, and the Brisbane and Clarence rivers.

tip of Fraser Island. Their aim was to determine the geological history of the sands and they were digging with great care, using only a hand-held auger.

When the auger reached a depth of 2.2 metres, Graham Roberts, who was turning it, felt it strike something solid. At first the researchers assumed this to be a lump of pumice, but when the sand was dumped into a sieve it turned out to be something very different. It was in fact a piece of sand-encrusted lead, about six centimetres by four centimetres and weighing almost 122 grams. Originally it had been a single sheet eleven centimetres wide, but it had been folded over once from each side and a hole had been drilled through all three folded layers near the top. This was the vital clue to its purpose: clearly it had been intended as a fishing sinker. To have reached the shore, it evidently must have been attached to a floating, buoyed net. How long had it been lying there, buried in the sand and pumice?

Detailed chemical analysis of the pumice brought an astonishing result—it indicated that the lead weight had apparently been cast up on this Fraser Island beach some time between 1410 and 1630 AD. But this was at least 140 years and possibly even as much as three and a half centuries before the arrival of the *Endeavour*, the first European ship known to have sailed in these waters! Any thought that something as technologically advanced as a lead sinker could have been manufactured by Aborigines was out of the question, so who could have left it there so long ago?

Intrigued by this question, the scientists began to pursue it further with the help of other specialist colleagues. By analysing the radioactive isotope ratios of the lead sinker and comparing these with the known 'fingerprints' of international lead sources, they were able to narrow down the possible sources of the lead to just three European countries: Portugal, Spain or southern France—with the latter region giving the best correlation. Homing in even further, they were finally able to arrive at a remarkable conclusion: the lead used to make the Fraser Island sinker most probably came from one particular mine in the mountainous Massif Central region of southern France—a mine from which lead had been extracted ever since the end of the Stone Age. The findings meant the sinker was by far the oldest European artefact ever discovered on the east coast of Australia.

In his earlier analysis of the small lumps of pumice among which the lead weight was found, one of the original three scientists, Dr Bill Ward, had been able to prove that some of the pumice had come from an eruption on Macauley Island, a volcano some 2500 kilometres east of Australia a little south of the latitude of Brisbane. Macauley is an uninhabited island in the Kermadec Group, a chain of volcanic peaks on the Pacific 'arc of fire' extending from New Zealand to the Tongan archipelago. Carried by ocean currents, floating pumice ejected by this volcano would have reached the Queensland coast a year or two later. The data was difficult to assess, but it was the identification of the Macauley Island pumice which put the approximate date for the lead sinker's stranding on the beach at somewhere between 1410 and 1630. The midpoint between these two extremes (the most likely date) is the year 1520—which, as it so happens, coincides almost exactly with the date of Mendonça's voyage.

But even though they had rigorously assessed their methodology to eliminate the possibility of error, Ward and his five colleagues encountered a puzzling reluctance by scientific journals to publish their findings. Time after time their research paper was rejected. Different reasons were given by various editors, but none of them seemed entirely convincing. Ward eventually drew his own conclusions: 'With rejection slip after rejection slip it became more and more obvious that some critics were reluctant to believe that some European visitor might have encountered the east Australian coast before Cook in 1770.' It was not until 1999—more than two decades after the original discovery—that the research findings were finally published, in the authoritative scientific journal *Archaeology in Oceania*.[2]

Even then Ward was unaware that thanks to an extraordinary coincidence, independent confirmation was awaiting recognition in the form of two words on the Vallard map—a document which he had not even seen. The only possible explanation for the strange word 'pomezita' in the map is that off this part of the Queensland coast Mendonça's ships found themselves sailing through floating pumice. And not just any pumice, but specifically small pumice, just like that discovered with the lead sinker. And then, a little further down the coast, the ships' crews sighted *camonron*—shoals of prawns.

Surprising as it might seem, many Portuguese mariners of this period would have been familiar with the phenomenon of floating pumice. During the sixteenth century the island of Fogo (meaning 'Fire') in the Portuguese-ruled Cape Verde Islands was in frequent eruption, spewing out clouds of pumice which rained down onto the surrounding sea. A spectacular eruption of this sort was witnessed by the English captain Sir Francis Drake and his crew as they sailed past Fogo in the *Golden Hind* at the beginning of their epic round-the-world voyage in 1578.

As for Mendonça's voyage, we now have enough information to picture what evidently happened. Portuguese seamen by tradition and upbringing were expert fishermen, and on a long voyage of exploration with fresh food scarce they would have been ready to exploit any bounty the sea might offer. So we can well believe that in the calm, crystal-clear waters off Fraser Island they quickly unrolled their shrimp nets and began floating them out. Juicy, cooked prawns would have been on the menu that day. But while they were fishing, one of the nets, or perhaps part of a net, must have broken off and drifted away. Eventually it was washed up on the beach at Fraser Island, along with some of the same pumice the ships had passed through just a short time before. And if any further confirmation of all this were needed, the name *camonron* appears in the Vallard map on the Hervey Bay coast at the location of the present-day Queensland town of Burrum Heads, while *taragontray* is shown at the position of Burnett Heads, a little further north at the mouth of the Burnett River. What are these two places notable for? Prawn fishing, as we might well guess!

As a sequel to the twenty-year delay which Ward and his team encountered in getting their work into print, it has to be said that the eventual publication of their remarkable findings has been met with deafening silence from Australia's academic community. Not one expert in maritime archaeology has displayed any interest in following up the team's scientific conclusions. As for the University of Queensland, despite the fact that its own museum today houses the sixteenth century Portuguese lead sinker discovered on nearby Fraser Island, influential academics at the university still choose to ignore this crucial evidence, dismissing the whole idea of a Portuguese discovery of Australia as 'speculative'.

We noted earlier a third strange name on the Fraser Island section of the Vallard map—*taragontray*. This is difficult to interpret and has almost certainly become garbled in the process of copying. The first part of the name, 'tara', could be a misreading of the familiar word 'terra', meaning 'land' but if so, 'gontray' defies translation. Another possibility lies in the fact that the only Portuguese word which 'taragon' even faintly resembles is 'tartaruga' meaning 'turtle'. Given Joseph Banks' reference in his journal to the sighting of a large turtle near Fraser Island, it would not be at all surprising if this is what it indicates.

North of the names 'taragontray' and 'pomezita' the sixteenth century map depicts a river with the apparent name of *Rio Andrea*. It is difficult to be absolutely certain about this name because it falls exactly between the two facing pages of the original Vallard Atlas, with the result that the two central letters of the name are indistinct. It could possibly be 'Andria' but 'Andrea' seems more likely, since this would be the equivalent of the English name 'Andrew'. So this is probably the 'River of St Andrew'. The feast day of St Andrew is 30 November, which could fit the likely progress of Mendonça's ships were it not that it seems to conflict with the dates of other saints' days further down the coast.

There is in fact one other very good reason why a river in this locality might have been given such a name. St Andrew was a patron saint of fishermen. Here, one might think, would have been a suitable occasion to offer thanks to this Saint for a bountiful catch. And as we will see later, a river with the name

'St Andrew' is actually referred to in a seventeenth century English geographic treatise which lists a number of Portuguese place names on the coast of what it calls 'Terra Australis Incognita'. In present-day terms, the River of St Andrew would appear to be the inlet known as the Port of Baffle Creek.

While in the vicinity of *Illa Plata*, Mendonça evidently stopped for some time so that his smaller ships could explore the great sheltered bay up to seventy kilometres wide which lies on the landward side of Fraser Island and which Cook was to call 'Herveys Bay'. This is apparent from the fact that the Vallard map depicts headlands, sandbanks and shoals in these waters. Big Woody Island, lying at the southern end of the bay some seventy-five kilometres from its northern entrance is also shown, suggesting that Mendonça must have dispatched a shallow-draught vessel to investigate Great Sandy Strait—the narrow passage between this part of Fraser Island and the mainland.

At the southern end of *Illa Plata*, one coastal feature is initially puzzling. This is the depiction in the Vallard map of a wide gap between what is obviously Hook Point, at the southern tip of Fraser Island (where the lead sinker was found) and the mainland coast. In actual fact the gap between Hook Point and Inskip Point on the mainland is quite narrow—hardly more than a kilometre. The sixteenth century map also shows a large river with the name *Rio Merico* flowing into the sea at this point.

An explanation for these anomalies soon becomes apparent. Inskip Point is the northern end of a sandspit some ten kilometres long, and near Fraser Island the spit becomes narrow and low-lying. As Mendonça's ships sailed toward the southern end of Fraser Island they would have had to stand out to sea in order to keep well clear of the potentially dangerous Double Island Point and its outlying Wolf Rock. From this distance the low-lying sandspit near Inskip Point probably would not have been visible, thus giving the false impression of a wide river mouth.

At this same spot, and no doubt for the same reasons, Captain Cook's chart also erroneously depicts a broad inlet opening to the sea. The British captain even called it 'Wide Bay'—the name which the harbour behind Inskip Point retains to this day. As for the Vallard map's name for this inlet, *Rio Merico* is most probably a French misreading of the Portuguese words 'Rio Mérito', meaning 'River of Merit'.

Finally, what of the name *Illa Plata*? This could mean one of two things. The ambiguity exists because on the original Portuguese chart the name could have been either 'Ilha Plana' (Flat Island) or 'Ilha Prata' (Silver Island). Either would fit, but 'Ilha Prata' is perhaps the more likely because in certain lights the sands of Fraser Island are said to shimmer like sheets of silver.

If the outline of Fraser Island on the Vallard map is unmistakable, so too, and for similar reasons, are those of the next two large islands which Mendonça's flotilla came to and charted as they voyaged south. Both are long and narrow; the northernmost island is the smaller of the two, and a narrow channel is shown running between them. These are exactly the shapes and parameters of the two other big sand islands off the Queensland coast—Moreton Island and North Stradbroke Island.

Before reaching this spot, though, Mendonça's ships had first to traverse a 120-kilometre stretch of coastal waters beginning at the southern end of *Illa Plata*. The first significant headland they noted, shown in their chart as an unnamed broad arrowhead at the southern end of a long curving beach, can hardly be other than the prominent Noosa Heads. Both its shape and location are correct, jutting out into the Coral Sea. *Rio Sorigno* (possibly *Rio Sereno* or 'Serene River'), drawn a short distance further south, appears to be the Noosa River. It is shown entering the sea on the wrong side of Noosa Head, probably because Mendonça's lookout mistook the river mouth for what is actually a lake several kilometres wide lying just south of the headland.

Now Mendonça's chart depicts two rivers close together with the names *R (Rio) S Caterini* and *R S Thomis*. This is an interesting nomenclature, reflecting again the importance in the sixteenth century of the Christian calendar of saints' days. The original Portuguese versions of these names would have been 'Santa Caterini' (St Catherine) and 'São Thomas' (St Thomas). St Catherine of Alexandria, the legendary Christian saint martyred on a spiked wheel (the origin of the Catherine Wheel), was especially revered in Portugal. The river named after her is evidently the large Maroochy River, today also the base for a prawn-fishing fleet. Was this river perhaps named for St Catherine because the ships were here on her feast day, 25 November? That would be in the early part of the southern summer, the most favourable time for ships to be heading toward southern latitudes in these unknown seas. This date would also accord with some other saints' days which we will come across on the coast further south.

As for St Thomas, his feast day is given in some calendars as 21 December, and in others as 3 July. Neither date fits at all comfortably with St Catherine's day, or with several other dates we will come to.[3] The 'St Thomas River' itself seems to equate with a large inlet close to the town of Caloundra, north of Brisbane; this inlet looks like a river mouth, but it is in fact the northern entrance to a strait called the Pumicestone Channel running between Bribie Island and the mainland.

This strait was given its name by Matthew Flinders on his exploratory voyage in July 1799, because of the 'quantities of pumice stone' he saw washed up on either side of its entrance. Flinders described the channel as more than a mile in width. At this spot the English navigator also had an unfortunate encounter with a party of Aborigines, one of whom, according to the captain's journal, was shot and wounded after the Aborigines made an unprovoked attack. One possible clue why Mendonça apparently named this same waterway after St Thomas might relate to the fact that the Saint is said to have preached and been martyred in south India, the region from which the Portuguese fleet had sailed.

This now brings us, in the Vallard map as in a modern atlas, to Moreton Island. We can now look closely at the detail. The first large headland shown in the sixteenth century map south of 'St Thomas River' and facing Moreton Island appears to represent South Point, at the southern end of the Pumicestone Channel. Immediately south of this headland is depicted a large semicircular bay which is exactly the shape of Deception Bay, and then a cape with a distinctive, broad arrowhead shape. This feature matches the headland of Redcliffe—which has Reef Point at its northern end, Redcliffe Point in the centre, and Woody Point at its southern tip.

The name opposite this headland in the Vallard map is one of the most astonishing we will come across in this great Portuguese voyage of exploration. It is written *Cap Bon Espoir*, which is clearly a French translation. In the original Portuguese it would have been Cabo da Boa Esperança. Where have we seen this name before? At the southern tip of Africa, of course, for it means 'Cape of Good Hope'!

Mendonça, it hardly needs to be said, would have been under no illusion that he was off Southern Africa. He had rounded the African Cape of Good Hope on his outward voyage from Lisbon two years earlier and would no doubt have been aware that Bartolmeu Dias, its Portuguese discoverer in 1488, had with some justification called that headland the 'Cape of Storms'.

On this part of Australia's coast the name 'Cape of Good Hope' must surely have been intended to mean exactly what it says, so we can believe that at this point the morale of Mendonça's expedition had taken a major turn for the better. Behind them lay the dangers of the Barrier Reef and the oppressive heat of the tropics; now they were sailing in a pleasant, temperate region abounding in marine wildlife that included not only whales, turtles and prawns but also dugongs, those strange, rare sea mammals that inhabit the waters of Moreton Bay. And having got safely through the dangerous reefs, shoals and currents of which stories concerning the Island of Gold had warned, Mendonça's men might well have speculated that finally they were nearing their promised land—the land of King Solomon's Mines. Perhaps they would yet return laden with gold. Good Hope indeed!

It seems open to question, though, whether in Mendonça's original chart the name *Cap Bon Espoir* was intended to refer to the cape of Redcliffe, as it appears to do in the Vallard map (where most place names are written on the mainland). In reality Redcliffe is a low-lying and fairly unimpressive headland—today one of Brisbane's more affluent residential areas. It is hard to imagine that it would have seemed particularly noteworthy when viewed from a ship approaching from the open sea. And quite unlike the original Cape of Good Hope it lies in sheltered waters, with the long expanse of Moreton Island shielding it from the ocean. In the sixteenth century map the actual shape of the headland has been quite accurately drawn, but for some reason it is shown jutting out much too far from the shore. Did the Vallard cartographer perhaps take the liberty of enlarging it to match the portentous name?

A likely alternative explanation is that the name in the original Portuguese chart was intended to refer not to Redcliffe but to Point Lookout, at the north-eastern tip of North Stradbroke Island. This headland is also more or less in line with the legend *Cap Bon Espoir*, although further away. One of just two headlands sighted and named by Cook on this section of the coast, the rocky, bush-covered Point Lookout makes a spectacular sight, jutting out into the vivid blue waters of the Coral Sea. It would have been a significant turning point for the ships of Mendonça's flotilla when they resumed their voyage southward and a more worthy recipient of this evocative name.

Whatever the truth about the precise identity of the Australian 'Cape of Good Hope', the unmistakable depiction of Redcliffe headland and other nearby features leaves no doubt that Mendonça's ships must have sailed south through Moreton Bay, the twenty-five kilometre-wide passage separating Moreton Island from the mainland. To the groups of Aborigines who would have been watching from vantage points on shore, Mendonça's flagship must have been an awe-inspiring sight as it sailed majestically down the bay with its great sails billowing in the breeze. And strangely enough, a scene rather like this actually appears in the Vallard map. Out in the Coral Sea just south of the Tropic of Capricorn the cartographer depicts a three-masted carrack with flags flying and all sails set, leaving a foaming wake as it runs swiftly before the wind toward the coast of Queensland. The ship's huge mainsail, high forecastle and towering stern structure decked with heraldic shields and rows of gun ports all confirm that this is a Portuguese carrack of the early sixteenth century. And it seems to be heading straight for *Cap Bon Espoir*!

As for Mendonça and his officers, they must surely have found it perplexing that although this vast country was obviously inhabited there was still no sign of a town, a port, or even the humblest village. Where did these black people have their homes? Were there other more civilised people living in the interior or perhaps further south? Where, indeed, was the hoped-for Land of Gold?

On the mainland coast today opposite the southern end of Moreton Island is the site of Brisbane International Airport, and next to it lies the entrance to the Brisbane River. Did Mendonça's people sight the river also? The answer is that unquestionably they did and not only that, we can be reasonably sure that one or more of his ships entered it. The outer section of what can only be the Brisbane River, shown correctly south of the equivalent of Redcliffe, has been drawn in the Vallard map in bold, stylised form with the name *Bonno Porto* meaning 'Good Port'. Mendonça could hardly have chosen a more appropriate name for what is today one of Australia's major international shipping terminals.

Although in the early decades following European settlement, larger ships entering the Brisbane River had problems with a sand bar caused by soil erosion, a British naval party which made the first survey of the river mouth in 1824 found an entrance channel with a depth of nearly four metres (twelve feet) at high tide. This would have been ample for the average Portuguese caravel, which drew only six feet. Inside the bar, the navy surveyors reported finding deep water of approximately twenty-four feet (seven metres) for a distance of some three kilometres upriver. The dimensions and draught of Mendonça's flagship are not known, but it would almost certainly have been quite a lot bigger than Christopher Columbus's flagship *Santa María*—a small carrack which also had a draught of just six feet.

Suggestions have been made that 'Bonno Porto' was intended to refer to Moreton Bay rather than the Brisbane River, but this seems highly unlikely given that Moreton Bay in the nineteenth century had a reputation for being anything but a 'good port'. The Brisbane Roads, where bigger ships had to unload in Moreton Bay, were an exposed anchorage prone to the sudden and dangerous development of high winds and rough seas.[4]

What the Brisbane river mouth was like for navigation 300 years before the arrival of British settlers we can only guess, but the name 'Bonno Porto' strongly suggests that the Portuguese fleet anchored for a while in this sheltered haven. While they were there, Mendonça would no doubt have dispatched a small boat to explore the river for some distance.

Apart from the possible *Cap Bon Espoir*, neither Moreton nor North Stradbroke Island has been given a name in the Vallard map. Both are positioned slightly further north in relation the mainland than is actually the case. Whether this misalignment was in Mendonça's portolan or whether it is the result of an error by the Vallard cartographer when he copied the coastline from the original Portuguese chart is a matter for conjecture, but at all events the southern end of North Stradbroke Island appears to terminate at a 158-metre-high hill today called Mount Scott. None of the numerous low, sandy islands immediately south of this point, including the smaller South Stradbroke Island, are shown. The southern tip of North Stradbroke Island has also been drawn further out from the mainland than is actually the case and here again one might wonder whether the Vallard cartographer may have exaggerated the size of this gap to better fit his assumption that Redcliffe was 'Cap Bon Espoir' and that the Portuguese voyagers must have exited at the southern end of North Stradbroke Island.

In reality it seems clear enough that Mendonça wisely decided against attempting to proceed south through the maze of winding, shallow channels and islets which lie between North Stradbroke Island and the mainland, otherwise some of these features would surely have been shown on the map. And indeed there can be no real doubt that his ships exited from their 'Good Port' to the open sea by sailing eastward through the passage between North Stradbroke and Moreton islands, turning abruptly south as they rounded

Point Lookout—their likely 'Cape of Good Hope'.

One reason we can be confident they sailed this way is because just off this cape the Vallard map depicts three small islands. At first this might appear to be a problem, for no islands are shown in this vicinity on any popular maps of the Brisbane region. Nautical charts, however, present a different picture. These show the presence of potentially dangerous rocky islets off the approaches to the North Stradbroke-Moreton passage and it so happens that there are exactly three of them! The largest, Flat Rock, is some 200 metres across, with its top rearing five metres above the sea at low tide. The Vallard map's three islets are correctly shown lying off the north-eastern end of North Stradbroke Island, where the cartographer has also drawn a dotted line depicting a shoal or sandbank. This may well be an indication of the treacherous sandspit on which today lie the remains of the World War II Liberty Ship *Rufus King* which ran aground and broke in two while heading for Brisbane in July 1942. Captain Cook, we might add, also drew three rocks or islets on his chart in exactly the same position.

And here we might pause for a moment to compare again the quality of Mendonça's charting with that of his British counterpart. On this part of the coast Cook failed to discover either Moreton or North Stradbroke islands: the nearest he came to it was naming Point Lookout, and a second more northerly headland 'Cape Morton' but he wrongly assumed that both these capes were part of the mainland. This mistake caused him to miss charting both Moreton and North Stradbroke islands and of course he did not see either Moreton Bay or the Brisbane River. Mendonça's charts, as copied by the French, may have their shortcomings but here, as at Fraser Island, it can hardly be denied that they provide far more detail than those of the illustrious Cook.

Leaving the region of *Cap Bon Espoir* and the Brisbane River behind them, Mendonça's ships turned south-east once more and soon sighted another wedge-shaped headland where the coast began to trend even more to the south. From its appearance and position on the sixteenth century chart this headland resembles Cape Byron, the most easterly point of mainland

Australia. This cape, too, was named by Cook, who described it as 'tolerable high' and with 'a remarkable sharp peaked Mountain lying inland from it'. Seen from the sea, as Joseph Banks noted on board the *Endeavour*, Cape Byron does indeed have the shape of a wedge.

Further south again the Portuguese ships sighted a group of offshore islands, shown on the Vallard map with the name *Illes Basses* (Low Islands). Four of them are depicted. These are clearly the Solitary Isles, a collection of small, low, rocky islands whose very existence would be unknown to most Australians but which happen also to be the next feature named on Cook's chart heading south.

Mendonça's 'Solitary Isles' have been placed closer to Cape Byron than is actually the case, but nevertheless there can be no doubt about their identity. There are indeed four main Solitary Islands (if we exclude one very small islet lying close up against the shore) but more importantly, if we look closely at the Vallard chart we can see that the two northernmost of these islands, coloured blue, have been drawn so close together that their shores are touching. Carelessness on the part of the cartographer perhaps? The answer is no—this is absolutely correct! In reality these two islands do indeed lie within just a stone's throw of each other, unlike the remainder of the group (coloured red on the Vallard map) which are spaced a considerable distance apart. The impressive accuracy of the original Portuguese cartography is confirmed by the fact that a present-day admiralty chart also depicts the two northern islands as conjoined at low tide.

Further out to sea the Vallard map features a drawing of another sea monster, this time a grotesque-looking fish resembling a dinner plate with staring eyes. This is without doubt a giant sunfish a rare marine monster which can attain the astonishing weight of almost half a tonne. Apart from its tail, which should be wider and shorter, the detail of the Vallard sunfish is strikingly realistic, a whitish fish which appears to consist of just a great head with a small fin on either side, a small beak-like mouth and two prominent round eyes.

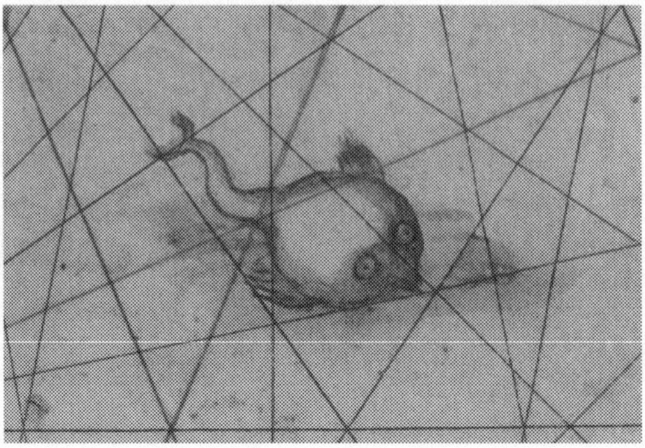

Sunfish pictured in the Vallard map off Australia's east coast.

Is it just coincidence that this drawing of a sunfish has been placed in the waters of eastern Australia? Sunfish are known to be in the habit of resting on the surface in fair weather and, as it happens, in December 2004 one of the leading boats in the Sydney-Hobart yacht race had to be abandoned a day after colliding off the New South Wales coast with a huge basking sunfish, estimated by the yacht's crew to have weighed more than 300 kilograms.

As for Mendonça's voyage, sightings of exotic fishes in the region of the Solitary Isles would have been well within the bounds of possibility. The seas in the vicinity of these islands—today a protected marine park—are known to provide a habitat for no fewer than 280 species of fish, four species of turtles and ninety species of corals.

The next river the Portuguese ships sighted on their voyage south is depicted in the Vallard map with name *Rio Grand*. This is the third river on Australia's east coast to have been given this name and in this case it is well deserved. This 'Rio Grand' can be identified with confidence as the Clarence River, in northern New South Wales. It is shown further north than it should be in relation to the map's Solitary Isles but then, as we have seen, it is the placement of these islands which is at fault. The fact of major significance here is that the Clarence is by far the largest river system in this entire region, draining an area of more than two million hectares. Its original English name was even 'Big River'—the exact equivalent of 'Rio Grand'!

A prominent cape shown in the sixteenth century map at the northern headland of 'Rio Grand' fits the location of the Clarence River's Wooded Bluff, today a World Heritage listed nature reserve containing the region's last remaining area of coastal rainforest. As we might expect, fishing is an important industry for the two towns located on either side of the mouth of the 'Big River'. The small port of Iluka, on the edge of the rainforest, is the base for a deepsea fishing fleet, while the town of Yamba on the opposite shore boasts rich prawn-fishing.

The Clarence is navigable by small craft for nearly 100 kilometres, so if Mendonça's flotilla had entered this river's estuary one of his smaller ships would have had no difficulty sailing far upstream. And the Vallard map suggests that this is exactly what they did, for unlike most of the rivers shown on this coast 'Rio Grand' is drawn not in the usual symbolic style but with detail, as a real river. Moreover, judging by the scale of the rest of the map, the distance the river is shown winding inland seems to be about 100 kilometres—exactly the distance for which it was navigable!

Most intriguing of all, there exists a mysterious oral tradition attributed to Aboriginal people of this region which could well refer to such an event. The tradition tells of a 'great canoe with sails' which voyaged up the river from the coast many centuries ago. This exotic vessel must have predated the arrival of the British, because the Aboriginal account speaks of its white-skinned crew being clad in 'garments of stone', a description strikingly reminiscent of the Kimberley Aboriginals' account of invaders—apparently armoured Portuguese soldiers—with 'skins like crocodiles and turtles'.

A second version of the Clarence River tradition describes the men in the 'sailing canoe' carrying what would seem to be iron weapons and recounts that they had come 'from the centre of the world'.

Conservative historians tend to dismiss these oral legends as lacking credibility. As for the people who are custodians of the legends, it can sometimes lead to personal embarrassment even to put their stories forward. Ron Heron, an Aboriginal elder living in the Clarence valley, says the second oral tradition was passed on to his father by an old friend, the late Alex Vesper. He adds that Vesper was not believed, and even ridiculed, because he insisted that the strangers' boat had sails and that they carried iron weapons. If we assume the legends to have a basis in fact, though, they would clearly point to one of the first encounters between Europeans and Aboriginal people on the east coast of Australia.[5]

Rio Pescadores, shown on the early chart some distance south of 'Rio Grand', is also intriguing because this name means 'River of Fishermen'. The wide estuary, depicted with a small island inside it, is strongly suggestive of Port Macquarie at the mouth of the Hastings River—the Vallard's island then equating with Port Macquarie's Pelican Island. At this harbour a particularly harsh prison colony was established in 1821 for the incarceration of convicts who had fallen foul of the British penal regime at Botany Bay.

Why did the Portuguese voyagers give it the name 'Rio Pescadores'? The most obvious explanation would be that Mendonça's men saw Aborigines fishing here in canoes. Again we might wonder whether they were able to make contact of some sort.

Not far south again from this 'Fishermen's River' is another estuary which has been given the name *Rio S Ambrozio*. This is 'St Ambrose River', probably the entrance to the substantial Manning River. Shown on the chart after St Ambrose River, but not named, are coastal headlands corresponding to Cape Hawke (described by Cook as 'a high point of land which made in two hillocks') and the headland of Seal Rocks, which Cook did not see because he passed by it at night.

The feast day of St Ambrose is 7 December, twelve days after that of St Catherine (25 November), after whom Mendonça named the Maroochy River, north of Brisbane. As we have seen, saints' days cannot always be a relied on as an indicator of progress and should be interpreted with some caution. This sequence could, however, provide a realistic time frame for a voyage of about 400 nautical miles down the coast, assuming an average speed of about five knots in favourable winds and allowing for several days spent at anchor. Earlier in the voyage the date of 4 October, the feast day of St Francis, apparently marked the fleet's arrival at the Fitzroy River on the Tropic of Capricorn. We cannot tell how long the voyagers stayed at the Fitzroy River but it seems that in any event they would have had ample time to sail the distance of approximately 300 nautical miles to the Maroochy River. This would suggest that Mendonça's crews may well have enjoyed a fairly lengthy stopover in the sheltered bay behind Fraser Island, replenishing their fresh water casks, cleaning and making repairs to their ships—and of course, netting more of those succulent prawns!

The Portuguese mapping of this section of Australia's east coast may have its shortcomings by modern standards, yet it has to be said that not only is it more accurate in some important respects than Cook's chart, it also bears comparison with the results achieved twenty-nine years after Cook by a second famous English navigator, Matthew Flinders. Flinders made a detailed survey of the coast from Sydney to Hervey Bay in 1799, yet his chart derived from that voyage, like that of Cook, shows both North Stradbroke and Fraser islands as part of the mainland. And despite the fact that the King's representative, Governor Hunter, had specifically instructed him to find out if there were any rivers of importance on the New South Wales coast that might allow access to the interior, Flinders returned to tell the Governor that no such rivers existed.[6]

5

In search of Botany Bay: cracking the Mendonça code

See from the Bear to the Pole opposite
Vast countries stretching on without a let.

CAMOENS, *THE LUSIADS*

Retracing Mendonça's voyage down the east coast of Australia, we now come to the great enigma of all the maps of eastern Terra Java produced by the Dieppe cartographers—a huge wedge-shaped promontory jutting out almost due east into what is now called the Tasman Sea for a distance (reckoned on the general scale of the Vallard map) of at least 500 kilometres. In reality, nothing even remotely resembling this strange feature actually exists. It is the main reason why orthodox historians have invariably rejected the idea that the Dieppe maps are intended as a representation of the east coast of Australia.

What explanation can there be for this peculiar anomaly? There would seem to be two possibilities. The first is that it is simply fiction—a mythical coastline and landscape. However, if this is the case then it is very odd indeed that Mendonça's chart-maker should have delineated the Australian coast with such great fidelity up to this point and then suddenly, for no apparent reason, taken off on a flight of fantasy. It simply does not make sense.

The alternative explanation—and surely the more realistic—is that this promontory is a *real* coastline. But if so, what country or region does it represent?

Before attempting to answer this question we need to look back at the coastline on which we have been up to now following the course of Mendonça's voyage.

The region of eastern Australia depicted thus far in the Vallard map has a coastline trending steadily south-east on a compass bearing of about 130 degrees until it reaches Fraser Island. Here the Terra Java coast begins to turn more southerly and once past North Stradbroke Island it turns again to run for a considerable distance almost due south. Interestingly enough, the direction of this coastline consistently trends some ten degrees further to the east than the *actual* direction of the east coast of Australia on a modern map. This discrepancy is not accidental. On the contrary, it was an inevitable outcome of the technical limitations of navigation and cartography in the early sixteenth century.

Map-making errors of this kind arose out of the fact that although Portuguese and French navigators of the early Renaissance period were well aware that the earth was round, this was a relatively new concept and they still used the traditional method of plane, or 'flat-earth', geometry to calculate their position and course at sea. They had no choice, as it was the only method available. What they did not know was that it was seriously inadequate for coping with the complex mathematical problems of spherical geometry, arising from the fact that the earth is not flat but a globe.

In higher latitudes in particular (latitudes closer to the poles) the 'flat earth' methodology tended to cause significant chart distortions known as errations. These could be in either an easterly or westerly direction, depending on the geographical circumstances. In the case of Mendonça's expedition setting out from the Portuguese base at Malacca and using that location as their longitudinal reference point, the distortion as they sailed down the east coast of Australia would have been always toward the east—just as depicted in the Vallard chart.

The concept is not all that easy to envisage. One can perhaps best imagine an elastic globe stretched out to become a flat chart. The need to expand the areas closer to the poles distorts them in an easterly direction if, as was the case with eastern Australia, they lie well to the east of the navigator's equatorial reference point.

Added to this distortion would have been errors caused by compass variation: the difference between true north and magnetic north. Magnetic rather than geographic north was the reference point toward which ships' charts were aligned in the sixteenth century and much later. Captain Cook, master cartographer that he was, still used magnetic north as his reference point 250 years after Mendonça, which explains why in his charts, too, the line of the eastern coast of Australia is shown trending too far toward the east as one proceeds in a southerly direction.

Unlike Mendonça, Cook was aware of the problem of compass variation, which on several occasions on the Australian coast he calculated at about eight degrees, effectively to the east. The variation is a few degrees more today, due to the fact that the magnetic pole is constantly moving, but in the sixteenth century it is thought to have been about the same as in Cook's time. This, added to an erration error of some two degrees, gives a built-in eastward error for the Australian coastline of around ten degrees. And this is exactly what we see in the Vallard map.

No amount of erration or compass variation, however, can explain the huge eastern promontory. So how can it be accounted for?

Over the years several diverse theories, some of them bordering on the bizarre, have been put forward in an attempt to provide an answer to this baffling question. One of the more imaginative suggested that the promontory is in reality the East Cape of New Zealand's North Island—but it is difficult to take this seriously when even the New Zealand historian proposing it had to concede that the adjoining sections of the map depict the coasts not of

New Zealand but Australia.

One retired Australian academic has cast his net even wider, claiming to have proved that the entire map of Terra Java represents the coast of Vietnam. This imaginative theory has met with acclaim in orthodox academic circles—but the truth is that it cannot stand up to even the most elementary scrutiny.

Its most obvious flaw is that Vietnam happens to be situated in the northern hemisphere, not the southern. To imply that leading cartographers of the Renaissance—the age of intellectual giants like Michelangelo and Leonardo da Vinci—were so ignorant that they could not even tell the difference between the northern and southern hemispheres seems a somewhat dubious proposition. Furthermore, in order to try and make this theory seem even remotely plausible, its proponent had to postulate that for some inexplicable reason a number of islands, and even some place names, were drawn and written upside-down.

And if all that were not enough, the Vietnam theory ignores the very awkward fact that in the Vallard Atlas a companion page to the Terra Java map clearly depicts northern Australia, the Indonesian archipelago and the *actual* coasts of Cambodia, Vietnam and south China—all instantly recognisable, in their correct position in the northern hemisphere, adorned with numerous identifiable place names (for instance *Quinguapura* (Singapore), *Prignant* (Hainan Island) and *Chinna*) and with no islands or names upside-down! And finally, the Vallard map of South-East Asia actually *overlaps* the northern section of the atlas's Terra Java map—where, as we have seen the cartographer correctly identifies and names the Tropic of *Capricorn*, not *Cancer*. So there can really be no question but that Vietnam and Terra Java are two entirely different places.[1]

On this point, there is also one fundamental difference between the manner in which Terra Java is depicted in the Vallard and other Dieppe maps and the way in which other countries are shown. This difference has been pointed out by no less an authority than Helen Wallis, until recently Curator of Maps at the British Library. In a commentary in the library's elegant facsimile edition of the sixteenth century Dieppe atlas known as the *Rotz Boke*, Wallis firmly rejects the notion that the southern continent of Terra Java which the cartographer Jean Rotz drew (and which he styled *Java-la-Grande*) could represent coastlines of Malaysia or other South-East Asian countries. According to Wallis, the key lies in the nature of the place names. They show that these are two completely different regions. She writes:

> It is clear that the land [*Java-la-Grande*] represents a discovery made on a European voyage of coastal exploration. Secondly, it was apparently not a region of settled and civilised populations whose people would tell a visiting expedition the names of towns and other places—in fact there is no sign of towns. Thirdly, the number of Portuguese names suggests that the voyage was made, or at least recorded, by the Portuguese. The flags on Deslien's charts of 1561 and 1566 are more explicit, indicating a Portuguese discovery.[2]

Returning to the puzzle of the strange eastern promontory of this new land which Helen Wallis suggested the Portuguese must have discovered, other controversial interpretations include that of Melbourne historian Kenneth McIntyre, who claimed that it was supposed to be Cape Howe, on Australia's east coast. McIntyre put forward an elaborate theory to support his contention, despite the total lack of any resemblance between Cape Howe and the promontory of Terra Java.[3] And finally, in what is perhaps the most bizarre interpretation of all, a British writer has claimed in a recent popular book (apparently in all seriousness) that this jutting-out coastline represents no less than the edge of the Antarctic ice sheet, as discovered in the early fifteenth century by the crew of a Chinese junk.[4]

Leaving aside such flights of imagination, the huge eastward-pointing land mass depicted in the Vallard and other Dieppe maps could surely not have been the result of a navigational error. Given the impressive accuracy of the Vallard map up to this point it is inconceivable that an obviously skilful navigator who had thus far accurately charted the progress of his ship southward for many hundreds of miles through the perils of the Great Barrier Reef, with the sun rising every day from the ocean on his ship's port side and setting over hills of the unknown land to starboard, should suddenly and for no apparent reason mistakenly begin to draw this same south-trending coast as turning abruptly east—directly toward the rising sun! The idea is simply not credible.

So if the Vallard and the other Dieppe maps are to be interpreted as an attempt at a true representation of Australia there has to be another explanation for the mysterious eastern promontory. And sure enough, there is. The answer to the conundrum, as we shall see, is that the coastline depicted here is real enough, but it has been incorrectly aligned in relation to the rest of the map. And this error can be shown to have been an outcome of the circumstances in which the Dieppe maps originated.

It needs to be emphasised again that these elegant maps of the Dieppe school are not mariners' charts. They were painstakingly produced at great expense by the best French cartographers and illustrators of their day, not for rough use on board ships but to ornament the walls and libraries of the nobility and the wealthy bourgeoisie. The famous 'Dauphin' map, as its name implies, was presented to the young Dauphin (Crown Prince) of France; the 1542 world atlas of Jean Rotz was begun as a commission for King Francis I of France and completed for England's Henry VIII after Rotz crossed the Channel to take up a position in Henry's court. The Vallard Atlas was made for a wealthy Dieppe businessman.

But the original mariners' charts, or portolans, drawn on board ships during voyages of exploration in the early sixteenth century were a different proposition entirely to these expensive works of art. In Portuguese ships, as we saw earlier, the portolans were drawn on a kind of parchment made from dried animal skin, usually goat skin, stretched over a wooden frame. Animal skin was preferred to paper largely because it was much more durable: it could withstand far better the rigours of exposure to humidity, heat and sea water encountered during a long sea voyage.

An outstanding example of the art of Portuguese chart-making is the series of portolans of the East Indies sea routes drawn by Francisco Rodrigues in 1512. Rodrigues sailed as a pilot with the very first Portuguese expedition to the Spice Islands, commanded by Antonio de Abreu. His ship, a caravel, was one of three vessels which set sail from Malacca in December 1511, only a matter of months after this strategic Muslim port city had been captured from its Malay ruler by the redoubtable Afonso de Albuquerque. Rodrigues, on the title page of the book of charts and sailing directions which he subsequently produced, styles himself 'Pilot-major of the first Armada that discovered Banda and the Moluccas'.

One ship of Abreu's flotilla was lost on the outward voyage, but Rodrigues's caravel and the remaining carrack safely reached Banda island, the source of nutmeg and mace, some 500 nautical miles north of Australia. On the return journey Rodrigues drew a series of six portolans covering almost the entire length of what is today the Indonesian archipelago and parts of Malaysia. He also produced a companion set of sixty-eight exquisitely drawn sketches depicting all of the main coastal features along the way.

Rodrigues' portolans and sketches were thought to have been irretrievably lost until copies of them unexpectedly appeared in a French atlas in 1849. However the publisher of this atlas, the Viscount de Santarém, did not divulge the whereabouts of the originals and it was not until 1937 that they were finally tracked down by a Portuguese historian, Armando Cortesão, in a most unexpected place—the archives of the French Chamber of Deputies. Rodrigues' portolans provide the vital key to understanding not only the method of compilation of sixteenth century Portuguese charts but also the reason why the Vallard and other Norman French Dieppe maps of the east coast of Australia seem suddenly to go wildly astray.

Out of necessity, Rodrigues charted his voyage from the Moluccas in six separate sections, each taking up one of his animal hide portolans. This point is of crucial importance. Cortesão, after rediscovering the lost charts and recognising them to be separate pieces of a jigsaw, tried joining them together. Thanks to the elongated shape of the Indonesian archipelago with its many islands each with its own distinctive shape (not to mention the fact that he could refer to modern maps), this proved a fairly straightforward task. Cortesão then found he was looking at an elongated, composite chart covering the entire region.

We can be quite certain that Mendonça's chart of the east coast of Australia would have been produced in a similar fashion, because this coast was far too long to fit onto a single portolan. Mendonça's chart-maker, of course, had no means of knowing how far this unknown coastline would stretch, or in what direction it would lead. Once the line of the coast reached the edge of the goat skin he was working on, this piece had to be taken off the frame and a new one substituted. But just where on the new portolan should the cartographer recommence his line? As no one could predict in which direction the coastline might go beyond the horizon, the best bet would probably have been to begin drawing it near the mid-point at the top of the frame. Of course we cannot be certain, and ultimately it does not greatly matter, but for the sake of argument let us assume that this is what happened when Mendonça's flotilla reached the point at which the hitherto accurate coastline of the Vallard map suddenly veers off toward the east.

Mendonça or his cartographer obviously would not have had the slightest difficulty joining their portolans together in correct order and correct alignment to each other later on if they had wanted to make a composite map after returning to their home port. But the situation would have been very different for the French cartographers in Dieppe when the charts, or copies of them, fell into their hands. If there were no clear compass directions on some of the portolans, putting the jigsaw together correctly could have posed a real dilemma.[5]

If we look closely at the Vallard map's coastline we can see that it turns eastward just south of an inlet called *Rio Pator*. There is no word 'pator' in Portuguese, but the name brings to mind one which we came across on the Queensland coast at what is probably Bustard Bay. Could this be another

Joining portolan charts to make a composite map: Portuguese historian Armando Cortesão's assemblage of four portolan charts drawn by Francisco Rodrigues on the first Portuguese expedition from Malacca to the Spice Islands in 1512. On this voyage Rodrigues drew a series of separate portolans, covering between them most of the Indonesian archipelago. The portolans had been thought lost until Cortesão discovered them in the archives of the French Chamber of Deputies in 1937. The dotted lines indicate where Cortesão joined the portolans to produce a composite map of the East Indies. A similar process would have occurred when French cartographers tried to assemble their composite maps of Australia from a series of portolan charts. (*Armando Cortesão, Lisbon, 1939*)

'Rio Patos' (River of Ducks)? A second possible meaning could relate to the fact that in Portuguese, *pastor* means 'shepherd'(though it is hard to imagine hunter-gatherer Aborigines being mistaken for shepherds). One other possible interpretation is that 'pator' might conceivably be a misreading of 'paulo', meaning 'paul'—hence 'River of [Saint] Paul'.

Regardless of the name's meaning there are two possible present-day identifications for this river estuary. The first is Port Stephens, 220 kilometres north of Sydney, into which flow several rivers. The other possibility would be the Hunter River, some fifty kilometres further south and today the site of the steel manufacturing city of Newcastle. However, when we consider that even with the aid of a telescope Captain Cook failed to spot the concealed entrance to the Hunter River, it seems highly unlikely that Mendonça, relying on the naked eye, would have seen and charted the Hunter rather than the far more obvious entrance to Port Stephens, which Cook both sighted and named.

Assuming *Rio Pator* to be Port Stephens, until now the line of the long Terra Java coast had trended steadily south-east: on the goat skin charts always toward the right (assuming Mendonça's portolans, unlike the Vallard maps, were drawn with north at the top). Hence each section of coastline is likely to have ended somewhere above the lower right-hand corner of the chart rather than running off the bottom edge. The next portolan chart could then be perceived as joining in one of two possible ways: either edge to edge, meaning that the new stretch of coastline would trend east, or overlapping at the bottom, in which case the coastline would continue south. Which to choose? This dilemma is illustrated on page 150.

It seems that the Dieppe cartographers took the eastern option—and the geography of Port Stephens offers a reason why. Port Stephens is the only harbour on this part of the New South Wales coast which has a north-facing shore of any significant length. This shoreline runs east toward the Tasman Sea for about twenty-five kilometres, and this is the very direction in which the north-facing shore of the strange eastern promontory now leads. Is this

How the error in the Dieppe maps came about:

Portolan chart No. 1 (below left) represents the central coast of eastern Australia. How should portolan No. 2 be attached?

Chart 1 Chart 2

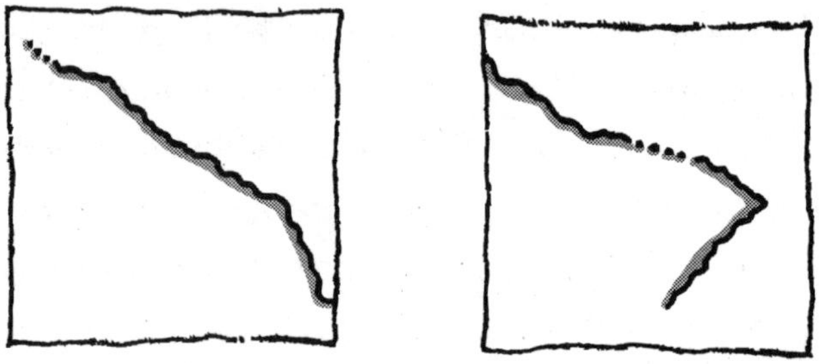

There are two options

It can be joined either this way...

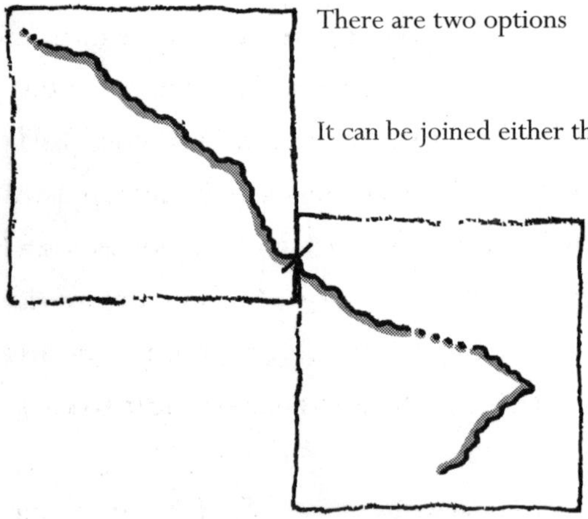

...or this way.

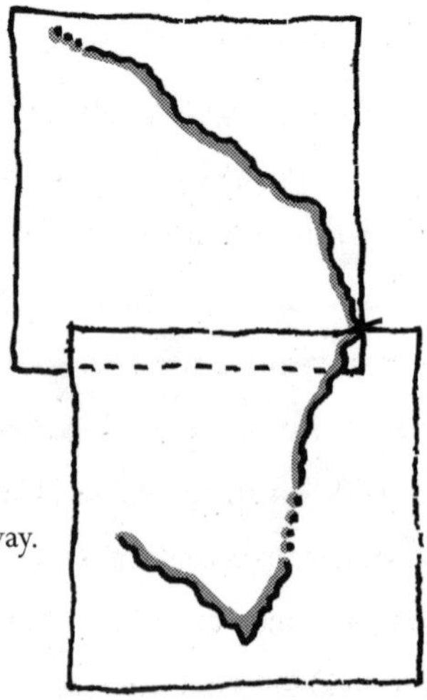

the key to the cartographers' error? Did they mistakenly assume that the Port Stephens shoreline was the beginning of a long, new stretch of coastline heading east?

The theory of a misaligned portolan can be put to the test by taking this section of the Vallard map and swivelling it clockwise through about ninety degrees around the point where the coastline goes wrong. By doing this we are, in effect, taking the second option which would have been open to the Dieppe cartographers. This will be the acid test on which the theory of realignment either stands or falls. If the theory is wrong, then the new coast created by this manipulation will also be fictitious and can be expected to make no more sense than did the coastline of the fanciful eastern promontory. If, on the other hand, the theory is correct it should now enable us to match the coastal features on the realigned Vallard map with the *real* geographical features of the Australian coast from Port Stephens southward.

The result of realigning the 'eastern promontory' in this way is shown on page 152. What is immediately apparent is that far from creating a fictional nonsense, the southern part of the Vallard map now bears an unmistakable resemblance to the outline of the south-eastern corner of mainland Australia. This is encouraging—but it is only the start. If we now take a closer look and compare the individual coastal features of the realigned Vallard chart with those shown in a modern map, there can be no disputing the fact that the two coastlines match to an astonishing degree. More detailed comparison will bring some extraordinary discoveries.

The first name we come to in the Vallard map south of its equivalent of Port Stephens is a headland called whose name is written as *Cap amato*. In Portuguese 'amado' means 'lover' but this does not make much sense; almost certainly the explanation is that this is a garbled version of 'Cabo Mato'—another 'Forest Cape'.

Just south of this cape we come to a very large bay with seven offshore islands. This same bay also features prominently in the Dauphin, Rotz and Henry II maps, so it was evidently considered important. In the Vallard chart

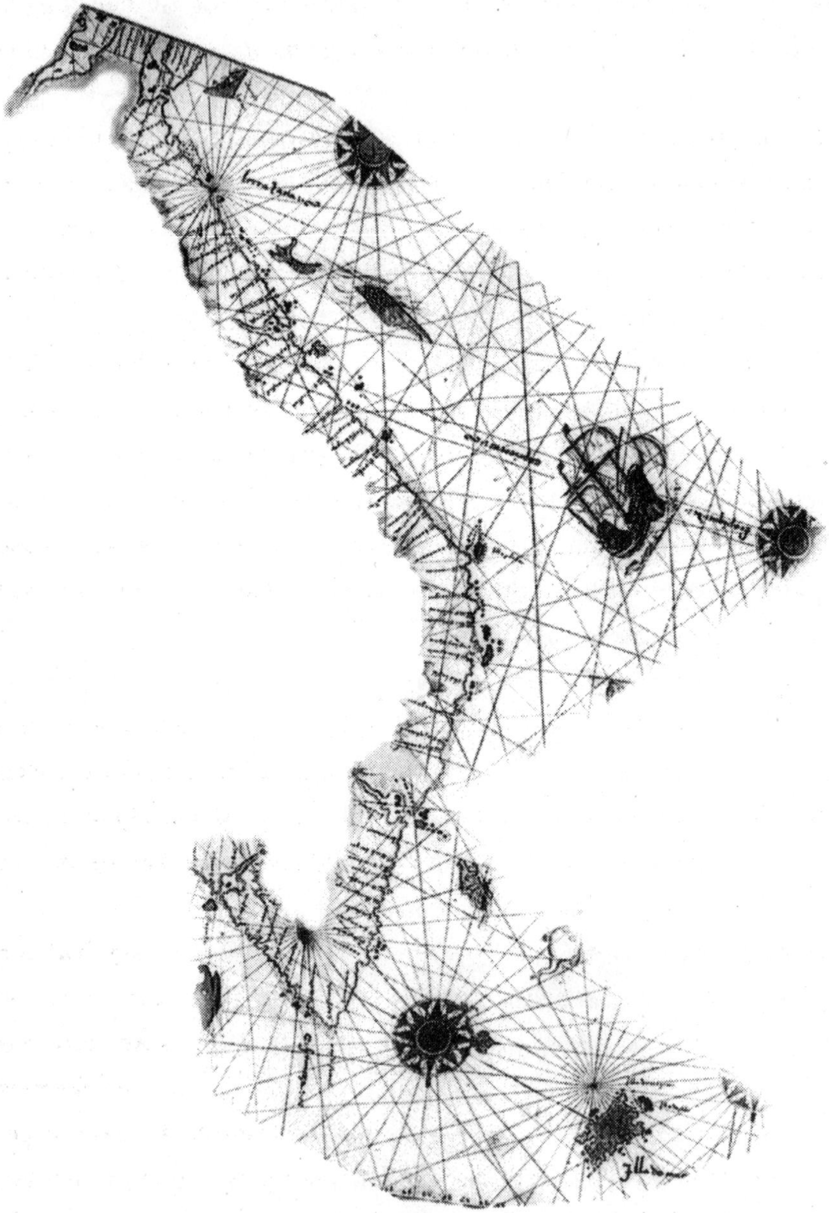

Outline of the Vallard map of Australia's east coast after realignment by computer. The puzzling 'eastern promontory' has been swivelled through ninety degrees at what was assumed to be the original break point between two separate portolans. Now, this previously nonsensical feature of the map portrays in detail the entire south-eastern sector of the Australian coast. (See also colour plate section)

it is boldly drawn with much distinctive detail, all of which strongly suggests that this is a real coastal feature, not something imaginary, as it might at first seem.

Is there anything to which this apparently capacious harbour could correspond on a modern map? There is, and the answer to this question brings us to one of the most extraordinary achievements of Mendonça's voyage—one which confounds accepted history. The answer is that this harbour matches none other than the birthplace of modern Australia—Botany Bay.

This conclusion, of course, brings with it the inevitable corollary that Captain Cook was not, after all, the European discoverer of Botany Bay! Shocking though this conclusion may seem, the facts are there to be assessed. A detailed comparison of the prominent harbour of the Vallard map with modern maps of Botany Bay will show that despite its small scale, the sixteenth century map delineates most of the bay's main features with astonishing accuracy.

As a starting point we can look first at the distinctive bulbous shape of the harbour shown in the Vallard map, resembling, one might think, a stomach with an intestine attached. The same outline can be seen on modern satellite weather maps of eastern Australia, with the 'stomach' representing Botany Bay and the sinuous 'intestine' of the Georges River flowing into its western end.

Geologically, Botany Bay is quite different to the other large harbours on this section of the Australian east coast. Port Hacking to the south, Port Jackson (Sydney harbour) and Broken Bay, to the north, all have deepwater confined in relatively narrow, irregularly shaped channels bound for the most part by steep sandstone outcrops. In contrast, the distinctive oval-shaped shoreline of Botany Bay was created by ocean waters flooding a tectonic basin at the end of the last Ice Age. Botany Bay has relatively shallow water and a mainly low shore, except at the ocean heads.

Before going on to examine the finer detail we need to pause for a moment to look at the question of scale. In relation to the Vallard map as a whole, the mystery harbour would be too large to be Botany Bay or for that matter any other bay on the New South Wales coast. This is partly why it is puzzling. But this is no longer a problem when we realise that it was not all that unusual for sixteenth century map-makers to enlarge the detail of important sections of their charts, such as useful harbours. We usually do this today by drawing these features as larger-scale insets, each defined by its own border. But as other examples show, Portuguese and French map-makers of the sixteenth century would, where they felt it appropriate, simply draw the enlarged section at its correct spot without any border to differentiate it from the rest of the map. This, we can reasonably assume, is what was done in the case of Botany Bay. It is also possible that the Vallard cartographer may have been working from a separate Portuguese chart showing this part of the coast on a bigger scale since we have the prior example of the cartographer Rodrigues sometimes drawing larger-scale portolans in his charting of South-East Asian coasts.

At all events, the amount of detail in the Vallard map demonstrates that Mendonça was every bit as quick as Captain Cook to recognise the strategic importance of what the English captain termed this 'capacious, safe and convenient' harbour.

Starting from just inside the southern headland, the first small bay, drawn neatly and very distinctly on the map, equates to what is now Cook's Landing Place and Kurnell Beach, behind which there stands today a large and ugly oil refinery. Next comes a larger bay which corresponds to Quibray Bay and then we come to a third inlet, the present-day Woolooware Bay. Correctly again, the sixteenth century map shows the western tip of this bay forming the southern point of the narrows at the entrance to the Georges River, present-day Taren Point. On the opposite side of the river mouth is shown another headland corresponding to Rocky Point, where nowadays a major Sydney motorway crosses the river over the Captain Cook Bridge.

The northern shore of Botany Bay has been drastically altered in recent times by container port development and extensive reclamations for the twin runways of Sydney International Airport. Despite this, the bold northern arc

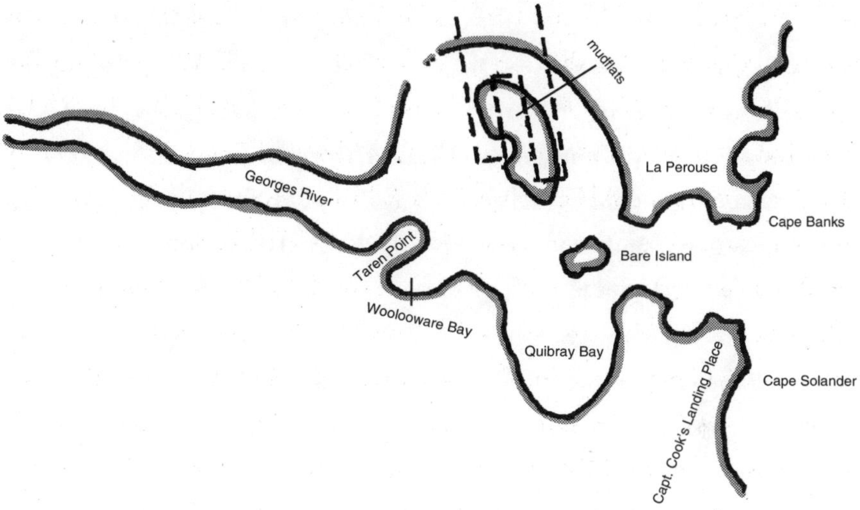

The Vallard map of Botany Bay, with identifiable present-day features indicated. Much of the area taken up by the map's large mudbank 'island' is now covered by the two main runways of Sydney international airport. The runways are indicated by dotted lines, drawn approximately to scale.

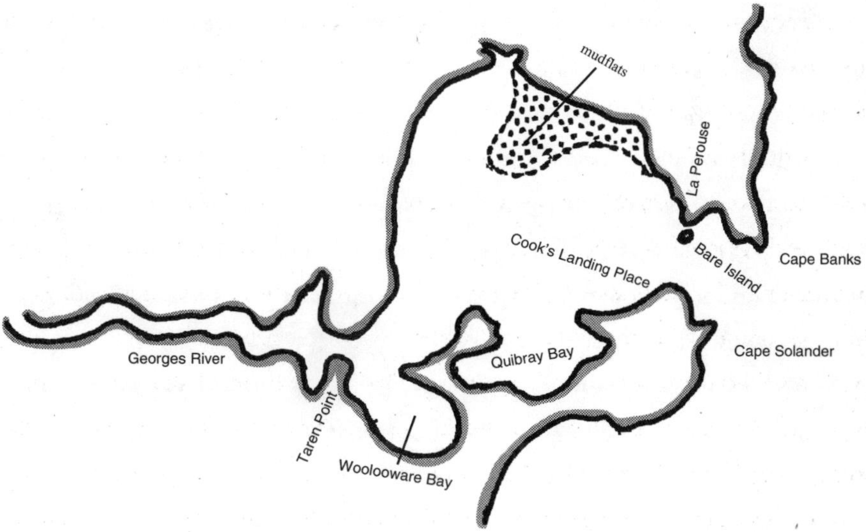

Botany Bay as it was before its shape was drastically altered by reclamations for shipping wharves, an oil refinery and Sydney international airport. Dotted in is the extensive mudbank off the north-eastern shore, drying out at low tide, which appears in the chart published with John Hawkesworth's account of 1773. Its position corresponds closely to that of the large 'island' depicted in the harbour of the Vallard map.

of the bay drawn by Mendonça still bears an unmistakable resemblance to this shore of Botany Bay on a modern map once we mentally delete the airport runways. A small opening appearing on this northern shoreline is in exactly the right place to be the entrance to Cooks River. Further on, the sixteenth century chart clearly depicts La Perouse bay and just off this bay's inner headland we can see an island which corresponds to Bare Island, drawn larger than life, to be sure, but again in the right place and correctly shown dominating the entrance to the harbour. (It was precisely because of its strategic importance that Bare Island was fortified by the New South Wales Government in 1881 due to fears of a Russian invasion; the Portuguese, past masters in fort construction, no doubt immediately recognised its potential.)

There is just one slightly puzzling feature of the bay depicted in the Vallard map: a larger island shown off the northern shore, west of Bare Island. Today there is nothing resembling an island of any size in this area—but then we have to remember that twentieth century dredging and reclamations have altered this northern part of Botany Bay out of all recognition. A far better point of reference is the detailed chart based on Captain Cook's survey which was published by John Hawkesworth in 1773. This chart shows in this same part of the bay an extensive mud or sand bank, drying out at low tide. This provides an obvious explanation for the 'island' of the sixteenth century chart, the more so in view of the fact that this unstable marine environment might well have undergone significant change in the 250 years since Mendonça's ships were at anchor there.

A more meaningful comparison would be to put the harbour shown in the sixteenth century map alongside the small-scale sketch of Botany Bay which appears in Cook's coastal chart of 1770. When we do this, it becomes immediately evident that the Vallard map's harbour bears a far greater resemblance to Botany Bay than does Cook's version. In Cook's chart, indeed, the shape of Botany Bay is barely recognisable.

The similarity of the Vallard map's harbour to Botany Bay is all the more impressive when we remind ourselves that this is not even the original chart—only a copy, hand-drawn on a very small scale, which in turn has had to be magnified several times over in order to reveal for us all the details of the bay's shape. And surely most remarkable of all is the fact that using this enlargement of a map drawn nearly five centuries ago, one can take a pencil and sketch in the modern airport runways in the right place and on approximately the correct scale without any difficulty whatever!

Mendonça's original chart must have emphasised the importance of Botany Bay, because along with the Great Barrier Reef coast (Costa Dangerosa) this is one of the very few recognisable features to appear in other surviving sixteenth century Dieppe maps of Terra Java. The names given to the bay may vary a little from one map to another, but they are all clearly derived from the same source. Three of the other versions (known as the Dauphin, Henry II, and Desceliers maps) have the Gallicised name *Baye Nevsne*. The later cartographer Guillaume Le Testu calls it *Baie de Nevs*. These have no obvious meaning in either Portuguese or French. The name in the Vallard map, *Baia Neve*, is not just closer to the original—it actually *is* Portuguese.

'Baia', as we have already seen, is the word for 'bay'. What about the second word, 'neve'? In Portuguese this means 'snow'. But snow at Sydney? The very notion seems preposterous: an absurd mistake. Before dismissing it out of hand, however, we need to consider a clue. Far behind the southern headland of Botany Bay lies a line of high, white sand dunes and with the sun reflecting off them these dunes shimmer for all the world like snowfields. What can be seen of these snow-white dunes today, however, is but a tiny fraction of what once existed. Sadly, over the past half-century the once-extensive and spectacular dunes behind the southern shore of Botany Bay have all but vanished under the impact of large-scale commercial mining to feed the City of Sydney's rapacious construction industry. At least 70 million tonnes are estimated to have been removed. Government preservation orders introduced in May 2001 were only just in time to save the very last remnants. This is a far cry from the 1940s and 1950s, when this dune country was still so unspoiled that it was chosen as the location to film desert scenes for major

Hollywood and Australian movies with titles like *Forty Thousand Horsemen, The Rats of Tobruk* and *Thunder in the Desert*.

Other historical records confirm that when the first British settlers arrived, undulating snow-white dunes capped all of the high ridges and cliff tops on both sides of the entrance to Botany Bay.

Cook observed in his journal in May 1770 that the sea coast at the entrance to the bay resembled a long, moderately high island and that the soil everywhere was 'a light, white sand'. Eighteen years later, Watkin Tench, the perceptive Captain of Marines serving with the First Fleet which transported the first British convicts to New South Wales, noted as his ship sailed up the Australian coast toward Botany Bay that the hills were 'clothed with trees, to which some little patches of white sandstone gave the appearance of being covered with snow'. Even more interesting is a pen and watercolour sketch of the entrance to Botany Bay drawn by George Raper, a midshipman and unofficial artist with the same fleet. The most notable distinguishing features in Raper's sketch, made from the deck of a ship well offshore, are two prominent white cliffs or headlands which stand out like beacons on either side of the bay's entrance.

The white sand dunes which formerly marked the northern side of the entrance to Botany Bay have now also disappeared—in this case, under the green grass of a golf course. It calls for some imagination to envisage today how in earlier times the snow-like dunes and white cliff faces would have been an arresting landmark for mariners seeking the partly concealed entrance to the wide bay. But it is now clear enough that this is the reason why Mendonça gave this inviting harbour a name which has come down to us as 'Baia Neve'. The name could be translated literally as 'Bay of the Snow', but a better rendition would probably be 'Bay of the Snowfield', bearing in mind that in Portuguese 'neve' can mean either 'snow' or 'snow-*field*'. The Dieppe cartographers evidently attempted to copy the Portuguese name without realising its meaning. The plural versions in the other Dieppe maps suggest that in his original chart Mendonça probably wrote it in the plural—*Baia das Neves* (Bay of Snowfields).[6]

One other conundrum remains to be addressed. If this really is Botany Bay, charted 250 years before the arrival of Cook and the *Endeavour*, what about the peculiar cluster of islands shown in the Vallard map just off the bay's northern headland? There are seven islands in all—two of them larger than the rest. How are they to be explained when there is no such group of islands at this point of the coast? Surely, one might think, this at least has to be a mistake on the map-maker's part! And if so, does it not cast some doubt on the map's reliability? These are not unreasonable questions. The explanation is not obvious—but once again the clues are there, waiting to be discovered. All is not quite what it seems.

To try to solve the riddle we might reasonably start from the premise that the islands are part of the same enlarged section of the map as Botany Bay. That being so, rather than trying to view them realistically, should they perhaps be treated as navigational symbols—in other words, some kind of nautical code? If this is the case, then it should be possible to work out their significance. Let us see what happens when we try this out.

The first detail to examine is the group of small islets. These are denoted by circles, all about the same size and all drawn in red ink as if to differentiate them from the two others, coloured respectively gold and blue. The islets are clustered together and there are five of them. Why five? Is this just chance—or could it be a clue?

The number begins to acquire possible significance when we find that there is in fact a cluster of five small islands just off the coast near the present-day city of Wollongong, some fifty nautical miles south of Botany Bay. These islets are even called the Five Islands Group. The Portuguese could hardly have failed to notice this distinctive little archipelago. Its sighting would be a clear sign for the captains of ships who might be sailing along this coast from the south in later years that they were nearing the haven of Mendonça's 'Bay of Snowfields'.

This is a promising start. But what of the much larger trapezoidal-shaped island drawn in gold-coloured ink close to the presumed north head of Botany Bay? There is no such island anywhere in the vicinity. Surely this must be a mistake?

Once again there is an unlikely, but ultimately simple, explanation. To arrive at it, we have to recall again the difficulties faced by early sixteenth century voyagers in charting unknown coasts from the deck of a small ship without even such a basic item of equipment as a telescope.

There is no large offshore island in the vicinity of Botany Bay, but there is something which comes very near to being an island—the North Head peninsula of Sydney Harbour. Thanks to its preservation as a military reserve since the early years of British settlement, North Head still looks much as it would have done 500 years ago, when Mendonça's ships sailed by. Great ocean swells from the Pacific crash against the peninsula's steep sandstone cliffs, above which a scrub-covered, hog's back of a hill rises to a height of 100 metres. The peninsula is over two kilometres long and almost as broad. Only at its northern tip is it joined to the mainland—and there only by a low-lying sandy spit some 500 metres wide, rising at its highest point to just a few metres above sea level. Across this land-bridge nowadays, throngs of holiday-makers from the Sydney ferries stroll along a popular promenade to the world-renowned ocean beach at Manly. Today the skyline here is dominated by high-rise apartment blocks and towering Norfolk Island pine trees, but in pre-European times the low-lying spit would have been well-nigh invisible from a ship well out at sea.

Confirming this identification of the mystery island is the fact that when we enlarge the fine detail, the shape of the island in the Vallard map resembles that of North Head peninsula to quite an uncanny degree. This similarity of detail is the more remarkable when we recall again that the Vallard map is not even the original chart, but only a copy. Bays on the northern, eastern and southern sides of North Head peninsula can all be identified on the Vallard island, notwithstanding its tiny scale. Also recognisable are North Point, Quarantine Head, and the actual North Head itself.

Mistaking a headland for an island, and vice versa, was not unusual in voyages of exploration even in much later times. We have already seen that Captain Cook mistook Fraser Island for a headland. In his circumnavigation of the South Island of New Zealand this distinguished navigator also drew Banks Peninsula incorrectly as an island and mistook Stewart Island, in the far south, for a peninsula.

The fact that the Portuguese evidently did not sight the land-bridge to North Head also suggests that they probably did not enter Sydney harbour. The manner in which this part of the map has been drawn is consistent with the sightings that would be taken from a ship sailing along the coast in a southerly direction. From this viewpoint the North Head peninsula for a time obscures the entrance to Sydney harbour and appears to overlap the South Head, so that for observers on the ship the harbour entrance would not have come into view until the vessel had actually passed North Head. By then, if wind and currents were unfavourable, it would have been impracticable to turn back and investigate.

In this situation the Portuguese captain would have been able to catch only a brief glimpse of the concealed entrance to the main arm of Sydney harbour, Port Jackson, which is shown on the Vallard map as a deep semicircular indentation (similar to the way Cook was later to draw it on his chart). Looking back over the stern, however, Mendonça would have had a clear view up the wide reach of North Harbour extending beyond North Head peninsula on its sheltered, inland side. This would have confirmed his belief that North Head was an island. Continuing south, a short time later the ship's company would have found themselves off the entrance to Botany Bay.[7]

North Head has been drawn on an even larger scale than Botany Bay. In considering the likely reasons for this we have to remember that the Vallard map has been redrawn from a set of portolans. Each of these sectional charts was probably about the same size as the entire Vallard map which in turn is roughly the size of a modern A3 sheet. Since at least three portolans seem to have been combined to make the Vallard map, the scale would have had to be reduced to possibly about one-third that of the originals. The problem with drastically reducing the size of objects like small islands is that a point is eventually reached when it becomes no longer possible to show any meaningful features: either the island becomes an anonymous dot, or the scale has to be kept large enough for the shape to remain identifiable. In the case of North Head and a few other important islands the latter is the course that has evidently been taken.

Still remaining to be solved in Mendonça's depiction of the approaches to his 'Bay of Snowfields' is one final puzzle—the strange seventh and last island shown. Coloured blue, it is drawn very distinctively in a shape resembling a crude letter 'Y' or perhaps a three-armed starfish. It is the most northerly of this odd group of islands and is located in front of the headland 'Cap Amato' which we saw earlier. What possible explanation can there be for depicting such a peculiar, three-pointed island at a place where no such island exists?

For a long time this proved the most difficult riddle of all to solve. Yet it seemed there had to be a solution and it finally appeared in a completely unexpected place.

The answer was waiting to be discovered in none other than Cook's *Endeavour* journal. In the British captain's entry for 7 May 1770 one name fairly leaps out of the page: 'Cape Three Points'! This was one day after the *Endeavour* had passed the entrance to Sydney harbour (Port Jackson). Sailing north, Cook sighted a distinctive landmark five leagues south-west of the *Endeavour's* position. 'At Noon,' he wrote, 'some pritty high land which projected out in three bluff points and occasioned my calling it *Cape Three Points*.'

Unfortunately the historic name Cape Three Points has disappeared from most modern maps. In recent times, it has been replaced with an Aboriginal name, Boudi.[8] Cape Three Points, alias Boudi, is to be found just north of Broken Bay (which Cook also named), seventy kilometres north of Sydney. The three prominent points of this large headland are Tudibaring Head (also marked on some maps as 'First Point'), Mourawaring Point and Boudi Point. Much of the headland has been preserved as a national park notable for its dense cover of old-growth native bush, a fact which even today amply justifies its Portuguese name of *Cabo Mato*—'Forest Cape'.

The answer to the final part of the riddle of the islands is now clear: Mendonça sighted and drew on his chart a prominent three-pointed 'island'; Cook, sailing along the identical stretch of coast, sighted and named a prominent three-pointed cape. On any reasonable interpretation, both captains were looking at one and the same thing.

Having cracked Mendonça's code we can now read the instructions for locating the entrance to his secure 'Bay of Snowfields'—alias Botany Bay. Coming from the north a pilot should look for a three-pointed island (or headland), followed by a prominent trapezoidal-shaped island (North Head). To the south, he should keep a lookout for a group of five small islands. While this would not be immediately obvious to a present-day reader of the chart, it makes complete sense when one considers that the original portolan would no doubt have been augmented by a *roteiro* (in English 'rutter', or log of sailing directions) explaining the significance of important landmarks of this kind. This might even have explained that the three-pointed 'island' was actually an enlargement of the headland *Cabo Mato*.

In summary, then, what we have found is that a puzzling group of seemingly non-existent islands which at first appeared to be a serious defect in the Vallard map turns out to be the very opposite—powerful evidence of the map's authenticity. And this evidence is the more convincing for being totally unexpected.

Cook's journal also offers a compelling reason why Mendonça was at such pains to set out for other, later mariners the early warning signs that they were nearing the important harbour of the Bay of Snowfields. In his journal entry for 6 May, Cook wrote of the Botany Bay entrance: 'in coming from the Southward it is discovered before you are abreast of it, which you cannot do in coming from the northward'. Cook came from the south but Mendonça approached from the north as subsequent Portuguese voyagers could have been expected to do. It would have been important to provide them with

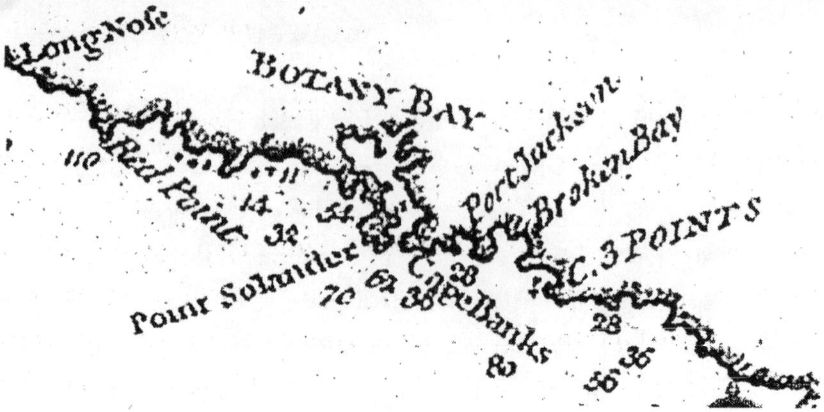

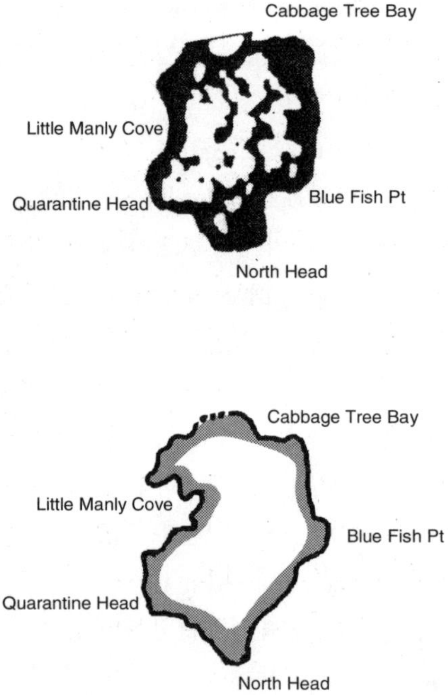

Top: Botany Bay, as it appears in Cook's chart of the coast of New South Wales: it is far less recognisable than the same bay depicted in the 16th century Vallard map.

Bottom left: Mendonça's code for the approaches to Botany Bay (clockwise from right), the Five Islands; North Head, and Cape Three Points (the Vallard map's *Cap amato*). North Head and Cape Three Points are both drawn as islands.

Bottom right: An enlargement of the Vallard chart's mystery island off Botany Bay (top) compared with Sydney harbour's North Head peninsula. The similarity of shape is unmistakable, even though the original island in the Vallard map is a mere 3 mm long. Recognisable in the 16th century map are North Head, Blue Fish Point on the seaward side, and Quarantine Head and Little Manly Cove on the opposite shore facing North Harbour. The dotted line on the North Head peninsula map indicates the point where this 'island' is joined to the mainland by a low sandspit.

easily recognisable landmarks, so that with a following wind they would not overshoot the harbour entrance and then have to try and beat laboriously back to it.

So many are the similarities of detail between the shape of Mendonça's harbour and Botany Bay that there really can be no doubting the fact that the Portuguese sailed into Botany Bay and explored as far as its upper reaches. Apart from the obvious conclusion that Cook and his men were not after all the first, this raises other intriguing questions. Did the Portuguese also land here? If so, where?

It would be most surprising if they had *not* gone ashore, if only to take on essential fresh water and firewood. Assuming they did land, the most likely place would have been the one chosen by Cook 250 years later, the eastern end of Kurnell beach, in the shelter of Sutherland Point, where water could be obtained. Another possible landing point would be on the opposite side of the bay, on the northern shore near Bare Island, where Cook also noted the presence of fresh water streams.

At this point, one unfounded rumour reflecting adversely on the English captain should be laid to rest. One or two commentators have claimed that a name which appears on the sixteenth century French 'Dauphin' map, *Coste des Herbaiges* (Coast of Pastures), is the source of the name given by Cook to 'Botany Bay' and that Cook therefore must secretly have been carrying a copy of the Dauphin map with him. This conspiratorial theory is completely at odds with the facts. In the first place, the *Coste des Herbaiges* is in entirely the wrong location to be Botany Bay—it appears on the coast of Queensland, north of Fraser Island. And if that were not enough, Cook's Journal makes it abundantly clear that he gave Botany Bay its name for an entirely different reason. He originally intended to call it 'String-rays Harbour', but changed the name to 'Botanist Harbour' to mark the fact that this was where his ship's naturalists, led by the eminent Joseph Banks, had spent many fruitful days ashore 'botanising'. Later Cook changed the name again to 'Botanist Bay', and finally 'Botany Bay'.

For Cook and his crew, Botany Bay was their first landing on Australian soil. Not so, we can safely assume, for Mendonça's men. On their long and arduous voyage down this unknown coast from Torres Strait they would by now have seen the smoke from countless Aboriginal fires and would without doubt have landed at one harbour or another to find out what they could about this land and its people. So the scene at their bay of the snow-white dunes would by then have become familiar: flocks of brilliantly coloured parrots and cockatoos wheeling overhead, a backdrop of grey-green eucalyptus forests, strange trees whose flowers looked like hairy candles, and exotic animals which bounded on their hind legs instead of running on all fours. A country sparsely inhabited by an elusive black people who went about naked, knew nothing of agriculture and seemed to have no permanent habitation.

Most of the illustrations in the Vallard map of Australia's eastern coast are misleading because it is scenes from Sumatra rather than Australia which mainly illuminate the interior of Terra Java. Much of this pictorial material, as we shall see in a later chapter, derives from a French voyage to southern Sumatra in 1529. Pictured prominently in the Vallard map are four Sumatran houses on stilts, family scenes and plantations of coconut palms, with one palm tree apparently about to be harvested by a climber. In the foreground is a rajah's procession. The rajah is mounted on a horse and shaded by a retainer on foot holding a parasol; following him are troops carrying lances and scimitars. Difficult to see at first glance is the outline of a recumbent lion, partly hidden by undergrowth.

Possibly of more relevance to Australia are quite different scenes in the far distance depicting groups of men with spears, mostly either naked or semi-naked. Some are mounted on horses and camels, but most are on foot. One group at the top left appears to be bending over a camp fire. These people, some of them coloured black and others reddish, are shown moving through a misty blue landscape of cliffs, rock pinnacles and mountains. The gaunt rock formations are not unlike those seen in parts of Australia. And without jumping to any conclusions, the pale blue colour in which the mountains in this part of the map are painted could bring to mind the fact that it was the hazy colour of the mountains visible on the horizon inland from Botany Bay which led the first European settlers to call them the *Blue* Mountains.

And amid the other exotica there are individual drawings which could fit Australia. At the northern end of Terra Java's east coast we can see a turtle, as well as a reptile which looks like a goanna. A lean dog (which might or might not be a dingo) is shown bounding in the far distance alongside a group of black men.

It seems that some of Mendonça's crew must have taken a small vessel a considerable distance up the Georges River, since in the Vallard map the river is shown winding sinuously inland, just as it does in reality. The river has been roughly mapped for about the same distance as the length of Botany Bay from Cape Banks to the river entrance that is, for about ten kilometres. Beyond that no further identifiable detail is shown.

Surprisingly, despite the fact that it loops through the densely populated western suburbs of Sydney, this section of the Georges River even today retains something of its pre-European character. For part of its journey it winds between yellow-grey sandstone cliffs where dry eucalypt forest reaches unbroken to the water's edge, just as it would have done when Mendonça's men sailed by those many centuries ago. Beaches of yellow sand alternate with muddier bays supporting dense mangrove thickets. Here and there, high-rise apartment blocks and million-dollar mansions shoulder the bush aside—but even so, the quiet river still seems a world away from the frenetic bustle of urban Sydney.

In Europe or Asia in the sixteenth century a river like this would often have been the highway to an important inland town or trading centre. But this was Australia and here no such settlements existed. The Portuguese explorers evidently gave up when the river narrowed to the point where it would have no longer been navigable for anything much bigger than a rowing boat.

An intriguing but apocryphal story exists of an antique iron helmet, described as 'Spanish', being found in a swampy area of the Georges River in the locality of East Hills about the year 1936. Could this provide a possible link with the chart? East Hills would have been near the upper point of the river depicted in the Vallard map, so the discovery here of an Iberian-type military helmet (for 'Spanish' we might read 'Portuguese') would have been a key piece of evidence. Unfortunately, if this helmet ever existed it seems to have disappeared without trace.[9]

Assuming that the Portuguese landed at Botany Bay, we can reasonably suppose that they would have made some attempt to trade with the indigenous people—hoping in particular, no doubt, for signs of gold. What sort of reception would they have got? A useful parallel here is the arrival of Cook 250 years later. Here is how Joseph Banks recorded the scene at Botany Bay on the morning of 28 April 1770:

> A few of the Indians who had not followed the [ship's] boat remaind on the rocks opposite the ship, threatning and menacing with their pikes and swords—two in particular who were painted with white, their faces seemingly only dusted over with it, their bodies painted with broad strokes drawn over their breasts and backs resembling much a soldier's cross belts, and their legs and thighs also with such like broad strokes drawn round them which imitated broad garters or bracelets. Each of these held in his hand a wooden weapon about 2 1/2 feet long, in shape much resembling a scymeter.[10]

Banks also noted that near the harbour's south head were four small canoes, in each of which was a man fishing with a long spear. On the shore nearby was a 'small village consisting of about 6 or 8 houses'. An older woman and several small children appeared, and the woman lit a cooking fire. 'Of all these people we had seen so distinctly through our glasses we had not been able to observe the least signs of Cloathing,' Banks added. 'Myself to the best of my judgment plainly discernd that the woman did not copy our mother Eve even in the fig leaf.'

The Englishmen's first attempt to row to the beach bearing trade goods was fiercely opposed by spear-brandishing Aboriginal warriors.

In Banks's words:

> As soon as we approached the rocks two of the men came down upon them, each armd with a lance of about 10 feet long and a short stick which he seemed to handle as if it was a machine to throw the lance. They calld to us very loud in a harsh sounding Language of which neither us or Tupia [the Tahitian interpreter] understood a word, shaking their lances and menacing, in all appearance resolvd to dispute our landing to the utmost tho they were but two and we 30 or 40 at least.

So resolute was the Aborigines' defiance that Cook had to resort to demonstrating the power of firearms before he and his officers felt secure enough to step ashore.

The Portuguese voyagers, for their part, would very soon have discovered that these black people knew nothing of gold—or any other metal for that matter. They had no towns and nothing to trade. For Mendonça's men there could have been only one disappointing but inescapable conclusion: if the promised land of gold did lie somewhere in this direction, they would have to voyage still further to find it.

6

South to Cape Beautiful and a Royal River

You should not marvel at all these accounts, for things are written as I who have made the voyages have seen them, and those who have made them or read them in books will know if it is true.

JEAN ALFONSE, *VOYAGES AVANTUREUX*, 1559

By now, many months would have passed since Mendonça's flotilla set sail from the Portuguese fortress at Malacca. Assuming that it was indeed the feast day of St Ambrose, 7 December when his ships passed the entrance to the Manning River north of Sydney, they should have arrived at Botany Bay no later than the middle of December. Did these staunchly Christian mariners pause to celebrate the festival of the 'Christ-mass' here or did they press on? There is no evidence either way. What we can be sure of, though, is that the discovery of the capacious 'Bay of Snowfields', welcome though it would have been, could have offered no useful clues as to the whereabouts of either their rival Magellan, or their ultimate objective, the Island of Gold.

So after leaving the white dunes of Botany Bay astern, Mendonça ordered his ships to turn once more to starboard, following the coast of the great unknown continent as it led still further south. We can say this with certainty because the coastline of the former eastern promontory, now realigned as the southern section of the Vallard map, portrays the *actual* coastline of this part of Australia with remarkable precision.

How far did the Portuguese ships venture and what did they discover? The answers will bring some more surprises.

To begin with, there can be no doubt about the identity of the coastline shown immediately south of the map's equivalent of Botany Bay. This has been given the name *Teramor*. In the original Portuguese it would have been *Terra mor*, 'mor' being old Portuguese for 'greater' or 'higher' (the modern word is *maior*). This, then, is the 'Higher Land', or perhaps 'Highlands', a fitting description for the great bush-clad rampart of the Illawarra Escarpment, which on this part of the coast rises abruptly to a height of more than 700 metres, providing a dramatic backdrop for the port city of Wollongong, some eighty kilometres south of Sydney.

Next on the map come a series of river mouths and headlands depicted in simplified, stylised form. The identity of these will become clearer if for the moment we pass them by and move further down the coast to a very distinctive feature which appears to have been drawn on the sixteenth century map with particular care. The feature in question is a large bay which, viewed from south to north, ends in a curve the shape of a fishhook. Is there anything comparable to it on a modern map? The answer is an emphatic yes. It is instantly recognisable as Disaster Bay, south of the present-day fishing port of Eden.

The bay itself has not been given a name on the Vallard chart, but just north of it is a prominent headland with the name *Cap Groca* (*Cabo Grosso* or Great Cape). This clearly is Green Cape, site of a lighthouse since the nineteenth century and one of the most prominent headlands on the southern New South Wales coast. We will look at this region in more detail later, it will prove to be of crucial importance in piecing together dramatic events which were soon to befall Mendonça's expedition.

The Portuguese navigators knew what they were about when they drew this section of coast carefully for the benefit of others who might follow.

The fishhook of Disaster Bay is a potentially lethal trap for northbound ships. As the name suggests it has been the scene of numerous disasters over the past two centuries, the worst of which was the wreck of the passenger steamer *Ly-ee-moon*, bound from Melbourne to Sydney, which ran at full speed onto the end of Green Cape in 1886. Seventy-two people died in that disaster.

With Green Cape and Disaster Bay firmly identified we can now retrace our steps and fill in the intervening coastline south of *Teramor*—the Illawarra Escarpment.

The first major river south of Wollongong is the Shoalhaven. On the Vallard map it has been given the name *Rio S Tarqno*. The 'S' is the map's customary abbreviation for 'Saint'—the saint in this case evidently being Iago (James) whose name in the early sixteenth century could be written either 'Santiago' or (as it is in Portugal today) 'São Tiago'.[1] A short distance further on we come to the name *P S Antonio*, an abbreviated version of *Ponto Santo Antonio*: in English 'Point St Anthony'. Here the Saint's name denotes a prominent headland.

Both St James and St Anthony were held in particularly high regard by Portuguese seamen, St Anthony because he was the patron saint of Lisbon, where he was born. This can be confusing because he is also known as St Anthony of Padua, the Italian town where he died. In Portuguese legend, St Anthony had the power to rescue shipwrecked sailors and provide favourable winds. The apostle St James (São Tiago) had a sterner aspect, reputedly taking an important role in the wars to free Portugal from Moorish occupation. On one such occasion in 1139, not long before the liberation of Lisbon, soldiers claimed to have seen the Saint leading the Christian forces into battle.

St James's feast day is 3 May and that of St Anthony 13 June, dates at the beginning of winter which seem unlikely to have any relevance in terms of the ships' progress. Probably these two saints were chosen for commemoration at this stage of the voyage simply because of their importance.

The Vallard map's 'River of St James', as we saw, is today's Shoalhaven. As for 'Point St Anthony', this prominent headland can be confidently identified

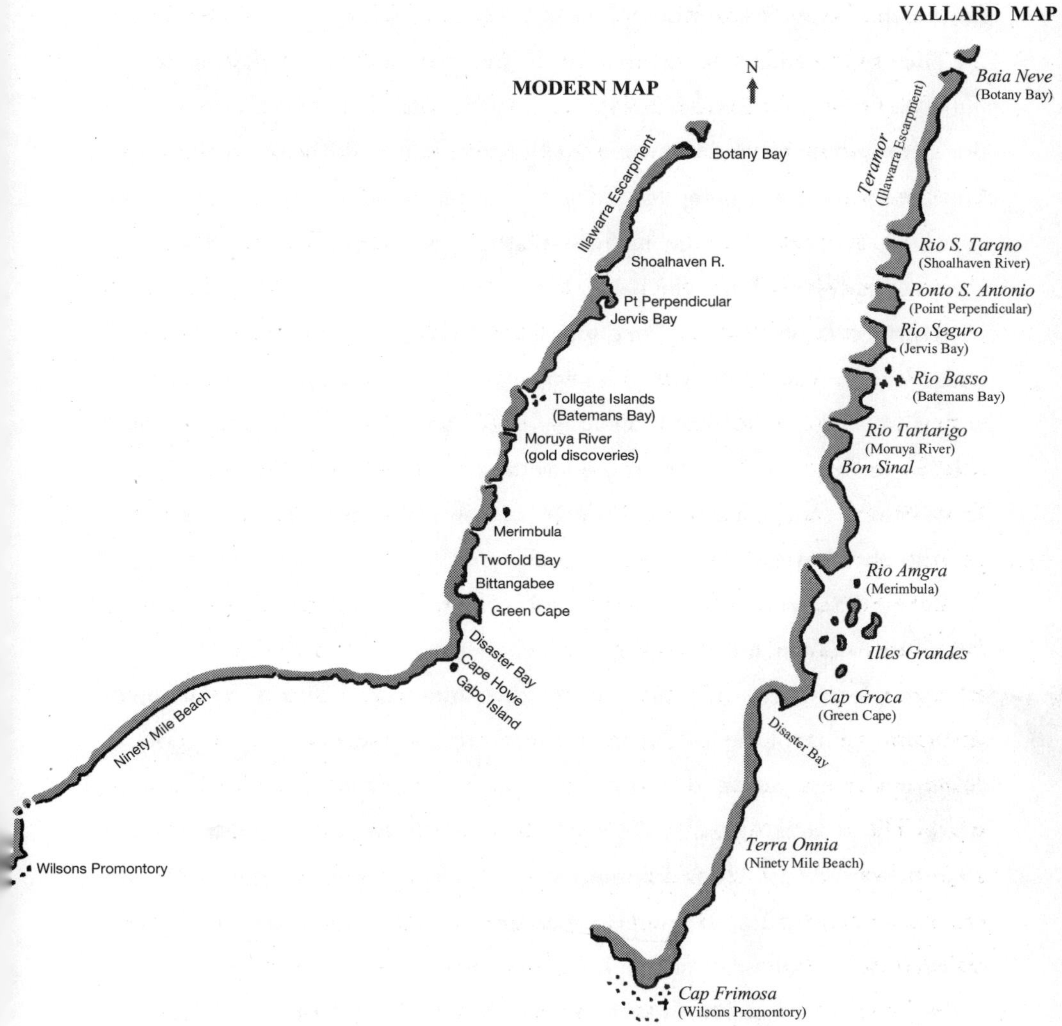

The Vallard map's coastline from Botany Bay south to Wilsons Promontory, compared with a modern map of the same coast.

as Point Perpendicular, at the northern end of Jervis Bay. This headland's towering cliffs, dropping almost sheer to the water's edge, make an awe-inspiring sight. Captain Cook, too, noted the striking appearance of this headland, naming it Point Long Nose.

The next feature on the sixteenth century map, *Rio Seguro* (Secure River) equates with Jervis Bay itself. The bay is well named. Jervis Bay is indeed a fine, secure bay and claimed to be the deepest sheltered harbour in Australia. Until recently it was being touted as a location for the country's main naval base. The Portuguese name suggests that one or more of Mendonça's ships may well have sailed into the harbour and taken soundings.

The next large river the Portuguese flotilla passed has the name *Rio Basso*. This name has occurred already—on the Queensland coast south of Cairns—and as we saw then it means 'Shoal River'. It would be a neat conjunction if this 'Shoal River' could be identified as the present-day Shoalhaven River but its position is too far south and in any case we have already found a match for the Shoalhaven. Mendonça's 'Shoal River' is in fact the Clyde River at Batemans Bay, today a popular seaside resort favoured by holiday-makers from Australia's inland 'Bush Capital', Canberra.

This identification can be made with confidence thanks to yet another intriguing point of detail. Off the entrance to his 'Shoal River', Mendonça's chart-maker has indicated the presence of a group of three very small islands, marked by crosses. These correspond exactly with the position of the Tollgate Islands, lying some three kilometres off Batemans Bay. As seen from the shore, though, the Tollgates appear to be just two islands, so does this indicate an error in Mendonça's chart? In fact the answer is just the opposite. The group does in fact include a third much smaller island but this is visible only from the ocean side: the side from which, naturally enough, the Portuguese mariners would have been viewing it.

Interestingly, this little collection of islets was also noted by Captain Cook who, in naming the estuary 'Bateman Bay', described it as 'an open bay wherein lay 3 or 4 small islands'. As for the number, Cook, with the aid of a telescope, was probably including Snapper Island, which lies closer to the shore. And if further evidence were needed that *Rio Basso* is indeed Batemans Bay it can be found in the fact that the sixteenth century map also depicts with remarkable clarity the distinctive, scallop-shaped Broulee Bay, a little further south.

Rio Tartarigo, the name which appears alongside the next river, is almost certainly a misreading from the original Portuguese of the word 'tartaruga', which as we saw earlier means 'turtle'. This 'Turtle River' is in the right place to be the substantial Moruya River, fed by numerous creeks and originating on the slopes of distant bush-clad ranges. And the fact is that both the Moruya River and its tributaries are indeed the haunt of freshwater turtles—a native species variously known as the Long-Necked, or Eastern Snake-Headed Turtle.

We will return to Mendonça's 'Turtle River' shortly. First, though, we need to look at the next feature on the Vallard map. It is something very puzzling: a wide bay behind which, on land, appears the legend *Bon Sinal*. This name, written on the map in correct Portuguese, means literally 'Good Sign' or 'Good Auspices'. What can it mean? Oddly enough, it has a counterpart far away on the western shore of the Indian Ocean. Here, an almost identical name was given by Vasco da Gama to the Zambezi River in southern Africa when his ships found refuge there on 24 January 1498 during their epoch-making voyage to India. Da Gama, with his crews beginning to suffer from scurvy, took the opportunity presented by the estuary of the great African river to replenish his water supplies, careen his ships and repair the damage caused by storms off the Cape of Good Hope. He named the Zambezi *Rio dos Bons Sinais*—'River of Good Signs' ('sinais' being the plural of 'sinal').

So what was the 'good sign' recorded by Mendonça on the south-east coast of Australia? It does not seem to be intended as the name of a river. The location of *Bon Sinal* would appear to be not far from the bay of Tuross, behind which lies the extensive Tuross Lake, where in pre-European times there was a large Aboriginal population. Even so, what possible connection could there be? Alternatively, might the name relate to nearby Mount Dromedary, whose peculiar camel-like hump was remarked on by Captain Cook (who also gave it its name)? Might the shape of this mountain have been seen by Portuguese seafarers as auspicious?

These may be possible explanations, but neither seems entirely convincing. Perhaps we need to think again. Why was this 'sign' so important that it had to be recorded on the map? What about the expedition's primary objective—the discovery of gold?

If we follow this lead we may well find ourselves on the right trail.

The first question that needs to be asked is whether there is any known link between this part of Australia and the discovery of gold. The answer turns out to be an emphatic yes. This may not be one of Australia's best-known goldfields, but it was in this area that one of the first major discoveries of the precious metal was made in 1851, in the early years of British colonisation. The discoveries were of fine alluvial gold, and they were made just a few kilometres inland from the coast south of Batemans Bay—an area which included, most significantly, the numerous small waterways leading to Mendonça's 'Turtle River' at Moruya. These finds were the trigger for one of Australia's first great gold rushes, with as many as 30 000 men reportedly panning the region's creeks for what was termed 'mustard gold' at the height of the rush. One gold-bearing quartz vein was even discovered on the sea shore.

The precious metal was mined commercially in the area for several decades. Today the field is recognised historically as Australia's largest alluvial goldfield—and even now fossickers still pan for gold in the local creeks. So it would hardly be surprising if some of Mendonça's gold seekers had found signs of it (not to mention turtles) in these very same waterways. Clearly it would have been an important discovery. To some extent, at least, the expedition's main goal might at last seem to have been realised.

As for the significance of the name on the map, given the Portuguese obsession with secrecy, and particularly in view of the fact that Mendonça would have been well aware that he was intruding into what was officially

Spanish territory, the cryptic reference 'Good Sign' would be enough to mark the place of the discovery—without giving away the secret of what the discovery actually was!

Leaving their 'Turtle River' astern (and quite possibly taking with them samples of their 'good sign'), Mendonça's ships would soon have been sailing past rocky Montague Island, more than a kilometre long and lying some ten kilometres from the coast. Nowadays a nature reserve and the haunt of seals and penguins, Montague Island is moderately high in places and visible from a considerable distance, so Mendonça's lookout would surely have had no difficulty sighting it (even though Cook for some reason apparently failed to see it). And in fact the hourglass shape of one of the first of a group of islands depicted in the Vallard map near this point does bear some resemblance to Montague Island.

Here, correctly, the map also shows the coast bulging for a short distance eastward toward a prominent headland corresponding to Cape Dromedary, lying in front of the mountain of the same name. Just past this headland the coast (again correctly) is shown turning south-west as far as a river called *Rio Amgra* ('amgra' probably being a misreading of the Portuguese word 'angra', meaning 'bay'). This bay corresponds with the entrance to Merimbula Lake, which would have been visible from some distance out at sea. Just south of the river the map depicts a double bay evidently intended to represent Twofold Bay, at Eden.

The island in the Vallard map which would seem to be Montague Island is one of the first of a cluster of seven with the name *Illes Grandes* (Big Islands) though none of them are shown as being particularly large. This island group is something of a puzzle. The smaller islands could possibly represent some or all of the islets that do exist on this stretch of coast among them Bournda Island, the Mimosa Rocks and Leonards Island just north of Twofold Bay. If this is the case, though, some of the islands are out of position. Alternatively, there might perhaps be a coded navigational meaning, as with Botany Bay—but if so, it is elusive.

We are now back at Green Cape—the Vallard map's *Cap Groca*. South beyond its clearly drawn Disaster Bay the map correctly depicts a long, slightly indented but otherwise unbroken coastline trending south and then west of south. Cape Howe, nowhere near as prominent as Green Cape, has not been given a name and is indicated (as in reality) by just a small bump on the coastline. The lack of detail on this part of the coast, together with the fact that the south-flowing ocean current sweeps strongly eastward off Green Cape, suggests that the Portuguese ships were far offshore when the coast of what is now Victoria began to turn westward. On the map this coastline has the name *Terra Onnia*. The word 'onnia' is puzzling, since there is no such word in either Portuguese of French. This is evidently a misspelling of 'Terra Omnia', meaning 'Featureless Land'. It was an appropriate enough name for the low-lying, featureless shoreline which the Portuguese ships were sailing past—today called Ninety Mile Beach.

Now we have arrived with Mendonça at what will turn out to be one of the most remarkable features of his extraordinary chart—a very prominent cape whose name has been written in the Vallard map in unusually large blue letters *C: Frimoza*. As if to further emphasise its importance the cape's name appears again in the map with slight variations no fewer than three times, in red ink as well as blue. The colours themselves do not seem to carry any particular meaning, but in one of the two red versions the name is given as *Cap Frimosa*, while the second red version calls it *Cap de Frimosa*.

This prominent headland appears in several other Dieppe maps. In the Jean Rotz world map of 1542 it has no name but in the Dauphin chart of about 1536 it is called *C: de Fremose*. In the 1546 Henry II map by Pierre Desceliers the name has been altered slightly to *Cap de Freinose*, while Guillaume Le Testu (in 1556) calls it *Cap de Fremoze*. In this instance the Dauphin is closest to the original: the Portuguese name would have been *Cabo Fremoso* (Cape Beautiful)—*fremoso* being the old spelling of the word for 'beautiful'. In modern Portuguese the word is *formoso* (or *formosa*, in the feminine form; the Portuguese name for Taiwan, used until last century, was *Ilha Formosa*, Beautiful Island).

Identifying *Cap Frimosa* has been a highly contentious issue. The most widely held view, that it represents Cape Howe, was put forward by its advocates because all the Dieppe maps show it as the most easterly point of the continent at the tip of the controversial great eastern promontory and therefore Cape Howe seemed the obvious candidate. But this theory ignored two important facts. Firstly, Cape Howe is not the prominent, wedge-shaped tip of a dramatic triangular land mass, as drawn on the Dieppe maps—on the contrary, it is a low and rather nondescript headland. And secondly, Cape Howe is not even the most easterly point on this part of the coast: that distinction belongs to the much more impressive Green Cape, which we have identified in the Vallard map as *Cap Groca*. It needs only to be added that the Cape Howe theory becomes redundant once we accept that on the original Portuguese chart there was no eastern promontory at all.

So where in fact is *Cap Frimosa*—Mendonça's 'Cape Beautiful'? It was not so very often that Portuguese mariners bestowed romantic names on features of the lands they discovered. The names that appear in the Vallard map tend to be practical, sometimes almost to the point of banality: *Rio Grande* (Big River) occurs time and again, as do *Terra Alta* (High Land), *Illa Grossa* (Big or Great Island) and *Bonno Porto* (Good Port). It must have been a spectacular sight indeed to have moved Christopher de Mendonça to christen this headland 'Cape Beautiful'. On the reoriented Vallard map there is really only one candidate for the honour and it fits the description perfectly—the cape now known as Wilsons Promontory.

Relatively few Australians are familiar with the appearance of Wilsons Promontory, particularly from out at sea. This southernmost headland of mainland Australia, known to local Aborigines as *Wamoom*, is distant from major shipping routes as well as from the paths of ocean yacht races. On land the cape can be reached only on foot through forest tracks. At sea, apart from local fishermen only a handful of people venture there in summer on the occasional seal-watching cruise.

But does Wilsons Promontory live up to the Portuguese description? The answer has to be an unqualified yes. Viewed from a small boat out at sea the promontory is a magnificent sight. Still looking much as it would have done when Mendonça's ships sailed by 500 years ago, it rises from the ocean against a backdrop of rugged, bush-clad peaks whose forest canopy sweeps down unbroken almost to the water's edge. These forests, now a nature reserve, are home to a rich variety of native wildlife including koalas, wallabies, wombats, emus and rainbow lorikeets.

Opposite the headland, out in Bass Strait, the spectacular red granite cliffs of conical Rodondo Island rise almost sheer from the ocean to a height of 450 metres. The shoreline of the promontory itself is fringed by huge, wave-sculpted granite boulders varying in colour from grey or cream to brilliant orange, with here and there a beach of dazzling white sand fringing a turquoise sea. The towering wedge-shaped headland itself was once the northern rise of a mountain ridge that extended all the way to Tasmania. Today crowned by a lighthouse, it breasts the great swells of the Southern Ocean like the bow of some majestic ocean liner. Well might Mendonça have christened it 'Cabo Fremoso'. 'Cape Beautiful' was a fitting name indeed![2]

In the Vallard map the eastern shore of the promontory looks a little unfamiliar because it would have been impossible for Mendonça's ships to have sighted the hidden expanse of Corner Inlet, the large, shallow gulf which cuts in behind Wilsons Promontory and the adjoining coastline on its eastern side. From out at sea the long, dark line of scrub-covered sand dunes fringing the eastern arm of Corner Inlet has the appearance of a continuous shoreline connecting the promontory with Ninety Mile Beach, exactly as drawn from Mendonça's chart.

To present a reasonable comparison, the modern map of the coast on page 173 shows only the entrance to Corner Inlet. The resemblance between the Vallard's *Cap Frimosa* and Wilsons Promontory is now abundantly clear. The shape of Mendonça's promontory is correct, even down to such detail as the depiction of a small bay near the promontory's south-eastern tip and just next to it a small but distinctive east-facing headland. Clusters of dots warn,

How Christopher de Mendonça's flagship would have looked: the carrack *Santa Catarina de Monte Sinai* carrying to France Princess Beatriz, second daughter of Portugal's King Manuel, for her marriage to Charles III, Duke of Savoy in 1521 – the same year Mendonça sailed on the first of his voyages to the 'Island of Gold' (Australia). This is one of the few authentic pictures of a carrack of Mendonça's day. One notable feature of ships of this kind was their huge mainsail, which could give surprisingly fast speeds in favourable winds. The *Santa Catarina* has been richly adorned for the royal occasion with flags and banners, and crewmen can be seen in the crow's nest and rigging. The painting is anonymous, but is ascribed to the early 16th century Flemish school of Joachim Patinir. (*National Maritime Museum, London*)

Right: This portolan chart of the North Atlantic was made by the Portuguese cartographer Lopo Homem about 1550; the rough natural shape of the animal hide parchment on which it was drawn is clearly evident. A series of animal hide parchments of this sort would have been used by Mendonça to map the coasts of Australia. (*Biblioteca Nacional, Lisbon*)

The Vallard map of the eastern section of 'Terra Java'. Does it depict the east coast of Australia? To make recognition easier, in this reproduction the map has been inverted because the cartographer invariably put south at the top. (*This item is reproduced by permission of the Huntington Library, San Marino, Calfornia HM29*)

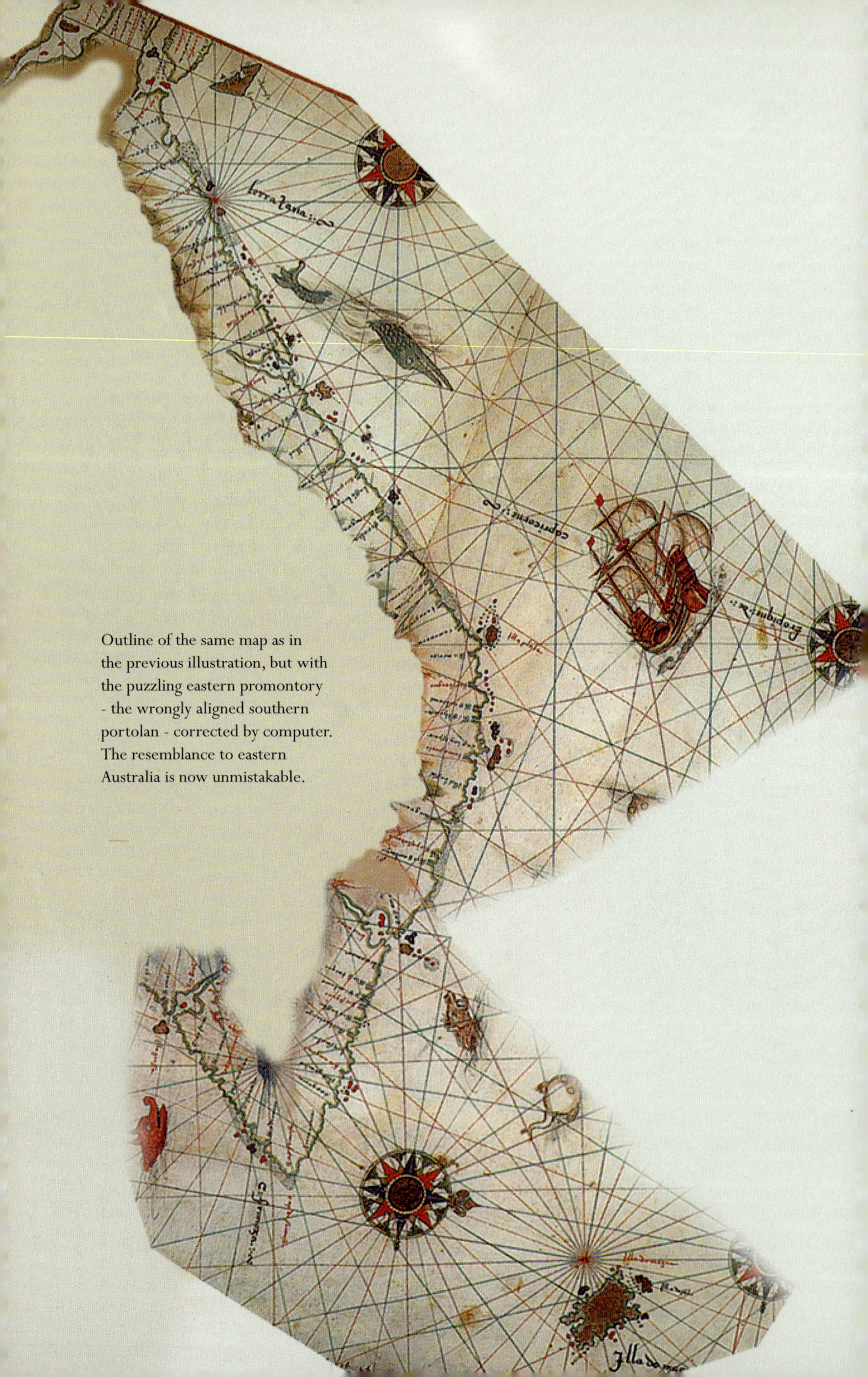

Outline of the same map as in the previous illustration, but with the puzzling eastern promontory - the wrongly aligned southern portolan - corrected by computer. The resemblance to eastern Australia is now unmistakable.

The Vallard map of western 'Terra Java'. Again for ease of recognition the map has been inverted, because the cartographer had south at the top. As with its eastern counterpart, realignment of one misplaced portolan will produce a map which is unmistakably that of the coast of Western Australia. (*This item is reproduced by permission of the Huntington Library, San Marino, Calfornia HM29*)

The decorated bronze cannon (above) found on an Aboriginal sacred site at Carronade Island on the Kimberley coast, compared with a replica sixteenth century Portuguese cannon (below) on display at the Lisbon Maritime Museum. The similarity between the two is self-evident. (*Photos: the author*).

Christopher de Mendonça's family home was this castle of Mourão commanded by his father, Pedro de Mendonça, who also had a distinguished career as a sea captain. The castle was one of a chain of fortresses guarding the turbulent Spanish border.
(*Photo: the author*)

Above: The snow-white cliffs and sand dunes which once marked both sides of the entrance to Botany Bay can be seen in this watercolour by George Raper, midshipman and artist with the first British convict fleet which arrived off Botany Bay on 19 January 1788. Raper painted this scene while the fleet was hove-to off the bay's entrance. (*The Land about Botany Bay*, National Library of Australia, an21511990-2). **Below:** A surfer prepares to practise snowboarding on the last remaining white sand dunes of Botany Bay, Mendonça's 'Bay of Snowfields'. (*Photo: the author*)

Below: A corner of the mysterious stone ruins at Bittangabee Bay: legend links them with the coming long ago of a Portuguese ship. (*Photo: the author*).

Wilson's Promontory, Mendonça's 'Cape Beautiful', at the southern extremity of mainland Australia. In the distance is Rodondo Island. (*Photo: the author*)

Above: This Portuguese-type pottery jar trawled up from the sea bed off Gabo Island has been scientifically dated to about the year 1500. It would probably have held either wine or olives.
Right: A similar pot trawled up in the same area more recently. It has yet to be dated. (*Photos: Dan Fitzhenry and the author*)

correctly, of the danger to passing ships from the numerous small islands and reefs lying off the southern and western shores of Cap Frimosa. The entrance to Corner Inlet is also hinted at, in the form of a bay.

If one were to go by the scale of latitude shown on the right-hand edge of the Vallard map, the global positioning of Cap Frimosa would appear to be misplaced. After correction for the original misalignment of the map's southern section, the scale would place Cap Frimosa at about latitude fifty-four degrees—half way between Tasmania and the coast of Antarctica! This error, however, is the fault of the Dieppe cartographer—not Mendonça or his navigator. It was in fact an inevitable outcome of the cartographer's original mistake when he decided that the north coast of Australia was the same as the north coast of Java. This put Cape York almost on the Equator, with the result that the Vallard cartographer's scale of degrees of latitude, based on the distance between this false 'Equator' and the accurately positioned Tropic of Capricorn, is far too small. The mistake is compounded the further south the map goes and this is why there is such a large error in the scale when we reach the latitude of Cap Frimosa.

One further piece of evidence will confirm that Cap Frimosa is indeed Wilsons Promontory. As we will see in detail in a later chapter, this cape's *actual* latitude and its bearings from other known landmarks are set out with astonishing accuracy in a sixteenth century book on world navigation written by one of the outstanding European mariners of that period.

In this southern section of the sixteenth century map, both the east coast of Terra Java from Cape Howe to Wilsons Promontory and the coastline extending westward from the promontory trend more steeply to the north than is actually the case. This is almost certainly due largely to the difficulty of determining longitude, particularly when combined with the imponderable effects of winds and ocean currents in distant, unknown seas. With Mendonça's fleet now heading west rather than south, progress could only be measured by dead reckoning, an estimate of the distance sailed each day and night. In these circumstances, errors of longitude were inevitable. Matthew Flinders found

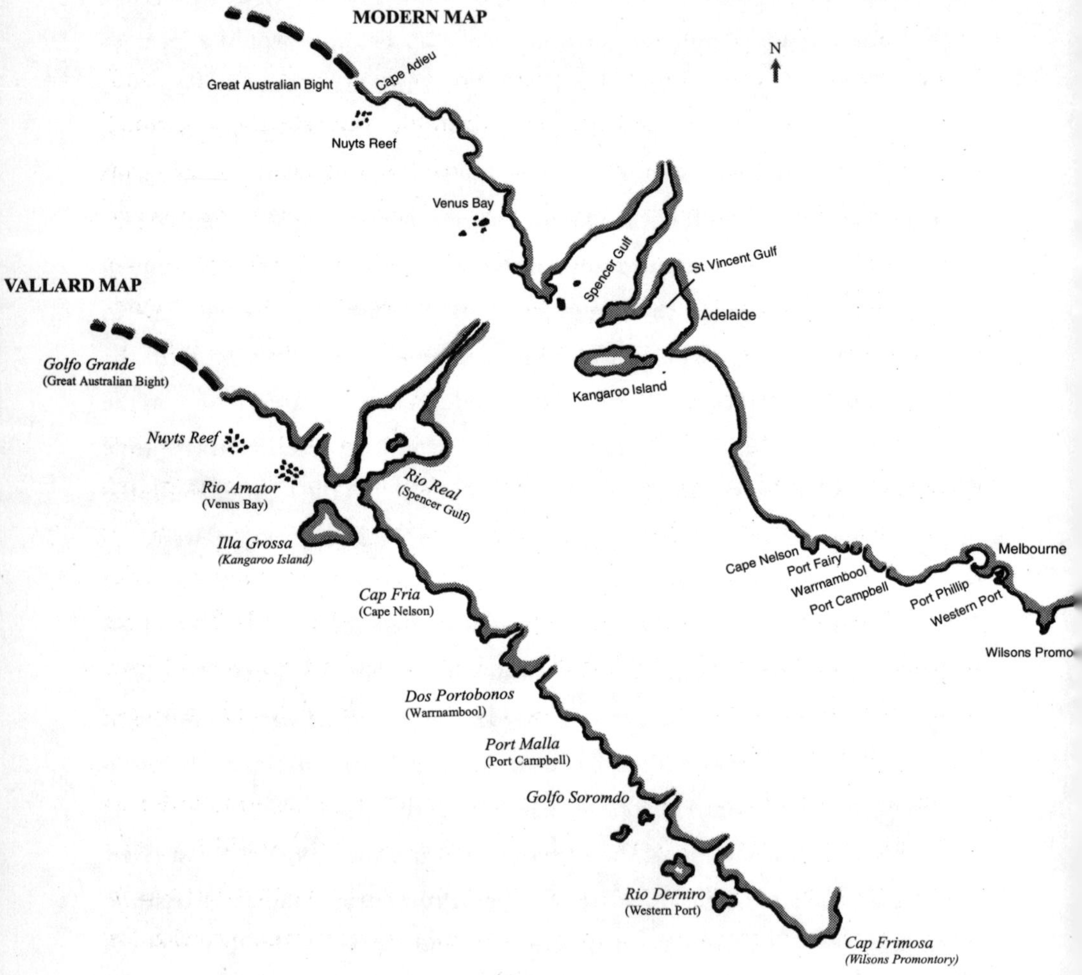

Australia's south coast as depicted in the Vallard map, alongside a modern map of the same coast. Clearly identifiable features in the sixteenth century map include Kangaroo Island and Spencer Gulf.

that even Cook, similarly sailing westward through Torres Strait, was over sixty nautical miles in error with some of his longitude reckonings. In the case of the Vallard coastline of southern Australia, the distortion indicates that the Portuguese ships travelled considerably further west than their navigator estimated.

Beyond Cap Frimosa, Mendonça's lookouts evidently sighted two prominent estuaries. The first of these has been given the name *Rio Derniro*, which looks as though it could be a corruption of 'Rio De [Ja]neiro', meaning (as with the well-known harbour in Brazil) 'River of January'. Such a date would be consistent with what we can determine about the fleet's progress after leaving Botany Bay, as well as with another name which we will encounter further along the coast. This likely new Rio de Janeiro would be Western Port, near the present-day city of Melbourne.

Two islands depicted on either side of this harbour entrance are particularly intriguing. The larger one, drawn with blue ink, has a rough yet undeniable resemblance to the shape of Western Port's French Island. But if so, why has it been placed off the coast when French Island lies plumb in the middle of Western Port? The answer (as with the islands depicted off the coast at Sydney) would seem to be that the scale of the Vallard map would have been too small for it to be drawn in its correct position while still retaining any meaningful detail.

But this proposition brings with it another arresting thought: in order to know the shape of French Island, one of Mendonça's ships would have had to have sailed around it—which in turn would mean that the Portuguese vessel must have entered the harbour. In that case the man credited with the European discovery of Western Port in 1798, George Bass, was 275 years too late.

On this interpretation the smaller island shown in red on the eastern side of the harbour entrance would have to be Phillip Island, and in fact Phillip Island does occupy approximately this position. As for the second opening shown a little further west on the map, this appears to be the entrance to

Melbourne's Port Phillip Bay. Mendonça's ships would seem not to have entered this bay, for otherwise its spacious harbour surely would have been depicted on the map. But this omission is not surprising because the harbour entrance would have been difficult to negotiate, and from out at sea the size of the bay inside would not have been apparent.

A little further west again we come to the name *G: [Golfo] Soromdo* which could possibly be a garbled reference to an Iberian saint called St Ordonho but is more likely a muddled version of 'Sereno' (Serene Gulf). This would seem to be the gulf which begins at Apollo Bay, east of Cape Otway, and sweeps in a gradual curve north-east to Cape Schanck, near the entrance to Port Phillip Bay. Two small islands drawn in red ink nearby could perhaps represent the more prominent of the spectacular Twelve Apostles islets which rear up steeply from the sea west of Cape Otway. Thanks to the rapid wave erosion which occurs in this stormy region of the Southern Ocean, the shape, size and even the number of these islets would probably have been very different 500 years ago.

On the map, a dotted zone close to the shore apparently indicating marine hazards is shown extending as far as a bay named *Port Malla* (Bad Port)—seemingly today's Port Campbell. The two final river inlets shown on this part of the coast, called *Dos Portobonos* (Two Good Ports), correspond to the fishing ports of Warrnambool and Port Fairy. The latter's Moyne River has been given the name *Rio Anno*, or perhaps *Anna*. There are two possible clues here. *Anno* (spelled in modern Portuguese with one 'n') means 'year' and this would make sense if it was intended to signify 'River of the [New] Year'. Alternatively the name could be an abbreviation of 'Rio Santa Anna', (River of Saint Anne), St Anne's day being 30 January. Either way, this name indicates that the ships were here during the southern hemisphere midsummer, the optimum time for a voyage of exploration in these unknown waters.

But summer notwithstanding, the name given to the next prominent headland speaks eloquently of the more extreme conditions Mendonça and his men had to be prepared to face now that they were heading out into the unknown expanse of the great Southern Ocean. This is *Cap Fria*—(Cape Cold). In the original Portuguese it would have been 'Cabo Frio'. We can find this identical name today on the coast of Brazil, marking a prominent headland not far from Rio de Janeiro exposed to storms from the South Atlantic. This Atlantic Cabo Frio even has a connection to Australian history. It was the first sighting of the Brazilian coast for England's infamous First Fleet, carrying its cargo of convicts to Botany Bay on 2 August 1787.

In Mendonça's chart of southern Australia the distinctive sickle-shaped bay shown leading up to Cap Fria is immediately recognisable as Portland Bay, while 'Cape Cold' itself is just as obviously the nearby Cape Nelson. Here, well beyond Bass Strait which separates mainland Australia from the unseen shore of far-off Tasmania, the Portuguese ships would have been exposed to the full ferocity of any southerly storms.

The map now depicts an anonymous, indented coast with not a single harbour or river estuary. This corresponds closely to the actual South Australian coast running from Mount Gambier north-west toward Adelaide. Flinders, on his surveying expedition in 1802, also noted that this stretch of coast appeared barren 'with neither river, inlet nor place of shelter'.

This brings us to yet another arresting feature in the sixteenth century map—a very large island lying off the entrance to a wide estuary. This will serve as a further crucial test of the map's accuracy. Does such an island exist at this location—and if so what is it?

The answer is clear enough: this is Kangaroo Island, Australia's third-largest offshore island. If the shape appears somewhat distorted, once again this can be accounted for mainly by the problem of reckoning longitudinal distance. The island is in the right place and Kangaroo Island is the *only* large island on this stretch of coast. On a map showing all of Australia it looks deceptively small; in actual fact it is nearly 150 kilometres long and up to fifty-five kilometres wide, almost as big as Corsica in the Mediterranean. On the Vallard map it bears the suitable if not particularly imaginative name of *Illa Grossa*—'Big Island'.

How close, we might wonder, did Mendonça sail to his Big Island? If he came at all near to its shores, he and his men could hardly have failed to catch sight of its lordly masters—huge kangaroos. In the absence of humans, dingoes or other significant predators before the coming of Europeans, giant marsupials on this island reigned supreme.

Flinders, regarded as the first European to land there, arrived in March 1802 in the *Investigator*. Scanning the island through their telescopes from out at sea, some of his junior officers reportedly thought they could see 'black lumps, like rocks' moving about and speculated that they might be looking at elephants. They soon realised their error and when Flinders went ashore he found the kangaroos so tame that they could be 'shot in the eyes with small shot, and in some cases knocked on the head with sticks'. Flinders killed ten himself with a double-barrelled shotgun, while the rest of his party easily slaughtered another twenty-nine. They represented a welcome supply of fresh provisions and it was Flinders who bestowed the name 'Kanguroo Island' on this southern outpost 'in gratitude for so seasonable a supply'. The following January, when the French captain Nicolas Baudin landed on the island from his ship the *Géographe*, there was still no shortage of pathetically trusting marsupials. Hunting was so easy that twenty-seven kangaroos were captured alive and taken on board in addition to a large but unspecified number said to have been killed by the crew without their having to fire a single shot.

If Mendonça's men had gone ashore here, they too would surely have made use of this ready supply of fresh, nutritious red meat. Did they in fact land? Of this we cannot be sure. There is, though, one tantalising possible clue to be found in the journal of Matthew Flinders. The English navigator noted that a long time before his visit a vast number of huge trees on the island had fallen, apparently all at the same time as a result of a great fire. Flinders wondered whether it could have been lit by humans but there were no Aborigines on the island and no sign of any ever having lived there.[3] Sceptics will of course attribute the fire to natural causes, and even if Mendonça's crew had landed and started a bush fire it seems improbable that it would have left many traces more than two and a half centuries later. Flinders, nevertheless, was sufficiently puzzled by what he had seen to muse in his journal: 'Can this part of Terra Australis have been visited before, unknown to the world?'[4]

Whether or not Mendonça's ships anchored for a while off Kangaroo Island, they eventually sailed on past the island's western headland and then turned north toward the mainland coast to enter a wide estuary which lay invitingly before them. As with 'Cape Beautiful' it is clear that they were greatly impressed by this majestic sea gulf, for the name Mendonça gave it was *Rio Real*: 'Royal River'.

The shape drawn on the Vallard map leaves no possible doubt about the identity of *Rio Real*. This is Spencer Gulf, the 300-kilometre-long waterway leading toward the Australian continent's vast interior. Just inside its entrance, Spencer Gulf widens significantly (to well over 100 kilometres), as does the gulf depicted in the Vallard chart. *Rio Real* is also shown as having a substantial island inside the heads. This could be either Thistle Island, slightly misplaced, or perhaps Spilsby Island in the Sir Joseph Banks Group. These islands, along with Kangaroo Island, are a haunt of Australian sea lions—large mammals of the seal family which spend most of their time in the ocean but which come up on land in the breeding season.

In Mendonça's day these creatures were known as 'sea wolves'—and in the Vallard map, a very strange animal can be seen emerging from the waves off the south-east coast of Australia. Whether by coincidence or design, its appearance perfectly fits the description of a 'sea wolf' provided by the Italian chronicler Pigafetta during Magellan's voyage down the coast of Patagonia: a creature 'the size and thickness of a calf...a head like that of a calf, small round ears and large teeth'. Another bizarre-looking sea beast with a red skin, pointed ears and long, up-curled snout—perhaps another attempt to depict a sea lion—is pictured in the Southern Ocean off *Illa Grossa*.

The detail in the sixteenth century chart leaves little doubt that Mendonça's ships must have sailed almost to the head of Spencer Gulf, since it records the fact that the gulf narrows significantly toward its northern end.

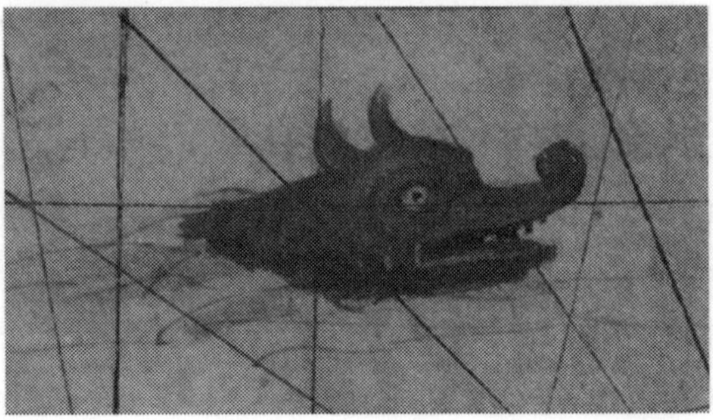

Vallard sea monster pictured off the coast of South Australia.

What would the feelings of Mendonça and his comrades have been as they made their way up this great waterway? At first their spirits would surely have lifted. Perhaps they were at last coming to the land of gold! Would this 'Royal River' also prove to be a sea route leading all the way back to the tropics?

In the latter belief they would not have been alone, for other later voyagers had a similar vision. Joseph Banks, after his return from the *Endeavour* journey, was convinced that Australia, 'being as large as all Europe, must have vast rivers capable of being navigated to the heart of the interior'. Baudin hoped to find a strait leading north, while Flinders believed the interior might be found to conceal a second Mediterranean Sea leading to the Gulf of Carpentaria. It was not an entirely unreasonable hope: geologically the Spencer Gulf is a drowned rift valley and north of it there was indeed long ago a great inland sea, some of whose remnants are today's vast salt pans of Lake Torrens and Lake Eyre.

We can perhaps capture some of the flavour of this part of Mendonça's voyage by turning to Flinders' account of his journey up Spencer Gulf in the *Investigator*, in March 1802. A rugged ridge, he noted, dominated the eastern skyline. They saw flights of black swans and smoke from fires but there was no sign of the human inhabitants who had lit the fires. In the crystal-clear air distances were deceptive. A mountain thought to be nine to eleven kilometres away proved, when Flinders and some companions trekked to it, to be twenty-four kilometres distant. This was a peak nearly 1000 metres high in what are now called the Flinders Ranges, an ancient, fractured mountain chain which ends at the Southern Ocean, only to reappear far away on the facing shore of Antarctica. From atop this peak Flinders, perhaps like Mendonça before him, saw his hopes cruelly dashed. The majestic gulf which had seemed to offer such fine possibilities tamely petered out into a swamp.

We can be quite sure that Rio Real is not the smaller Saint Vincent Gulf leading past the present-day city of Adelaide if only because, as we have seen, Mendonça's 'Royal River' is shown widening significantly just inside its entrance, something which the St Vincent Gulf does not do. The sister gulf does not appear at all on the Vallard map, an omission presumably due to the Portuguese ships having sailed outside Kangaroo Island, which would have blocked most of the St Vincent Gulf from view.

We have now almost come to the end of this part of Mendonça's epic voyage. After retracing their passage down the gulf his ships must have pressed on a little further toward the north-west, for two more names remain from his chart. The first is a small river called *R Amator*, seemingly a misreading of 'Rio Mato', meaning 'Bush River'. This best fits the small harbour of Venus Bay, at the head of Anxious Bay. The scattered Investigator Islands are indicated on the map by rows of dots offshore.

The second name is far more significant. On the Vallard map it is *G Grand*. In Mendonça's original chart it would have been *Golfo Grande* (Great Gulf). There can be no doubt about what this signifies. So portentous a name can refer to one thing and one only, the Great Australian Bight, that forbidding rampart of cliffs constantly pounded by the swell of the Southern Ocean which arches for 1000 kilometres toward the distant Indian Ocean. Here, where the Great Australian Bight begins and fittingly at about the location of Cape Adieu—named by France's Admiral d'Entrecasteaux in 1792—the Vallard coastline of Terra Java abruptly comes to an end. A reef shown here off what would be the tip of Cape Adieu exactly matches the position of Nuyts Reef, named after the Dutch voyager Pieter Nuyts in 1627.

Did Mendonça's ships press on toward the west, thus circumnavigating Australia? The answer almost certainly seems to be no, they did not. No Portuguese or Dieppe map showing anything resembling the 'Great Gulf' has ever come to light, which it surely would have done had the Bight been traversed.

Sailing conditions would have been one important deterrent. Although in summer the winds off the south coast of Australia at times reverse their normal direction and blow from east to west, Mendonça and his officers would by now have realised that if they attempted to continue sailing west they would for the most part have to beat against the prevailing winds, a task for which their ships were ill-suited. Added to this was the danger of a menacing lee shore offering neither refuge nor shelter.

We perhaps should remind ourselves, as Mendonça no doubt would have done, that he had ventured into a totally unknown sea in one of the most remote corners of the world where no Europeans had ever gone before. On perilous missions of this sort, ships had been known to vanish without trace. Had not a fate of just this kind overtaken even the great Bartolomeu Dias, pioneer of the sea route around the Cape of Good Hope, who had been swallowed up by the ocean somewhere off the Cape a few years later, with the loss of four great ships and their entire crews?

Mendonça and his men were in a situation not unlike that of twentieth century astronauts navigating on the far side of the moon; one mishap could spell catastrophe with absolutely no hope of rescue. And besides, Mendonça might well have reflected, was it not a vital part of his mission to intercept the turncoat Magellan, perhaps sailing westward even at that very moment from the coast of South America? Anyone could see that there was no realistic prospect of finding him here.

All the evidence, then, leads to the conclusion that at this point the fleet turned back. And in fact we shall soon discover other clues which will support that view. Those clues will also lead to yet another strange twist in the story of Mendonça's search for the Island of Gold.

7

A Ring of Stone at the Bay of the Portuguese

Impetuous winds from the W.S. W. blew with terrible violence. The sky was covered with thick black clouds; torrents of cold rain like melted snow fell, accompanied by flashes of terrific lightning; the ground swells were so violent and so sudden that we were obliged to tack continually till day appeared.

JOURNAL ENTRY BY FRANÇOIS PÉRON, NATURALIST WITH THE FRENCH SHIP *GÉOGRAPHE* OFF SOUTHERN AUSTRALIA, APRIL 1802

I'll never look at waves the same again. Those waves were out to kill you... You could see death working in the water.

TONY MOWBRAY, SKIPPER OF THE YACHT *SOLO GLOBE CHALLENGER*, DISMASTED NEAR BASS STRAIT IN THE 1998 SYDNEY-HOBART RACE BY A STORM WHICH CLAIMED SIX LIVES AND SANK OR BADLY DAMAGED A LARGE PART OF THE RACE FLEET.

Many puzzling questions surround the fate of Mendonça's ships after they evidently turned back at the beginning of the Great Australian Bight. No account of their return voyage along the southern coastline of Australia would be complete without mention of the wreck on this dangerous coast of the legendary 'Mahogany Ship'. This phantom-like vessel has figured prominently in local folklore. Could it be the wreck of a Portuguese ship, as some have suggested?

The legend of the Mahogany Ship surfaced in 1836 when two seamen searching for seals were shipwrecked on a remote part of the southern coast of what was to become the British colony of Victoria. Trekking toward a small European settlement called Port Fairy (the Vallard map's 'Rio Anno'), they stumbled upon the wreck of a wooden ship lying in a valley between two sandhills. The men reported their find but nothing more was heard of it until a decade later when it was rediscovered by John Mason, a sea captain from Port Fairy, while he was riding on horseback toward the neighbouring settlement of Warrnambool. Mason estimated the wreck to have been a ship of about 100 tons. Its bleached and weatherworn appearance convinced him that it must have been lying there for a great many years. Its masts and most of the decks were gone and the hull was full of drift sand. Its timbers, he believed, were mahogany—a hard, reddish-brown wood.

The Port Fairy harbour master, Captain John Mills, was the next to visit the wreck. He tried to cut a sample from the hull with his clasp knife but found that the blade glanced off 'as though I was trying to cut iron'. Captain Mills particularly noticed the redness of the timber. When he asked some Aboriginal elders about the wreck he was told that tribal knowledge of it extended 'from their earliest recollection'. According to another story attributed to local Aborigines, some survivors from the shipwreck managed to get ashore and these lighter-skinned people subsequently intermarried with the local tribespeople. And intriguingly, even as early as 1869 there was already documentary reference to a legend that the wreck was that of a vessel from a fleet of Portuguese or Spanish ships on a voyage of discovery which had been parted from the others in a storm. There were many other reports of people seeing the wreck up to the end of the nineteenth century, but after that time sightings abruptly ceased.

The Mahogany Ship legend was taken up with enthusiasm in the 1970s by the writer Kenneth Gordon McIntyre. McIntyre, who lived in this part of Australia, became convinced that the wreck must have been that of a Portuguese caravel. In support of his theory he concluded that the southern coastline shown on the Dieppe maps he had studied ended just off Warrnambool and that these two things were therefore obviously interconnected. A flotilla of Portuguese caravels, he theorised, were hit by a storm off this part of the coast, and one of them was driven ashore.

In recent decades several determined efforts have been made to find the elusive Mahogany Ship but all without success. On the latest occasion, in August 2000, a mechanical excavator was brought in to dig up the dunes at a spot where ground-penetrating radar had detected signs of buried timber—but again there was disappointment; nothing of any relevance was found.

McIntyre's theory about the Mahogany Ship was an attractive one, but it does not entirely match the facts as we now know them. In reality, as we have seen, the south coast of Australia as depicted on the Vallard and other comparable Dieppe maps does *not* end at Warrnambool but over a 1000 kilometres further west at the beginning of the Great Australian Bight. This in itself does not by any means exclude the possibility that a ship of Mendonça's expedition could have been driven ashore by a storm on the Warrnambool coast, but a more serious problem with McIntyre's hypothesis is that the Mahogany Ship does not seem to even remotely fit the description of a caravel. According to Captain Mason, an experienced seafarer and the man who seems to have examined it most closely, the general appearance of the wreck 'bespoke a very slight acquaintance of the builder with marine architecture and resembled more the outlines of our local lighters, though of greater dimensions'.

This is the exact opposite of what one would expect to read in the description of a Portuguese caravel, which far from being of crude vessel similar to a lighter or barge, was in its day arguably the world's most technologically advanced ocean-going ship. Caravels had graceful, curved lines and were eminently seaworthy, as they had to be in order to voyage safely halfway round the world, contending with wild storms and the roughest of seas.

But there is another possibility that must now be considered. In light of the discovery that Mendonça's flotilla did *not* consist of three caravels, as has been up to now believed, we need also to look at the possible fate of the two smallest vessels in his fleet—the prau and the brigantine. The description of the Mahogany Ship does not sound much like a Malay prau (even assuming that such a relatively flimsy ship designed to sail in tropical seas could actually get as far as the Southern Ocean), but what about the brigantine?

The most significant feature of a brigantine in this context is that it was of shallow draught and flat-bottomed, lacking the elegant curved shape of a caravel. Its hull embedded in sand could well have resembled an outsize barge. And on this point it is interesting to note that the Portuguese term bargantim, the equivalent of the English word 'brigantine', is related to barca, the Latin word for 'ship', which in Portuguese has a specific meaning of 'barge' or 'lighter'.

From these facts we might reasonably conclude that if the enigmatic Mahogany Ship did indeed have a Portuguese origin it would be as the wreck of Mendonça's brigantine, not that of his caravel. Because of its shallow draught and flat bottom the brigantine would have been the most likely of the Portuguese ships to have been driven ashore in a southerly storm. Furthermore, the fact that the brigantine would have been built in India or South-East Asia is also consistent with the hull of the wrecked ship in the Warrnambool sandhills being made from a hard, red mahogany-like timber.

Before leaving the subject of the Mahogany Ship, one more theory needs to be addressed. This is the claim put forward by the author of a recent popular book on fifteenth century Chinese voyaging that the wreck was that of a Chinese 'Treasure Junk'. Some of this vessel's crew, the book's author claims, struggled ashore and then rode on horseback to Cooktown in far north Queensland. On any objective evaluation this remarkable theory must be rejected as untenable, first and foremost among many reasons for this conclusion being the fact that (according to the author's own figures) these huge Treasure Junks were almost ten times bigger than the wrecked Mahogany Ship![1]

There can be no real doubt that the Mahogany Ship existed, at least until the end of the nineteenth century. The lack of any confirmed sightings since then suggests that it was either washed out to sea in a storm or has simply disappeared among the shifting dunes. If one day its remains were to be rediscovered, scientific analysis of its timbers it might well provide a definitive answer to the mystery of its origin. But unless or until that happens all theories must remain to some degree speculative.

Regardless of the identity of the Mahogany Ship, however, there is other dramatic evidence which supports the belief that Mendonça's fleet was overtaken by a storm somewhere off Australia's southern coast. To find this evidence we need to look not among the sandhills of Warrnambool but further east, on the southern coast of New South Wales.

Trawling off this Tasman Sea coast in 1963, Olaf Mannes, an Australian commercial fisherman of Norwegian origin, had an unexpected catch. Hauling in his net from a depth of fifty-six fathoms (about 100 metres) off Gabo Island, south of the little port of Eden, he found it contained pieces of pottery. Most were fragments, but one piece was almost intact. It was a marine-encrusted earthenware jar some thirty-one centimetres high, onion-shaped with a narrow neck and a ring pedestal on its base. The fisherman's immediate reaction was that it looked like an old wine jug.

After returning to port at Eden, Mannes showed his find to authorities at the local whaling museum, who put it on display. The following year, according to a local newspaper article, the barnacle-encrusted pot caught the eye of three visiting scientists, who after closer examination suggested to the museum's curator that it could be a thirteenth or fourteenth century Spanish or Portuguese wine pitcher. If this was the case, they added, it could possibly indicate the shipwreck of an ancient galleon lying in the area where the find was made.

It was an intriguing suggestion—but nothing came of it. The pot remained on display as a curiosity until 1981, when a television crew arrived to do a documentary feature about the museum. Intrigued by the strange pot, they commissioned scientific tests to try and determine its age. The tests, carried out at the Australian National University in Canberra, involved a technique called thermoluminescence—a complex but normally highly accurate process involving light absorption and emission.

The result was extraordinary in more ways than one. The scientist who carried out the experiment, Dr David Price, made two separate tests on small amounts of sample material taken from different parts of the pot. One of them produced no meaningful results, possibly because that sample had been contaminated. But the other gave a very specific date indeed—1500 AD, plus or minus twenty years.

The pot and its unexpected age made an interesting news item—but again there was no real follow-up. Seen in the context of a Portuguese voyage down the east coast of Australia about the year 1522, however, the date obtained by the successful thermoluminescence test was and still is, little short of sensational. It fits precisely the date of the Mendonça expedition, the more so bearing in mind the fact that the tests indicates the date when the piece of pottery was fired, not when it fell into the sea.

A year earlier, in 1980, a National Parks and Wildlife Service historian, Mike Pearson, had reported hearing of another intriguing find: a figurine, brought up in a trawl in the same general area. This object might possibly have provided further important evidence in view of the fact that sixteenth century Portuguese ships are quite likely to have carried religious figurines—but unfortunately, as Pearson noted, even by the time he wrote his report the figurine had disappeared and could no longer be studied, even in photographs.[2] Another local fisherman working in the same area at about the same time as Mannes also trawled up a large quantity of broken pottery—all of which he threw back into the sea.

Yet another twist to this strange story came recently with the trawling up of a second earthenware jar strikingly similar to the pot recovered by Olaf Mannes. This pot was also recovered off Gabo Island and has been undergoing desalination at the Eden whaling museum. Its finder has asked the museum to keep his identity confidential. Like the first pot, its spout has been broken off. Its clay is a brownish-yellow colour, with here and there some tinges of

The great Portuguese Armada of 1519 sets out for the East. In this contemporary drawing each ship is identified by its captain's name, and Christopher de Mendonça's carrack can be seen at the left-hand end of the middle row ('Medosa' is Old Portuguese spelling for Mendonça). A strip has been torn off the left-hand edge of the page. (Detail from the *The Book of Lisuarte de Abreu*, The Morgan Library and Museum, New York. MS M.525, F.37v, Livro de Lizuarte de Abreu Part 2)

This is the only known picture of the bronze swivel gun recovered by pearl fishermen from a reef at Napier Broome Bay, on Australia's Kimberley coast, in 1918. Posing with the cannon as 'buccaneers' are men at a local meat works. The photo was taken by one of the gun's finders, Cyril Davey.

The fortified Malay port of Malacca, at the time of its capture by a Portuguese armada in 1511. With its defences strengthened, Malacca became the main naval base for Portuguese expeditions to other parts of the Far East, including the 'Island of Gold' – Australia. The drawing is from *Lendas da India*, by the Portuguese chronicler Gaspar Correa. (*Australian National Library*)

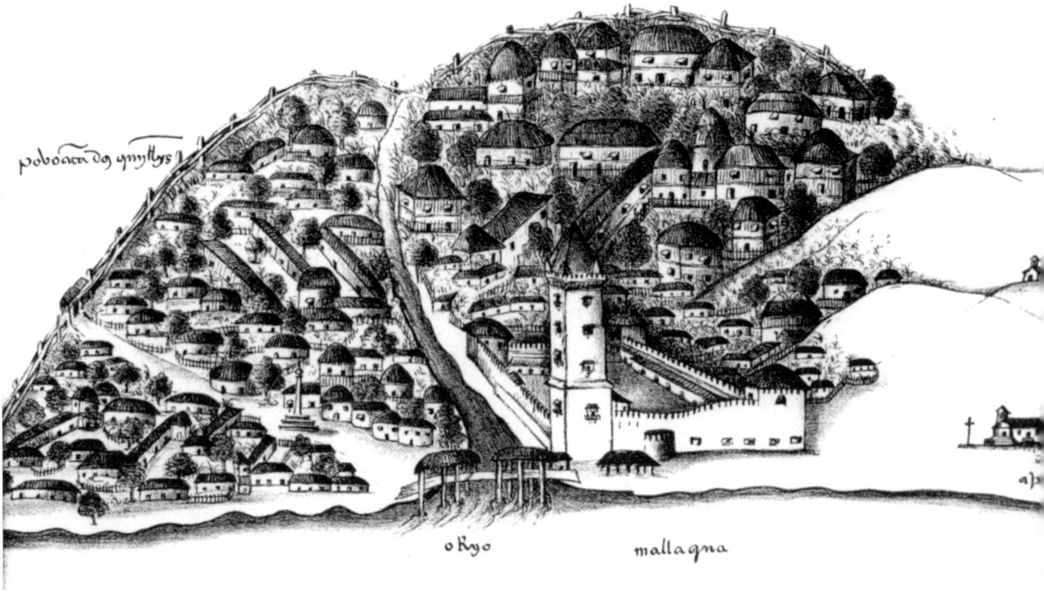

A Macassan prau. A shallow-draught vessel of this type was added to Mendonça's fleet before he set off on his main 'Island of Gold' expedition. (*Lithograph by Le Breton, 1839, in Atlas Pittoresque, National Library of Australia an20801273*)

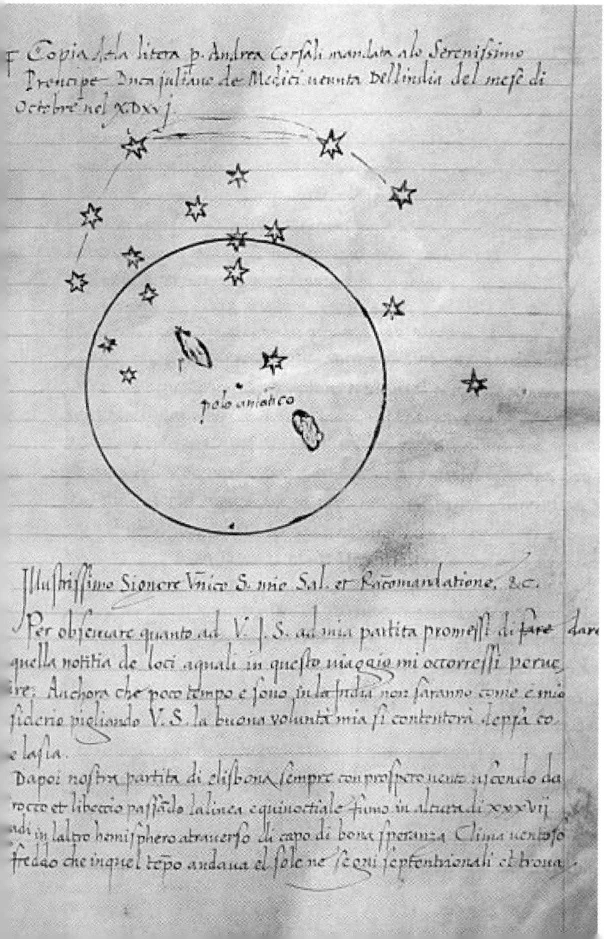

The first known European drawing of the Southern Cross. It appears in a letter to the Medici ruler of Florence by Andrea Corsali, who sailed in a Portuguese ship to Goa in 1516. By that time, though, the constellation was already well-known to Portuguese mariners. (*National Library of Australia MS7860*)

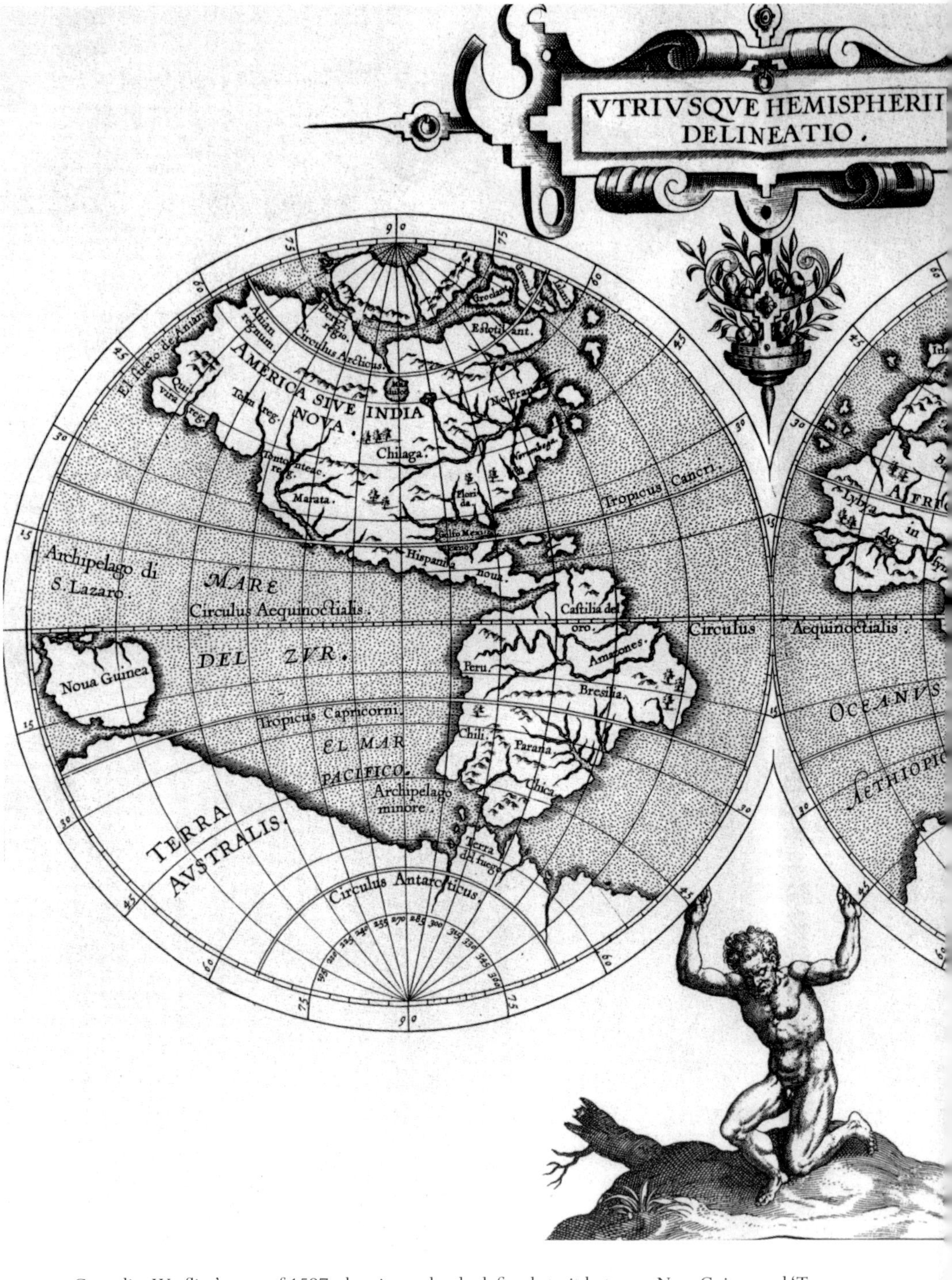

Cornelius Wytfliet's map of 1597, showing a clearly defined strait between New Guinea and 'Terra Australis' - nine years before Torres supposedly discovered it. In a note accompanying his map, Wytfliet revealed that there had already been two planned voyages to the coasts of 'Australis'.

The 500-year-old lead fishing net sinker found buried on a beach at Fraser Island, Queensland. It has been scientifically dated to about 1500, and the lead has been identified as having come from a mine in either Portugal or the south of France. (*Bill Ward*)

Aborigines spearing fish and diving for crayfish on the coast of New South Wales, early nineteenth century. In the background a group of people are gathered beside a fire, cooking fish. Could it have been a scene like this which inspired Mendonça to call Port Stephens *Rio Pescadores* - 'River of Fishermen'? (*Watercolour by Joseph Lycett, National Gallery of Australia an2962915*)

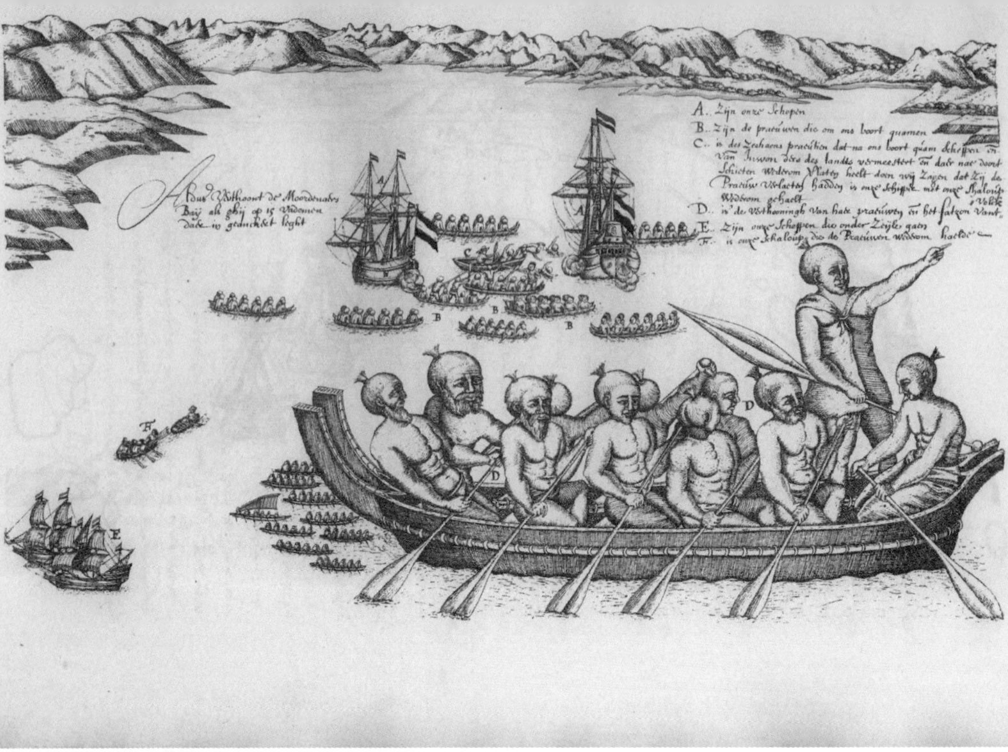

Maoris attack Abel Tasman's ships in 1642. This pen and ink drawing by Isaac Gilsemans shows Maori warriors intercepting the *Zeehaen's* cockboat and killing four members of his crew. A shocked Tasman called the place Murderers Bay. Did something similar occur when Mendonça's ship came into Wellington harbour? (*State Library of New South Wales*)

The Portuguese-type iron helmet recovered from New Zealand's Wellington harbour (right), the Tamil bell from South India (below), and the cannonball dredged up from the bed of Wellington Harbour. The helmet is of a similar type to sixteenth-century helmets on display in the Lisbon Military Museum. (*Museum of New Zealand, Te Papa Tongarewa, and Wellington Maritime Museum*)

The 'Mendonça Stone' unearthed by a gardener near the old shoreline at Cape Town in 1990. Its inscription, reconstructed by South African archaeologists, records that 'the ship Vitória, Captain Cristóvão de Mendonça, arrived here 26 Ma [March or May] 1524'. (*South African Cultural History Museum*)

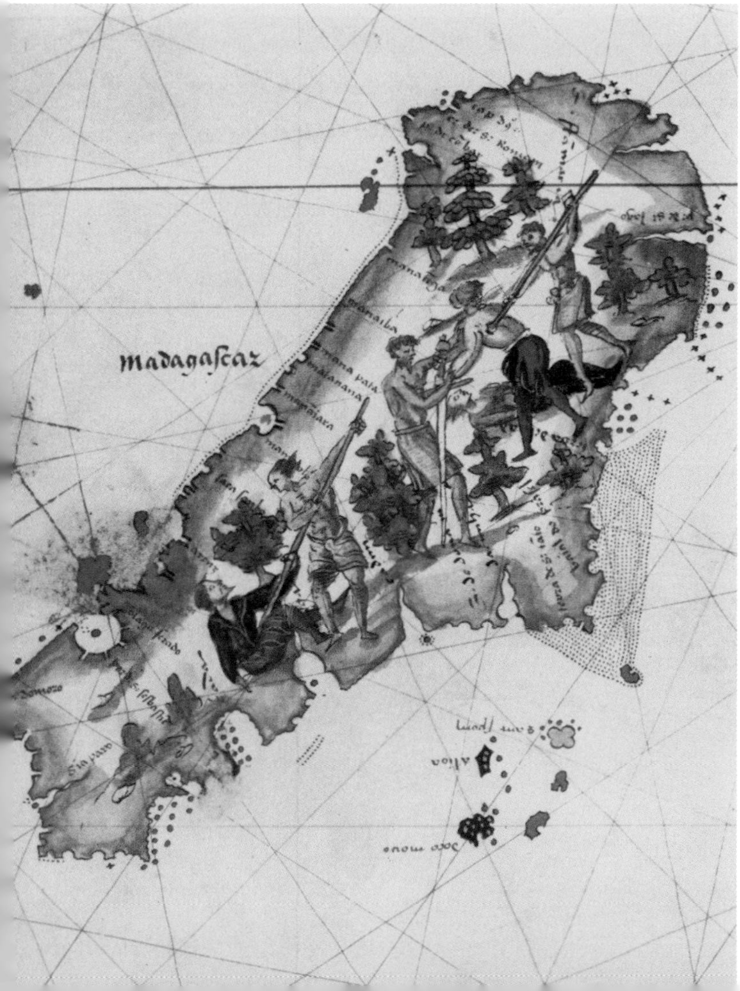

This sketch by the cartographer Jean Rotz vividly depicts the killing of two French sailors and a Scottish crewman by natives of Madagascar in 1529. It is believed Rotz may have witnessed this scene at first hand. (*Jean Rotz Boke of Idrography, 1542, facsimile edn., British Library*)

An unidentified man astride a sea turtle in the Buccaneer Archipelago. The photograph was taken during Bernard Ryder's expedition to the Kimberley Coast in the early 1930s, and the man's powerful body accords closely with Ryder's description of the 'magnificent physique' possessed by the 'Latin-speaking' people he encountered. The photograph was published in the French journal *La Geographie* in 1935.

One of the many mysterious beehive-shaped stone dwellings found by archaeologist Sue O'Connor in 1985 on High Cliffy Island, in the Buccaneer Archipelago. Who built these structures, and whether they had a connection with the 'Latin-speaking' island people encountered by Bernard Ryder, are questions which remain unanswered. (*Photo: Sue O'Connor*)

red. Some fine horizontal lines visible on the jar's surface appear to have been made when it was on the potter's wheel.

The original pot tested by David Price was eventually withdrawn from the museum by Mannes. It then came into the possession of a Sydney oceanographer and hydrographer, Captain Daniel Fitzhenry, who had it examined by experts in European archaeology at the Sydney Applied Arts and Science Museum. They confirmed that it was a storage jar of a type used in Portugal centuries ago, including the 1500s.

This brings us to the central issue: how are we to explain the presence of apparently 500-year-old Portuguese wine jars and other broken pottery on the sea bed in this remote spot near the southeastern extremity of mainland Australia when, if the accepted view of Australian history is correct, England's Captain Cook was the first European ever to have sailed in these waters in 1770? This confronting question perhaps explains why these extraordinary discoveries have generally been ignored, and sometimes even disparaged, by mainstream Australian maritime archaeologists and historians.

And what are we to make of the possibility, first suggested in 1964, that the pottery trawled up from the depts off Gabo Island could indicate the presence of a centuries-old Portuguese or Spanish shipwreck?

Olaf Mannes reasoned that it was unlikely there was a wreck lying on the sea floor because as far as he knew no ship's timbers had ever been trawled up in this area—something he supposed would have occurred if the remains of a wrecked ship were lying on the bottom. He therefore assumed that the pottery he and others had trawled up must originally have been thrown overboard. But Daniel Fitzhenry, managing director of a company which has been involved in major hydrographic projects including the Sydney Harbour Tunnel, disagrees with Mannes's conclusion—arguing that it is highly unlikely any loose timbers would be remaining on a wreck submerged for nearly 500 years. On the other hand, he says, local commercial fishermen have told him they avoid trawling a particularly notorious area where their nets get snagged—and this is exactly what he would expect if a ship's hull and its contents were lying partly embedded in sand on the sea floor. If there was such a wreck, it could contain a cargo which in historical terms would be of incalculable value.

As for the jars so far recovered from the sea bed, a comparison with pots salvaged from ships of the Spanish Armada wrecked on the coast of Ireland in 1588 suggest two possibilities. One is that the Gabo pots could be a type used for storing commodities such as olive oil and vinegar. Typical pots of this kind, known in Spanish and Portuguese as *botijas*, were mass-produced for the Armada with a capacity of 6.25 litres.

The second, and perhaps more likely possibility for the Gabo Island pots, given that they are considerably smaller than standard Armada botijas, is that they might have been wine jugs—used perhaps on the mess table. It is also possible that they might originally have had handles. Broken jugs of this sort have been retrieved from Armada wrecks, and reconstructed drawings of them look very similar to the Gabo pots. Originally the pottery handles on the Armada jugs extended from the widest part of the jug to the top of the spout. This could explain why the tops of both the Gabo pots have been damaged—caused, perhaps, by the handle breaking off at its weakest point. Daniel Fitzhenry thinks it is possible that his pot might originally have had a handle. Both the Gabo pots hold just over two litres—which would have been enough to accommodate a sailor's normal daily wine allocation in Portuguese and Spanish ships.

Another feature of the pots trawled up off Gabo Island which fits their presumed early sixteenth century origin is that on the available archaeological evidence they appear to be of a type predating the Spanish Armada. Very few of the pots recovered from Armada wrecks have a top similar to the elongated necks of the Gabo pots—and those samples that do exhibit such a feature are classed as being of an 'early' or pre-Armada style.

Closer to Australia, another interesting pottery vessel described as an 'olive jar' figures among a rich collection of artefacts salvaged from the wreck of a Spanish galleon sunk in a battle with Dutch ships in Philippine waters on 14 December 1600. Although the vast majority of the hundreds of pots recovered from the wreck of the *San Diego* are of Asian origin, and in appearance quite unlike the Gabo vessels, a few of the salvaged pots were found to have been made in Europe—and among these the little olive jar stands out. It has a green lead glaze, and although it does not have an elongated neck it is otherwise strikingly similar in both size and shape to the pots trawled up from the sea bed off Gabo Island.

As for Gabo Island, the seas in its vicinity are the sort of place one might well expect to find a shipwreck. The island lies a short distance off Cape Howe, at a point where the long south-trending Australian coastline finally begins to turn west. Here, the warm Queensland current flowing south meets the cold waters of the Southern Ocean—and here, too, sudden icy storms from the Antarctic hurl themselves into the funnel of the south Tasman Sea with an awesome ferocity, reaching as much as 150 kilometres an hour and whipping up immense seas. This is a danger which even today's sailors in the internationally-renowned Sydney–Hobart yacht race have good cause to respect. Over the years these storms have claimed dozens of lives and sunk many well-equipped ocean racing yachts.

Did a disaster of this kind overtake Mendonça's fleet on its way back from the beginning of the Great Australian Bight? Were one or more of his ships sunk, or at least badly damaged—and could the jars and the other pottery shards trawled up from the seabed off Gabo Island be evidence of such an event?

It is certainly not beyond the realms of possibility. Ferocious storms do sometimes sweep in from the Southern Ocean even in summer, and Gabo Island offers the first shelter of any kind for a sailing ship running before such a gale from the dangerous, turbulent waters of Bass Strait. We have to bear in mind that unlike François Péron's *Géographe* or other sailing ships of the late eighteenth century, Portuguese vessels in the sixteenth century were not designed to ride out a storm by beating into a gale. Their only recourse was to run before wind and seas and hope for the best.

So convinced is oceanographer Daniel Fitzhenry of the existence of a Portuguese shipwreck that he has drawn up a detailed plan for a preliminary underwater investigation. Central to this plan would be a modern, sixteen-metre oceanographic survey vessel which his company owns. This craft is fitted with sophisticated equipment enabling it to undertake a wide range of marine research tasks which previously would have required a much larger vessel.

The area of interest off Gabo Island is too deep for normal scuba diving, so Fitzhenry's project would involve first a side-scan sonar survey of the sea floor and then a video scan of the most likely locations. The requirements of the various authorities who superintend the investigation of shipwrecks would be fully respected, he emphasises. If the results were positive, this would be followed by the deployment of a Remote Operated Vehicle (ROV), an unmanned mini-submarine whose task would be to capture more detailed video images and still photographs of the possible wreck site. Using its strobe lights and mechanical arms, the ROV would also be able to bring to the surface any pottery or timber samples.

Because of the exposed position of the possible wreck site, the operation would have to be undertaken in summer coinciding with a period of calm weather and sea conditions. Even the initial survey up to the point where the existence of a wreck could be confirmed would be a high-cost operation. So far, Fitzhenry has been too busy with other work to take on the project—but it will happen at the first opportunity, he says. The crucial factors are time and funds.

If it were to turn out that there was no actual shipwreck associated with the two jars and other pottery shards, an alternative explanation for their presence on the sea floor could be that they were thrown overboard from a ship which had suffered damage in a storm. Gabo Island does not offer anything that could be called harbour or safe anchorage in a gale, but it lies close to the land and its lee shore might offer some shelter from the worst of the seas. Somewhere in this vicinity would be the place one might expect the crew of a storm-battered vessel to abandon ship if it was sinking—or alternatively, to clear the decks of broken gear and debris before running north again in search of more secure shelter.

We can trace a boat's likely course from Gabo Island today by climbing the high promontory of Green Cape, the easternmost point of this part of Australia (named, as we have seen, *Cap Groca* or 'Great Cape' in the Vallard map). Far below the crest of this high headland, the suitably named Disaster Bay stretches away south to the distant horizon—a crescent of yellow sand on which the great surge from the Southern Ocean thunders ceaselessly in a line of surf. Gabo Island lies just below the horizon. Many ships have come to grief on this treacherous stretch of coast, even after the building of a lighthouse on Green Cape in 1883.

Not until a ship is safely past Green Cape is there any real shelter from a southerly storm. But once in the lee of the cape, the breakers begin to subside. The first small harbour past Green Cape is called Bittangabee Bay. Inside this inlet, protected by a natural breakwater, the sea is smooth and placid even when the mightiest storm is raging out at sea. For a small vessel, or a ship's boats seeking urgent shelter, it would be a welcome refuge. Its gently shelving sandy beach would be an ideal spot for hauling a vessel out of the water for repairs.

Bittangabee (also spelt earlier with one 't', as 'Bitangabee') is the site of some strange and puzzling ruins. They consist of the stone foundations of a building, surrounded by a circular wall of boulders, in places nearly a metre high. The origin of these structures has long been the subject of debate and speculation. Particularly intriguing is an early legend linking them with the coming of a Portuguese ship. No one knows exactly how or when this story originated but it seems to have existed from the earliest days of known European contact.

Even today, Bittangabee Bay is one of the most isolated places on this part of the Australian coast. To reach it involves a twenty-seven kilometre drive after turning off the main coastal road near Twofold Bay. The first seven kilometres follow a sealed highway servicing a Japanese woodchip mill; the remainder of the journey is a bone-jarring ride along a narrow, rutted dirt road through dense eucalypt forest to within a short distance of the Green Cape lighthouse.

Bittangabee Bay, with its snug beach of white sand, is a place visited only by a few campers and hikers. The stone ruins lie on the far side of a stream, hidden behind a screen of native bush. The silence here is broken only by the pling-pling of bellbirds. In autumn the site is carpeted by yellow everlasting flowers. At the approach of a visitor, lizards scuttle among the rocks and a red-bellied black snake slithers into a hole among the stones.

The old legend that the ruins had a Portuguese connection was taken up by Kenneth McIntyre in the 1970s. He suggested that they were the remains of a blockhouse and stockade built by a ship's crew while they 'wintered over' at this spot, in the same way that Magellan wintered at St Julian's Bay, Patagonia, before finding a passage through to the Pacific. Sadly, it has to be said that McIntyre based his theory in part on some faulty interpretations of the evidence and this made it easy for his critics to discredit him. That was the more regrettable because essentially he was on the right track.

If we are to attempt to decipher the enigma of the Bittangabee ruins we have to look carefully at their structure and layout. They occupy a raised area of gently sloping ground facing the sea and are a few score metres from it. Today the site is hemmed in by a dense mass of trees and undergrowth, but in earlier times it would have had wide views across the bay and out toward the ocean. Close by are two streams of fresh water originating in dense bushland which never seem to run dry, even in the harshest of summer droughts.

The outer stone structure is an oval with its main axis across the slope, thirty metres in diameter at its widest point. It consists of a jumbled mass of boulders of varying sizes, the largest over a metre long and half a metre wide. None of them appear to have been worked in any way and none show any signs of mortar. Some are so large and heavy that it probably would have taken three or four men to lift and carry them any distance. The oval varies in height, parts of it evidently having been removed and used as construction material for the foundations of the partly built structure which lies within it.

What was the intended purpose of this inner building? McIntyre claimed that it was the remains of a blockhouse built by the Portuguese in the sixteenth century. However, after the publication of his theory the National Parks and Wildlife Service (NPWS), which administers the area, commissioned a survey of this and other nearby historic sites. The survey, published in 1988 concluded, inter alia, that the inner structure was in fact the beginning of a nineteenth century house for a shore-based whaling station.[3]

This finding, as far as it goes, is correct but it is only part of the truth. There was indeed a whaling station at Bittangabee Bay for a short period in the 1840s. It was owned by three brothers, George, Alexander and Peter Imlay, entrepreneurs who at the height of their fortunes operated several whaleboats in Twofold Bay, three of which were for a brief time stationed at Bittangabee. The foundation walls of the building inside the stone circle are clearly intended for a house of that period. They outline a rectangular building with seven rooms, an entrance verandah and a gable wing on each side facing the sea. This could not possibly have been a blockhouse, for apart from other considerations the building is the wrong shape and far too big for such a purpose.

In the latter fact, however, may lie a vital clue to the real nature of the site. A detailed survey of the site made in 1978 and republished in the official 1988 report noted a curious thing: that the outer circle of loose rocks apparently passed *underneath* a corner of the uncompleted north wall of the inner building. How could this be explained? The authors of the report speculated that there must have been a 'minor collapse' of the unfinished wall, thus accounting for the presence of loose rocks inside as well as outside it. But there is an immediate problem with this theory: the stones in the standing wall are all mortared into position, but apart from a few small rocks which have obviously been dislodged from the top of the wall in recent times the larger 'collapsed' stones lying embedded in the ground have no signs of mortar on them. This suggests that they were never part of the mortared wall in the first place.

The alternative explanation—and one which makes sense—is that part of the end wall of the house has actually been built *on top* of the original stone circle. And looking at the ruins this is visually apparent, for one can see that a corner of the building has passed through the curved line of the rock oval.

This raises some important questions. The official reports assumed that the stone circle must have been assembled by the builders as construction material before they began building the house. But now we can see a contradiction, for as any builder knows, the first thing to be done before beginning a new job is to clear the site and mark out the foundation lines. Would any builder in his right mind then proceed to dump large quantities of extremely heavy construction material right on top of his building line? So what can have happened here? The explanation which fits the facts is that the ring of stones was already in existence when the nineteenth century stonemason came on the scene. According to the 1988 NPWS report, the mason was probably working from a 'pattern book' plan with set dimensions. When the length of the house proved too long for the space inside the stone enclosure he had no alternative but to run one of his end walls right through the rock circle.

This interpretation makes sense for other reasons. Skilled labour in the Australian colonies at that time was expensive and in short supply. At such a remote spot as Bittangabee Bay, which could only be reached by boat or over a rough Aboriginal track, it seems unlikely that the work force employed in building the house would have been more than a stonemason and his mate.

But the amount of loose rock and boulders piled up in the stone circle is huge. Can we really believe that two men commissioned to build a house would have set out from scratch to gather, let alone 'quarry'—as the official reports suggest—this vast quantity of heavy rocks before they even began any construction work? And more puzzling still, why would they have placed all these rocks in a continuous ring? Normal and sensible building practice is to have one or more stockpiles of material placed in strategic positions which are handy, but for obvious reasons well clear of the actual building lines. Piling up a massive, continuous ring of rocks (even assuming a gap had been left somewhere to get inside!) would make no sense at all. And why a circle when the building lines were rectangular? In short, it would have been an arduous, impractical and completely unnecessary extra task.

And why, we might well wonder, was this house being built in stone at all? While in this region of New South Wales stone houses were occasionally constructed for the well-to-do, usually at major European settlements, the normal building material for whaling stations was timber. The Imlay brothers already had a whaling station at Eden, the main port of Twofold Bay and it was built not of stone but wood. This house has long since gone, but a well-preserved example of a whaling station of a similar kind can still be seen at an inlet further up Twofold Bay, not all that far from Bittangabee. This is the Davidson whaling station at the mouth of the Kiah River. It was more or less contemporary with the Bittangabee station, having been constructed by a Scottish immigrant, Alexander Davidson, in the late 1840s. And it was built entirely of timber, of which there was an abundance in the Twofold Bay area. The timber has proved so durable that the cottage was still being lived in at the beginning of the twenty-first century.

Is it really believable that such astute businessmen as the Imlay brothers would have employed a skilled stonemason to go Bittangabee Bay, the most distant inlet on Twofold Bay, merely on the remote chance that he *might*—somehow, somewhere—be able to find enough suitable building stone to construct a very substantial villa? On the other hand, of course, if a huge stockpile of stone was already available, then employing a tradesman to make use of it would make good sense. The fact that the house never got much above foundation level was certainly not due to any shortage of stone, because at the time work stopped there was an abundance of it remaining in the rock circle. The project was abandoned shortly after it began for two reasons: a drastic decline in the profitability of shore-based whaling, followed by the deaths of two of the Imlay brothers in 1846 and 1847.

George Augustus Robinson, Protector of Aborigines. Robinson was not as dandyish as this lithograph might suggest: in 1844 he made a spartan 3500-kilometre trek through largely unexplored regions of eastern Victoria and New South Wales - as he later wrote, 'tentless and without equipment, my chief shelter from the weather a blanket or a sheet of bark'. (*National Library of Australia, vn1191214*)

Just before the collapse of the Imlays' commercial empire an event occurred which will provide us with a crucial clue to the enigma of the Bittangabee Bay ruins. In 1844 the bay was visited by one of the very few Europeans who at that time took any active interest in the welfare of Aborigines. He was George Augustus Robinson, an educated, energetic English migrant who had been appointed by the government authorities in Victoria in 1838 to the high-sounding post of 'Chief Protector of Aborigines'.

After working among the Tasmanian Aborigines for some years, this extraordinary individual led a party of twelve, including six Aboriginal assistants, on a great circular trek of 3500 kilometres through Victoria and southern New South Wales. Their ostensible purpose was to check the condition of Aborigines employed in the whaling and pastoral industries.[4]

The journey took the party on foot around Victoria's Gippsland coast and up the coast of New South Wales beyond Twofold Bay, where they turned inland to trek back to Melbourne by way of the 'Limestone Plains', the site of Australia's present-day capital city, Canberra. It was a spartan expedition. Robinson noted in his diary that he went 'tentless and without equipment, my chief shelter from the weather a blanket or a sheet of bark'.

In addition to being a keen observer of all aspects of the countryside he travelled through, Robinson spoke several Aboriginal languages, an accomplishment which no other European of his day could even remotely match. Fortunately he kept detailed diaries of his travels, usually scribbled on scraps of paper in the light of a camp fire. Robinson's party arrived at Twofold Bay in July 1844 where he noted that 'three boats manned entirely with Aborigines were employed at the whaling station at the time of my visit'. On a page of his diary opposite a sketch of the mountains behind Twofold Bay, he wrote that what we now call Bittangabee Bay was a 'pretty place', with 'a stone house begun, bark huts, fruit trees, garden and waterfall near to a pretty neat bay' (we may note in passing that he does not say that building work at the stone house was actually going on: possibly it had already stopped).

But the really significant fact to emerge from Robinson's diary is that the Aboriginal name for this little bay was *not* Bittangabee at all—it was 'Pettungerbe'. This spelling, which the linguist Robinson would have written phonetically from the name he heard pronounced by the local Aborigines, can still be seen in the original of his diary now held in the early manuscript collection of the Mitchell Library, Sydney. The pages are stained and fire-singed at the edges but the scrawled name is absolutely clear: 'Pettungerbe'.[5]

The difference between 'Bittangabee' and 'Pettungerbe' might on the face of it not seem all that great but in the context of the oral tradition of a Portuguese ship having visited this place at some far-off time, Robinson's revelation of the true Aboriginal name for the bay is far-reaching in its

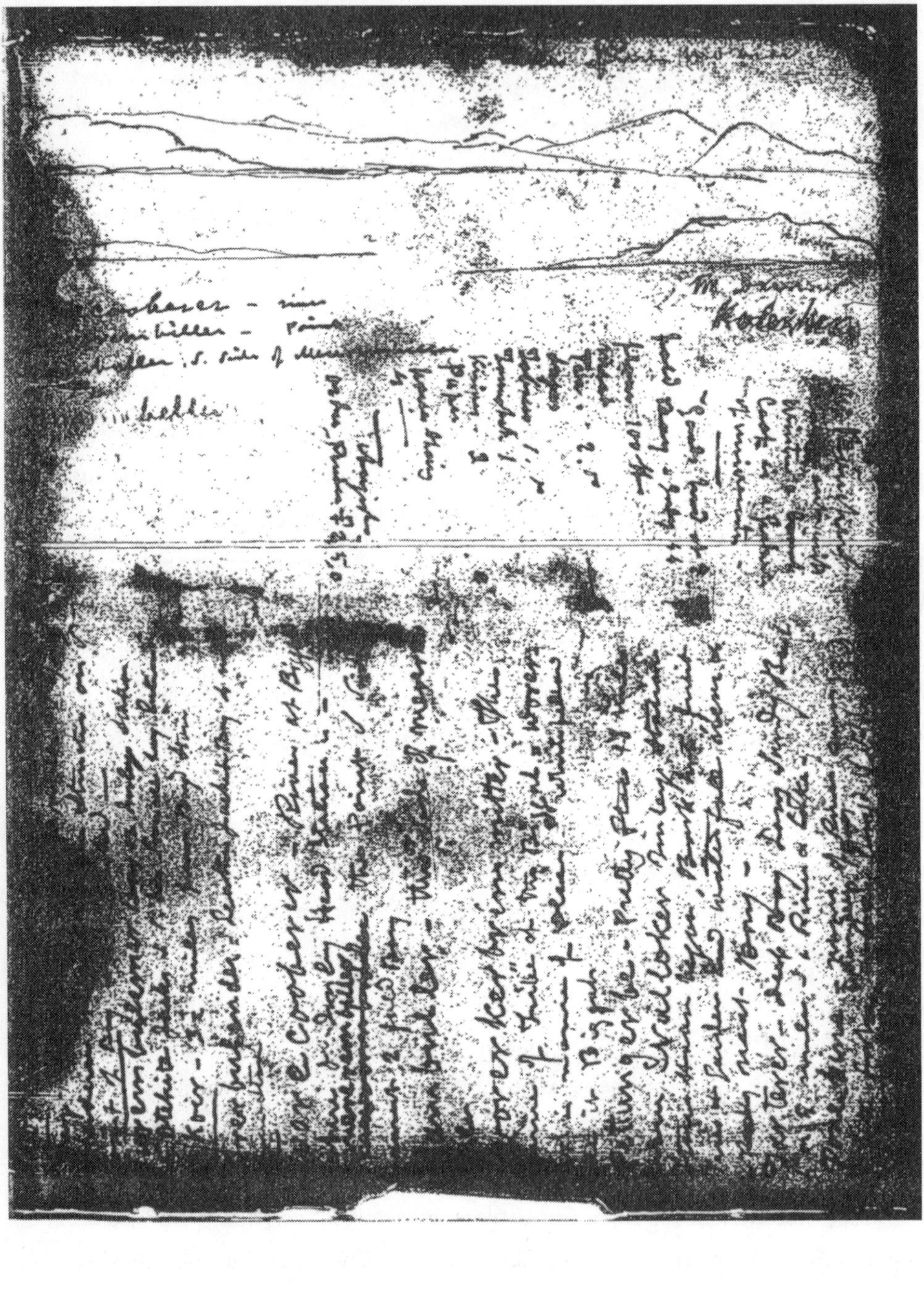

A page from Robinson's fire-singed diary, in which he records that the Aboriginal name for what is now known as Bittangabee was in fact Pettungerbe. (*National Library of Australia and State Library of New South Wales*)

implications. 'Bittangabee', no doubt derived from the rough pronunciation of European whalers, has no discernible Aboriginal meaning. And however carefully one examines or dissects it, the name provides not the slightest hint of any Portuguese link from the past.

'Pettungerbe' is a different matter.

In trying to discover the meaning of this name we have to allow for many uncertainties about the language details of the Aboriginal people who inhabited the 'Pettungerbe' area at the time of European settlement. The Twofold Bay people describe themselves as members of the Manaroo (today spelt 'Monaro') nation, whose territory extended from Eden all the way down the coast to the harbour of Orbost, where the Snowy River, which rises in Australia's highest mountain range, flows into the sea. Inland their territory extended west to the summit of the Great Dividing Range. Several alternative names recorded for the nation, or parts of it, include Thawa, Thaua, Thurga, Durga, Tadera-Manji and Gauanagal. The nation was also sub-divided into the Katungal (seacoast people) and the Paienbara (tomahawk—or forest people).

To complicate matters further the clan language, known variously as Thawa or Biddewal, was a sub-group of the Kuia language region extending as far north as Botany Bay. The clans of these coastal people were known collectively as the Yuin. Further inland the language was Wiradjuri, spoken over a greater territory than any other language in New South Wales.

Tragically, although some belated attempts have been made to record the linguistic remnants in two of the Kuia language sub-groups nearer Sydney, almost nothing is known about the Thawa (Durga) language. No one ever compiled a Thawa-English dictionary. An official report in 1976 concluded that the language was dead. Worse, it had vanished almost without trace, with no active speakers alive and only a few old people who remembered any words at all.[6]

The problem of resolving with any certainty how Aboriginal words of this region should be written or pronounced is further illustrated by the fact that g to the authoritative *Macquarie Dictionary of Aboriginal Words*, more than sixty different spellings have been recorded just for the language name 'Wiradjuri'. AW Reed, compiler of the dictionary *Aboriginal Place Names*, noted the difficulties caused by the fact that Aboriginal Australia had at least 500 languages or distinct dialects; he cautioned that a single word could have more than one meaning in one language and an entirely different meaning again in another.

In analysing the name 'Pettungerbe', therefore, a certain amount of intelligent deduction is necessary. To extract a possible meaning it is easiest to start with the last syllable, or suffix, '-be'. The Reed dictionary of Aboriginal place names and an Australian Museum dictionary both describe the suffix '-ba' as appearing at the end of many place names in New South Wales and meaning 'place of'—for instance Mulubin*ba*: 'place of ferns'. The Australian Museum dictionary says that '-dee' also means 'place', while Reed indicates that 'bee-' as a prefix can mean 'place of'—for example, *Bee*reegan: 'place of the quail'. The Aboriginal name for Cape Banks, the northern headland of Botany Bay, was Bunna*bee*—with 'bunna' meaning 'rain'. Inland in Thawa territory we find the place name, Norin*bee*. In the absence of evidence to the contrary, it seems a reasonable conclusion that '-be' (or '-bee') as a suffix also means 'place'.

Thus far then, the name means 'Pettunger place'. What then of 'Pettunger'? Robinson's double 't' spelling is optional, since a single 't' would give the same sound and as we have seen, in more recent times the name was sometimes spelt 'Bitangabee'. This is a minor point, but there is one other more important consonant to look at, and that is the final 'r'. One of the most common sound transpositions between different languages is from 'r' to 'l' and vice versa. Japanese and Chinese people, for instance, have difficulty with the 'r' in many English words, tending to pronounce a word like 'fresh' as 'flesh'. In Polynesian this even becomes a major factor differentiating one language or dialect from another: for example the word *aroha*, meaning 'love' in New Zealand Maori, becomes *aloha* in Hawaiian and *alofa* in Samoan.

In two relevant European languages, 'prata', meaning 'silver' in Portuguese, becomes 'plata' in Spanish.

If we look at the same possibility of transposition in the Thawa language, 'Petunger' could become 'Petungel', or perhaps 'Pertungel'. And now we can see a hidden meaning emerging. What we have here could be nothing less than an Aboriginal pronunciation of 'Portugal'! The similarity is even more striking if we compare 'Petungel' with the common sixteenth century and later English word for a Portuguese person—a 'Portingal'. This word was evidently an approximate phonetic rendition of the term which Portuguese people used to identify themselves, so it would be a fair assumption that something very like this is what Australian Aborigines would have heard.

The entirely unexpected fact that the bay known today as 'Bittangabee', when given its correct spelling and pronunciation of 'Pettungerbe' can be shown to contain within it the possible meaning 'Portugal' and this at the place where there is an existing legend of a Portuguese ship has to be more than yet another in a series of random coincidences. Statistically, the odds against this occurring simply by chance would be almost incalculable. But the coincidence is not at all strange if we assume that the legend and the names are linked.

And there is another curious fact. The Aboriginal clan living on the coast near 'Pettungerbe' were known in recent times as the Katungal. Whether by coincidence or not, in pronunciation and even in spelling, this name, too, has evident similarities to 'Portingal'.

Unfortunately, as with the Thawa language, the amount of traditional knowledge that has been preserved about the pre-European history of the Twofold Bay people is pathetically sparse, a mere fragment here and there, most of it describing minor encounters between Aborigines and European ships' crews in the nineteenth century, or an occasional doling out of government blankets. In 1990 Michael Organ, a researcher at the University of Wollongong, near Sydney, published a 520-page volume described as 'the first comprehensive work to explore the mostly unknown history of the

South Coast [Aboriginal] clans.'[7] But despite his extensive research, even Organ was unable to throw any light on the question of how the legend of a Portuguese connection came to be attached to 'Bittangabee'.

Are there any other clues as to how and when this legend originated? The last full-blooded Thawa people passed away a century or more ago. The few clan members of mixed-race descent still living in the Twofold Bay region are uncertain about the legend's origins, but they are able to recount information about some very curious traditions which have been handed down orally from one generation to the next.

The custodian of one such tradition is Ben Cruise, clan elder and chairman of the district's Aboriginal Land Council. He was told by his mother and grandfather that prior to the arrival of the British in the early nineteenth century his people inhabiting the coast south of Pettungerbe built sea-going outrigger canoes. This in itself would be quite remarkable, for Captain Cook, sailing north, only sighted his first outrigger canoe when the *Endeavour* was nearing the tropics off north Queensland. Cook had only a brief, distant glimpse of the coast south of Twofold Bay, but the Aboriginal canoes he saw at Botany Bay and other places further north were all flimsy craft made from a single piece of bark, totally unsuited in his opinion for use in anything but calm, sheltered waters.

In contrast, according to the clan tradition passed down to Cruise, not only did the Katungal people possess substantial outrigger canoes, these canoes even had sails made from possum skins sewn together. This would have been an impressive achievement, for the use of sails also implies the use of a rudder of some sort for steering. How could such a major technological advance have come about?

In those parts of the South Seas where sailing canoes are a long-established indigenous tradition, the sails are usually made of matting, woven by the women from plant fibres. Why were the Katungal people's sails not made in this way? The answer becomes apparent once we think about this question in the light of influence by foreign sailors. It is unlikely that men from a

Portuguese ship would have had any technical knowledge of weaving—but what they would have been experts on would have been sailmaking! And it is precisely the art of sailmaking that we apparently find so ingeniously applied to fashion canoe sails from the most suitable local material available—strong, light possum skins.

The oral tradition handed down to Cruise provides another curious fact—the women of the clan are said to have worn gowns made from possum skins and embroidered with what Cruise describes as Portuguese or Spanish-type flower symbols. Cruise was told it took fifty skins to make one gown. He recalls someone showing him an old photograph of some of these gowns many years ago, but he does not know where the photo would be now, or even if it still exists.

When we add to these Aboriginal traditions the finding out at sea of antique Portuguese jars and other broken pottery from an unknown ship, the mysterious stone circle begins to make sense. As we have seen, Pettungerbe—to give it its correct name from now on—provides the first real shelter for a small vessel running before a southerly storm from the turbulent region of Bass Strait. What more natural that one of Mendonça's ships, particularly a storm-battered smaller ship—or, if the vessel had foundered off Gabo Island, some ship's boats—would have put in here for shelter?

If it was a ship that sought refuge, the vessel in all probability would have been beached for hull cleaning and repairs—in the same way that Cook careened the damaged *Endeavour* after it was holed on the Great Barrier Reef. And whatever the exact circumstances, prudence would have called for the immediate building of some sort of fortification as a protection from possible attack by native people. There is an historical record of this actually happening on the coast of Madagascar when a Portuguese carrack was wrecked there about the year 1530. The one hundred or so survivors' first action was to build a makeshift fort—although notwithstanding this precaution, few of them survived to tell the tale. Another example is that of England's Sir Francis Drake on his epic voyage around the world in December

1579. With his ship the *Golden Hind* in urgent need of hull cleaning and repairs as he sailed through the East Indies, Drake came across a small wooded island which looked ideal for the purpose. But before bringing his cargo of captured Spanish treasure ashore, the English captain first cleared a campsite and then built a defensive wall which he fortified with his ship's cannons. Only then was the *Golden Hind* pulled onto the beach for scraping, caulking and a fresh coat of tar.

It is not hard to see how the stone stockade at Pettungerbe could have served a similar purpose. The stones may well have been carried up from the shore, where similar boulders can be seen today. It would have been a strenuous task, but nevertheless well within the capabilities of a crew of perhaps twenty to forty men working together.

In summary, then, there seems to be not one stone ruin at Pettungerbe but two, separated both in purpose and time. During the more than three centuries that elapsed up to the time of the whaling period, a roughly built Portuguese stockade of loose boulders would have gradually collapsed, before being partly plundered for the foundation walls of the whaling station house.

The wall of the stockade on the lower part of the slope facing the sea appears to be the one from which most of the building stone has been taken. Where the actual walls of the house have been begun, some of the boulders used have been cut to make them rectangular. The Melbourne writer Kenneth McIntyre claimed that it was possible to see a carved date '1524' on one of these walls, which he asserted was the front face of his 'blockhouse'. But this has proved to be a rash and unjustified claim. Quite apart from the fact that it is now apparent that there was no such blockhouse, McIntyre (now deceased) failed to produce any evidence of his claimed date, either photographic or otherwise, and no one else has ever been able to find it. Some of the squared stones do have eroded chisel or pick lines on them, and in one instance some of these lines might with vivid imagination be read as '15' but it seems far more likely that these are just random marks. In any case, as we shall see later when we look at a stone inscription which Mendonça undoubtedly *did* leave behind elsewhere, the Portuguese captain used a unique kind of shorthand for carving inscriptions. When he came to write dates he custom was to inscribed only the final three figures, omitting the initial figure '1' representing the millenium.[8] Thus 1524 was written as 524.

One other puzzling structure at Pettungerbe raises unanswered questions. On the slope above the ruins is—or rather, was—something known locally as the 'Slave's Grave'. It was described in the 1978 survey as being a stone platform about a metre high, one metre wide, and one and a half metres long. It had an indentation at each corner as if to house upright posts for a roof. Unfortunately it was destroyed in the 1980s by intruders apparently hoping to find buried treasure.

The 'slave's grave' legend referred to Melanesians brought to Twofold Bay as de facto slave labourers by another whaling entrepreneur, Ben Boyd, but this was not one of Boyd's whaling stations and so this legend appears to be without substance. As far as can be established now the structure does not seem to have been a grave at all. It would have been too small to hold anything but a child's body and in any case why would it have had a roof? Positioned as it was higher up the slope commanding a good view of the sea, it might perhaps have been a lookout post. The site's near-total destruction means that the answer may never be known.

Strangely, no archaeological work of any significance has been done at Pettungerbe to investigate the many possibilities. An official sign at the stone ruins states that 'this structure is undergoing study for its preservation' but the sign has stood there for many years, if not decades, and little has happened. The interior of the stone circle has been kept reasonably free of vegetation, but that is as far as the 'preservation' goes. The site is only about one hundred metres from the sea and would originally have commanded a sweeping view of the bay. To get to the water's edge today, however, one has to cut a path with a machete through the otherwise impenetrable undergrowth.

As for 'study', or archaeological investigation, the orthodox position has been to dismiss the tradition of a Portuguese connection out of hand, without any serious attempt at objective research. And even though it is accepted that this was the site of an early whaling station, still no field work has been done because the official view is that the ruins are merely the foundations of an unfinished nineteenth century building which was never inhabited—and therefore of only minor historical interest.[9]

If, however, we accept the evidence that a Portuguese ship's crew probably did land at this bay, perhaps at the beginning of the year 1523, the obvious question that follows is: what was the fate of the sailors? If their ship was still afloat, did they manage to repair it and attempt to return to Malacca by painstakingly retracing the course of their outward voyage along Australia's hazardous, reef-strewn east coast? What if their ship had sunk or was too badly damaged to continue? Did some or all of them stay on at Pettungerbe, either from choice or necessity, perhaps taking Aboriginal wives, as the clan traditions related by Ben Cruise might seem to suggest? In our present state of knowledge—or rather lack of it—we can only guess, but there are some tantalising clues.

The first of these is dated 1795. In February of that year, a mere seven years after the founding of the infamous British penal colony of Botany Bay, word circulated in Sydney about the arrival of a very unusual Australian character. A contemporary British colonial account described the event:

> The people who live about the south shore of Botany Bay brought with them a stranger of an extraordinary appearance and character; even his name had something extraordinary in the sound—Gome-boak. He had been several days on his journey from the place where he lived, which was far to the southward. In height he was not more than five foot two or three inches; but he was by far the most muscular, square and well-formed native we had ever seen.

The stranger entertained the European onlookers with mock fights against local champions. The account continues:

> He fought well: his spears were remarkably long and he defended himself with a shield that covered his whole body. We had the

satisfaction of seeing him engaged with some of our Sydney friends, and of observing that neither their persons nor their reputations suffered any thing in the contest.[10]

Two natives of New Holland advancing to combat: engraving from a sketch by Sydney Parkinson. The small, narrow shield carried by one of these combatants was the traditional type of Aboriginal shield. In Sydney in 1795, British observers expressed surprise at the 'extraordinary appearance' of Gome-boak, a muscular Aboriginal stranger from far to the south who had brought with him a shield which covered his entire body. (*National Library of Australia an9196443*)

Gome-boak evidently did not look like any other Aborigine the British had set eyes on at Sydney before. This 'extraordinary' stranger was 'muscular, square and well-formed'—meaning to the British, possibly, that his body had a build of European type. The lances he carried were much longer than the Aboriginal spears normally seen, and we are told that his shield covered his whole body. This latter point is particularly interesting. Gome-boak's shield was evidently unlike the normal Aboriginal shield, which was small, light and narrow. It would seem to have represented quite a major innovation for a traditional Aboriginal society, where tool and weapon design usually changed little over long periods of time.

Could this new design for a shield covering the whole body, coming from a clan on the far south-east coast, owe something to a European influence? What must also make us wonder if this could be so is the stranger's name, Gome-boak. The English narrator at Sydney instinctively felt that there was 'something extraordinary in the sound' but he was unable to say just what. A Portuguese-speaker might have been able to help him out. 'Gome' is virtually the same word as 'Gomes'—a very common Portuguese name. Assuming 'boak' to be an Aboriginal word (although 'boa' is also a Portuguese word, meaning 'good'), this 'extraordinary' stranger from the far south apparently had a Portuguese-Aboriginal name!

Who was this muscular, 'well-formed' man who had walked for many days up the coast (from the direction of Pettungerbe) to see for himself the white strangers from a far-off land? Did their coming evoke memories of things he had been told concerning his own distant ancestry?

Now a second clue. In 1798, three years after the visit of Gome-boak to Sydney, the English explorers Matthew Flinders and George Bass put in at Twofold Bay in the sloop *Norfolk* in the course of a voyage to determine whether or not a strait existed between Van Diemen's Land (Tasmania) and the mainland. In his log book entry for 11 October Flinders recorded that as they were setting up some surveying instruments on shore:

> A party of seven or eight natives broke out in exclamation upon the bank above us, holding up their open hands to show they were unarmed. We were three in number, and besides a pocket pistol, had two muskets. These they had no objection to our bringing, and we sat down in the midst of the party. It consisted entirely of young men, who were better made and cleaner in their persons than the natives of Port Jackson usually are; and their countenances bespoke both goodwill and curiosity, though mixed with some degree of apprehension.[11]

It is Flinders' last sentence which is particularly interesting. He, too, noticed something different about these Twofold Bay people. They were 'better made' (again, to English eyes, probably meaning more European in appearance) and 'cleaner'—which perhaps suggests that they did not follow the Sydney Aborigines' habit of rubbing rancid fish oil over their heads and bodies, a practice which the English found particularly repulsive. It is also curious that in contrast to the hostility or fear shown by most Aboriginal tribesmen, the Twofold Bay men displayed not the least sign of animosity toward the white strangers, greeting them instead with 'goodwill and curiosity'.

Yet another strange fact: a respected local historian, the late Henry Wellings, who diligently recorded as much as he could of the early history of the Twofold Bay area, found an intriguing entry in the diary of Oswald Brierly, a young Englishman who came to Twofold Bay with the entrepreneur Ben Boyd in 1842. It reads in part:

> The natives vary much in their features and appearance, some having quite the Hindoo features, others are more like negroes—curly hair, thick lips. One of them, called Toby, but whose native name is Buginboo, has a remarkably well-formed head...and is in reality a remarkably intelligent fellow...He described with considerable animation the first appearance of ships in the harbour [Twofold Bay]; how the blackfellows retreated to the hills round the bay; how a few first ventured out; how finally they began to carry for the white fellow.[12]

Two things in this account call for comment. Portuguese ships sailing in the East in the early fifteenth century were almost never crewed entirely by

Portuguese: while the officers and specialists such as gunners were invariably Portuguese, a chronic shortage of manpower meant that much of the rest of the crew would be a multiracial mix usually including Malays, Indians from Gujarat, and Africans (often slaves) obtained at the Portuguese-controlled harbours of west and east Africa. So the presence of Indian and African features among Aborigines at a locality where a Portuguese ship had stayed and some or all of its crew mingled with the local women would not be at all unexpected. The account of the Aborigines 'beginning to carry' for the first foreigners is interesting because it indicates that initially, at least, relations were on a friendly footing. Who were these first 'white fellows'? Probably this was a reference to the British, but could Toby have been recalling a much earlier event, relating perhaps to carrying stores from a beached ship, or ship's boats?

The diarist's observation that 'the natives vary much in their features and appearance' sits comfortably with Aboriginal elder Ben Cruise. In the past, he says, the people of his clan living further south were particularly noted for having lighter complexions and a more powerful build than other Aborigines. The men were said to be often big and square-shouldered, some standing over six feet tall. He says he has 'always believed there was some early introduction of European genes among our people'.

Oswald Brierly was an artist and one of the first Englishmen to make his home for a time at Twofold Bay. He soon struck up a close friendship with a young leader of the local Aboriginal people, evidently the 'Toby' referred to in his diary. It was an unusual friendship at a time when most Europeans tended to regard Aborigines in a somewhat patronising manner. Toby taught Brierly the correct pronunciation of important words in his language, Thawa, and it may well have been he who told George Augustus Robinson in 1844 that the bay used as a whaling station was called 'Pettungerbe'.

Could a distant Portuguese seafaring connection also help to account for the particular enthusiasm and skill which the Aborigines of Pettungerbe by all accounts brought to the new industry of whaling at Twofold Bay? Curiously enough, the white settlers' name for the coastal people of the Katungal clan was 'The Fishermen'.[13] Only a few decades after Flinders had met these people of 'goodwill and curiosity', men of this clan had earned universal acclaim for their outstanding abilities as rowers and harpooners in the European whaleboats.

Wellings' research convinced him that a European ship, or ships, had visited the bay 'long before' the coming of Flinders, and that their crew, or crews, had associated with the Aborigines. He based this belief on what he termed 'many Aboriginal statements' and a 'very careful survey' of Aboriginal records and stories. 'Long before' has to mean a date well prior to the arrival of the first convict fleet at Botany Bay, which occurred a mere ten years before Flinders' visit to Twofold Bay. It also indicates a date some time before Captain Cook, whose voyage along this coast preceded that of Flinders by only twenty-eight years and who in any case did not put in at Twofold Bay. Wellings' studies also led him to a conclusion that this first European arrival was not a whaling vessel.

What ship was it, then? Wellings found early nineteenth century newspaper clippings and other references speaking of a 'treasure ship from Cocos Island' and even earlier references to the visit of an unknown 'Portuguese vessel'. Various seventeenth century dates were assigned to this ship, all of them speculative. The 'treasure' was supposedly lost in a shipwreck on the Gippsland coast south of Twofold Bay. Wellings added that it was not only on the Gippsland coast that strange vessels were reported in Aboriginal stories, 'for there are in the same people's stories references to vessels being within the waters of what we know today as Twofold Bay'.[14]

The notion of a treasure ship may sound romantic, but might the fragmentary stories that Wellings stumbled across nonetheless be pointers to a hidden truth? Gabo Island lies at the northern end of the Gippsland coast where the treasure was said to have been lost. Could the 'treasure ship' legend in fact be a garbled reference to the Portuguese shipwreck believed lying off Gabo Island? Treasure ship or no, this wreck would be in about the right place—even if the only treasure it has so far yielded up has been antique pottery. Legends of an unknown 'Portuguese vessel' appearing in the waters of Twofold Bay are fully in accord with what we know about Mendonça's voyage. And the reference to this mysterious ship having come from the Cocos Islands is also curious because these islands do lie in the general region from which the Mendonça expedition would have originated.

Ben Cruise adds one other intriguing story. According to the oral tradition he inherited, at some time in the fairly distant past, warriors of his clan fought a battle against strangers whom they later described as 'men in bodies of stone', meaning that the Aborigines' spears glanced off them. This is immediately reminiscent of the Clarence River Aborigines' legend of a sailing boat manned by pale-skinned strangers with 'garments of stone' and it is surely likewise an attempt to describe European plate armour. There can be no question that this encounter must have occurred long before the coming of Captain Cook, for by Cook's time the development of deadly firearms had made body armour obsolete. Could this by any chance be a recollection of a battle that actually occurred between Aborigines and armour-clad Portuguese at the Pettungerbe stockade?

Finally, a fascinating linguistic clue. We have seen that the two 500-year-old Portuguese pottery jars were trawled up off Gabo Island, which lies just off the coast inhabited by the Katungal people—the clan whose name resembles 'Portingal', and who are said to have had outrigger sailing canoes. How the name of this island originated is something of a mystery. The island juts out from the coast near Cape Howe very much like a headland, and one commonly held explanation is that 'Gabo' was an Aboriginal attempt (though not a very successful one) to pronounce the English word 'cape'.

Any Portuguese-speaking person would immediately see a more convincing comparison. 'Gabo' is not just vaguely similar, but almost identical to the Portuguese term for a 'cape'–*cabo*.

Gabo: Cabo... The words speak for themselves.

8

An Island of Mahogany and the Vessel of the Gods

At 5 AM saw for a few Minutes the Top of a peaked Mountain above the Clowds...it is of a prodigious height and its top is cover'd with everlasting snow.

Captain Cook on his first sighting of the volcano of Taranaki, 13 January 1770

If a ship of Mendonça's fleet sought shelter in Pettungerbe bay after taking a battering from a Southern Ocean storm, it would not have been the flagship. We can be reasonably certain of this for two reasons. The first is that while the harbour at Pettungerbe would have been big enough to accommodate a smaller vessel, it would have been nowhere near adequate for a carrack. These big ships, while they had many excellent ocean-going characteristics, were somewhat clumsy in confined waters—hard to control and lacking in manoeuvrability. It would have been a desperate undertaking to try and pilot one through the narrow entrance to Pettungerbe bay: almost certainly the vessel would have been smashed on the rocks. And besides, for a large ship there is a much better harbour to be found further along Twofold Bay in Snug Cove at the town of Eden, nowadays the home port for a deepsea fishing fleet.

The second reason why we may deduce that if a ship from the fleet entered

Pettungerbe bay it was not the flagship is because the bay itself is not evident on the sixteenth century Vallard map derived from Mendonça's charts. There is a minor indentation on the coast at about the right place for Pettungerbe, opposite the legend *Cap Groca* (today's Green Cape), but the indentation is too vague to really signify anything. Pettungerbe is only a small bay, but had Mendonça's carrack by some miracle managed to enter it one would have expected the expedition's chart-maker to have drawn it with at least a little detail. A smaller auxiliary vessel, on the other hand, is unlikely to have carried a cartographer of its own.

If Mendonça in the expedition's flagship did not take shelter at Pettungerbe or nearby, then where did he go?

Portuguese ships of this period were not able to head into the wind or heave-to in bad weather, so the only option available for a vessel caught by a south-westerly storm in the region of Bass Strait would have been to run before wind and seas. Depending on the wind direction and other factors this would have carried the ship either northward along the Australian coast or toward the east into what is now called the Tasman Sea. Since it can hardly be disputed that Mendonça and at least his flagship returned safely from the voyage—for otherwise there would have been no portolan charts and no Dieppe maps of Australia—we might hope to find clues about what happened next by a careful examination of this part of the Vallard map. And when we do, some strange and unexpected things begin to emerge.

Far to the east of the Australian coast on the map is a large, odd-looking island with the apparent name *Illa do Magna*. This island also appears on some other Dieppe maps—notably the Rotz world map, the Dauphin, and the Henry II map. Rotz does not give the island a name, while in the other two maps it is called (with slight variations) *Ysles de Magna*. These names have invariably been translated as 'Great Island'. On the face of it the meaning does indeed seem obvious but as we have already seen elsewhere on this journey, first appearances can sometimes be deceptive.

In the original Vallard map (as with the other Dieppe maps which depict it), *Illa do Magna* is shown lying far out at sea, due east of the Queensland coast south of the Tropic of Capricorn at approximately the latitude of Fraser Island. In reality there is no island of anything like this size or shape in this part of the South Pacific Ocean. The only possible contender would be the former British penal colony of Norfolk Island, but quite apart from being a lot further south, Norfolk Island is the wrong shape and at five kilometres across far too tiny to be considered the 'Great Island'. New Caledonia would be nearer the right size in area, but it is long, narrow and totally different in appearance to *Illa do Magna*. Furthermore, it lies north of Capricorn, not south of it.

It is tempting to think that here we are indeed dealing here with the product of some sixteenth century cartographer's imagination. But before we reach that conclusion another possibility needs to be considered. On the Vallard map *Illa do Magna* is linked to the Australian mainland by a network of crisscrossing rhumb lines, or compass bearings. This suggests that whoever drew the map had reason to believe that there was a navigational linkage between this mystery island and the coast of Terra Java. Closer examination suggests that the island may in fact be aligned primarily to the section of coast which in the Vallard and other Dieppe maps became the northern shore of the enigmatic eastern promontory.

We have seen how reorienting this huge promontory so that it points south instead of east suddenly makes sense of this section of the map of southern Terra Java, alias Australia. Could *Illa do Magna* have been drawn on the same portolan chart as the promontory? If so, then it, too, needs to be reoriented—swivelled through the same ninety degree arc as the mainland coast to which it relates.

When we try this experiment it produces a remarkable result. Now, the mystery island occupies approximately the position of New Zealand on a modern map!

At first glance it certainly does not look much like New Zealand, but before making up our minds we need to take a closer look. To begin with, the coastline in the map contains a great amount of detail—so much so, that the overriding impression is of an attempt to depict a real island rather than one conjured from the imagination.

The second important point is that only one large island is shown. The shape seems entirely wrong for New Zealand's South Island, but what about the North Island? Do they match in any way?

One initial test of whether this island could be a depiction, however imperfect, of the North Island of New Zealand is to see whether the island's length and its distance from Australia approximate to the actual distances as we know them today. If the figures fail to correspond to any reasonable degree, then the case for this being the North Island will be greatly weakened.

The calculation is not difficult to make. The actual distance across the Tasman Sea from Australia to New Zealand measured from Cape Howe to Cape Egmont is about 2100 kilometres. As for the length of the North Island, the distance from Wellington in the south to just north of Auckland is approximately 600 kilometres (the reason for selecting the latter point for the measurement will become apparent later). The distance across the Tasman Sea is thus about three and a half times that between Auckland and Wellington.

How does the Vallard map compare? The same calculation here produces a very interesting outcome. Using the rhumb line already drawn on the chart from its equivalent of Cape Howe to the nearest point of the mystery island, the distance across the chart's unnamed sea turns out to be three and one-third times the longest north-south distance on *Illa do Magna*. In other words, the sixteenth century chart and a modern map provide close to identical results. The accuracy of the Vallard map is all the more impressive when we take into account the relatively primitive instruments which Mendonça and his navigator had to rely on when calculating such vital data as their ship's position, speed and distance travelled.

So far so good. Now comes the crucial test. What happens if we proceed to make a detailed comparison of the island depicted in the Vallard chart with the coasts of New Zealand's North Island?

A good place to start is a crescent-shaped island appearing just below the mid-point of *Illa do Magna's* west coast. At first this seems to present an awkward problem. The only island of any significance off the west coast of the North Island is Kapiti, former home of the feared warrior chief Te Rauparaha, but the island depicted in the sixteenth century map is the wrong shape, much too large and too far north for it to represent Kapiti.

There is, however, one important geographical feature on the North Island's west coast which is in the right place and about the right size to be a candidate for the Vallard island. This feature is not actually an island—but seen from out at sea it can easily be mistaken for one. It is the volcano of Mount Egmont, known to the Maori people as Taranaki. This graceful 2500-metre-high conical peak, usually snow-capped, was sighted above the clouds and given its English name in January 1770 by Captain Cook, who after noting its 'prodigious height' added in his journal that it could be compared with the peak of Tenerife in the Canary Islands.

It may well have been the first sight of New Zealand for Mendonça. If his ship was approaching the New Zealand coast in fine weather, his first view of Taranaki would have been its near-perfect cone rising as a lone sentinel just above the eastern horizon. Even at the height of the southern summer—in late January or February, the likely time of Mendonça's approach—it would in all probability still have been topped with snow, as Cook observed. If on the other hand the Portuguese carrack was approaching in partly overcast conditions the volcano would have come into view in an even more dramatic manner, sighted suddenly through a break in the clouds.

But what about the problem of its being shown in the sixteenth century map as an island? The answer is not difficult to find. The peninsula of Taranaki from which the volcano rises is semicircular, just as depicted in the Vallard map, and it is low-lying in all directions. Cook, sailing as close inshore as he dared, was in two minds for some time before deciding that the volcano was indeed attached to the mainland. It would hardly be surprising if Mendonça, lacking a telescope, reached the opposite conclusion.[1]

At this juncture Mendonça would have been faced with a critical decision. Which way should he turn his ship to explore the coast of this newly discovered land? Northward the coast would not have looked inviting. It runs more or less due north for over 150 kilometres before swinging toward the northwest. This is a lee coast pounded constantly by Tasman Sea breakers and offering no shelter, its few harbours lying behind dangerous sand bars. As Cook was to observe more than two centuries later, it was a coast that any sensible captain would do well to keep clear of.

What about the coast south of the volcano? Here the land trends for nearly 200 kilometres toward the east. This would offer plenty of sea room in which a sailing vessel could safely manoeuvre. It would seem the obvious way to go—and the Vallard map evidence suggests that this is indeed the course which Mendonça chose. The sixteenth century chart correctly shows this coastline running east and eventually turning back toward the southwest before ending abruptly at a prominent round headland. This headland is exactly the shape of Cape Terawhiti, at the extreme south-western point of the North Island. Behind this headland Mendonça's chart depicts a large, deeply indented bay running in a northerly direction. This is in exactly the right place to be Port Nicholson, the harbour of New Zealand's capital city, Wellington.

How do the details compare? The bay in the sixteenth century map is drawn in the cartographer's usual simplified form, but evidently on a larger scale than the chart as a whole, so that it provides a fair amount of detail. What it depicts is unusual: both an outer and an inner harbour—the latter with a river running into its upper basin. For Port Nicholson the broad details are absolutely correct. Wellington harbour has an outer bay which is large but does not give shelter from the frequent southerly gales. At its northern end this outer bay has a narrow entrance leading to a secure inner harbour, today an international container shipping port. At the furthest end of this inner harbour is a bay with the Maori name Petone, into which flows the Hutt River.

We will return to Wellington a little later to examine some fascinating evidence supporting the theory that a sixteenth century Portuguese ship did indeed enter this harbour, known to the Maori as Whanganui-a-Tara. But first, to confirm that this could be the North Island of New Zealand we should examine the rest of the island coastline depicted in the sixteenth century map.

Mendonça's flagship was evidently on its own now—as early Maori sightings of just a single exotic vessel with great white sails off the coast of Aotearoa (the Maori name for New Zealand) will eloquently testify. The wreck of one of the other ships of his fleet might well by now have been lying on the sea bed off southern Australia's Gabo Island. The fate of his other two ships is equally a matter for conjecture.

After leaving the double harbour of Whanganui-a-Tara, the map evidence indicates that the Portuguese carrack sailed south-east until it reached another prominent headland where the land abruptly turned north. This is evidently Cape Palliser (named by Cook) at the south-eastern extremity of the North Island.

Beyond Cape Palliser and now heading north, the Vallard map's east coast is foreshortened and it has errors of detail, perhaps because Mendonça's ship was a long way off the coast. Some coastal mountain peaks seem to have been mistaken for islands and the cartographer appears to have had second thoughts about the reality of the first deeply indented bay he has drawn, since he has cordoned off its entrance with a dotted line. A similar feature on this same stretch of coast—again a bay with a dotted line across its entrance—appears in a little-known French chart of New Zealand drawn in 1824 by Jules de Blosseville, an officer with the *Coquille* expedition under the command of Louis Isidore Duperrey.

Although Mendonça's charting of this coast appears to have been hurriedly done, the pronounced curve of Hawke Bay is not difficult to identify north of the sixteenth century map's equivalents of Cape Turnagain and Cape Kidnappers (both named by Cook: the former because this is where he turned back; the latter because here some Maoris attempted to kidnap a cabin boy from his ship). At the northern end of the Vallard map's equivalent of Hawke Bay the prominent Mahia Peninsula, which is only narrowly connected to the mainland, is drawn as a large island. A smaller island shown off its southern tip approximates Portland Island.

The final deeply indented bay shown on this coast would be Poverty Bay, whose distinctive horseshoe shape Cook remarked on. Accurately depicted on Mendonça's chart is Young Nicks Head named by the English captain after the ship's boy who sighted this promontory, the *Endeavour's* first New Zealand landfall. Next, Mendonça's chart depicts a very large headland jutting out toward the north-east. This can be seen to represent East Cape, with a mountain range at the cape's northern extremity mistaken for an island. Beyond this cape toward the north-west is a large semicircular bay corresponding in stylised form to the Bay of Plenty.

We will return to this area later to examine some fascinating evidence, in the form of an ancient Maori tradition, which could well be an eyewitness account of Mendonça's vessel sailing in these waters.

North of the Vallard map's 'Bay of Plenty' is another smaller promontory representing a truncated Coromandel Peninsula and then a long offshore island. This corresponds in both shape and position to the mountainous Great Barrier Island, guarding the entrance to the broad Hauraki Gulf. As with Wellington harbour, the island seems to have been drawn on a larger scale. This, together with the fact that two harbours on the island's sheltered western coast are shown in some detail, suggest that Mendonça probably put in here. The harbour of Tryphena, at the southern end of Great Barrier Island, can readily be recognised and possibly also Blind Bay, a little further north. The Hauraki Gulf and its approaches are littered with small islands, so the line of islets drawn on the chart between Great Barrier and the mainland may perhaps have been intended as a symbolic indication of this fact.

The Vallard map's *Illa do Magna* comes to an end just north of here, so on this interpretation the long Northland peninsula has scarcely been shown at all. Why would it have been omitted? It is not hard to see a likely explanation. Mendonça would in all probability have exited the Hauraki Gulf by sailing directly toward the east, back through the wide strait which he had already traversed between the northern tip of the Coromandel Peninsula and the southern end of Great Barrier Island. If, as seems likely, he changed course to the north after clearing the strait he would already have been at least 100 kilometres from the fast-receding mainland coast.

At about this point the Northland peninsula veers away toward the north-west, so the land in this direction would soon have been out of sight, particularly if Mendonça's ship had sailed on at night into the open waters of the Pacific. As for the long northernmost spit of the North Island, known as Ninety Mile Beach, this peninsula is low-lying and would not have been visible in any circumstances from far out at sea. Cook, we may note, ran into difficulties here. He had to battle against westerly gales and high, rolling seas for three weeks in order to round the northern tip of New Zealand. Only occasionally did he manage to glimpse the Northland peninsula, which he described in his Journal on 2 January 1770 as 'for the most part low and narrow' and 'everywhere a barren shore'. A little later he added:

> The land...wears a most desolate and inhospitable aspect, nothing is to be seen but large Sand hills with hardly any green thing upon them and the great sea which the prevailing westerly winds impell upon the Shore must render this a very dangerous Coast, this I am so fully sencible of that was we once clear of it I am determind not to come so near again if I can possibly avoide it unless we have a very favourable wind indeed.[2]

What other evidence do we have that the *Illa do Magna* is New Zealand's North Island?[3] In fact, the very name given to this island in the Vallard map will provide us with a clue as important as it is unexpected.

'Illa do Magna', as we have seen, has always been assumed to mean 'Great Island'. To a native Portuguese speaker, however, this translation is self-evidently wrong. Magna does indeed mean 'great', while *do* signifies 'of the'.

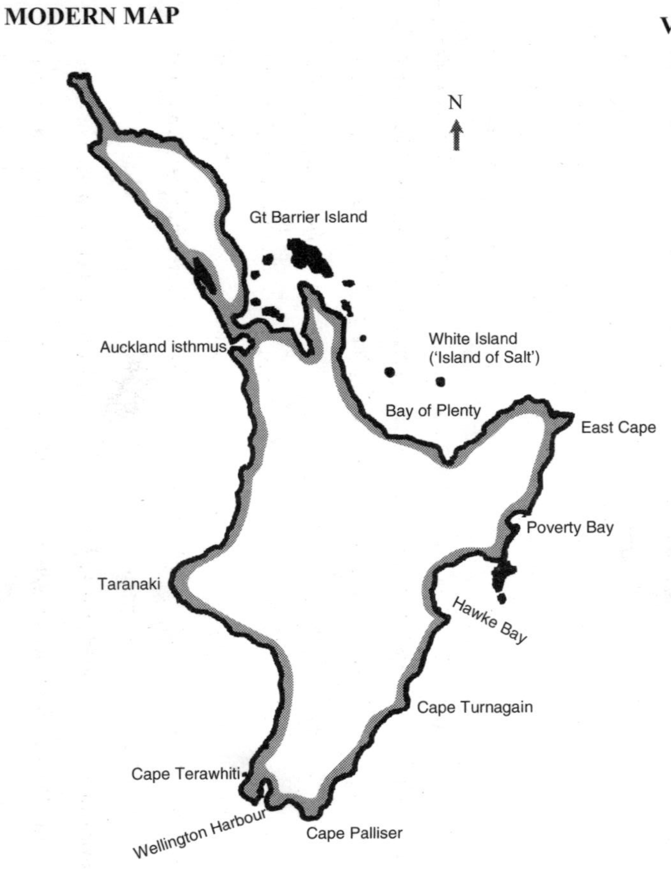

Not so different after all: New Zealand's North Island and (right) the Vallard chart's *Illa do Magna* on the same scale, with the crescent-shaped offshore island representing Mount Egmont (Taranaki) and the tip of East Cape now joined to the mainland, and the missing Northland peninsula sketched in.

But in Portuguese the gender must be the same for both the article 'do' and the word following it. In *do Magna* the word 'do' is masculine while 'Magna' is feminine, so something is amiss. It should be 'Ilha do magno'. But even now there would still be a problem because 'magno' is an adjective, not a noun, and so this name would literally translate as 'Island of Great'. The next question would be: 'great' what?

Moreover, if Mendonça had really intended to call this island 'Great Island' why would he not have used the name he had already bestowed on several other great (or big) islands on this voyage—'Illa Grossa'? As, for example, in the case of Kangaroo Island (Illa Grossa), off southern Australia.

In fact there is a quite different solution to the riddle. In Portuguese there is one word (and only one) which closely resembles *magno*, and that is *mogno*, the word for mahogany. Could this be the vital clue? In the Vallard map we can find numerous instances of words where the letter 'o' has been written so that it looks like an 'a'. Was the original Portuguese name for the Vallard island really *Ilha do Mogno*—'Island of Mahogany'?

It makes excellent sense, and not only in terms of linguistics. Always with an eye open for potential trade opportunities, the Portuguese would have been greatly impressed by this unknown island's magnificent native forests, containing towering mahogany-like trees such as the totara (fam. Podocarpus), a forest giant that grows up to thirty metres high, with a tall, straight trunk and reddish-brown timber. Joseph Banks, botanist on Captain Cook's voyage, described these huge New Zealand trees as the finest timber he had ever beheld. It was also from trees of this sort that the Maoris fashioned their mighty war canoes. Cook observed that the biggest of these canoes, carved from a single trunk, were up to fifteen metres long and capable of carrying 100 warriors.

Mendonça, too, would surely have seen some of these great canoes at close hand. The Portuguese were familiar with the value of mahogany, because species of it grew both on the Iberian peninsula and on their island colony of Madeira (*Ilha da Madeira*: the name actually means 'Island of Timber') not to mention in West Africa and India. This new island might be no El Dorado in terms of gold but clearly it offered plentiful supplies of excellent shipbuilding timber. It would be a potentially valuable find for the commander to include in the report he would have to compile for his King in Lisbon.

It is now time for us to return to the harbour of Wellington. Mendonça's chart leaves no doubt that his ship sailed in here because the inner harbour details—drawn on the enlarged scale customarily used to show important features—must have been recorded at first-hand. In pre-European times Wellington's spectacular harbour was forested to the water's edge. It would have been an ideal place for the Portuguese mariners to make running repairs and take on fresh water after an arduous ocean crossing from Australia. And there is evidence that this is indeed what happened.

The first clue to their presence came to light sixty or so years after the beginning of British settlement. Some time around the close of the nineteenth century, Wellington's Colonial Museum was presented with an iron helmet of medieval or Renaissance type which had been found in the harbour. Unfortunately, the date of its discovery, the name of the person who found it and the exact circumstances in which it was found are unknown. The item was first recorded in 1904 by the museum's then director, Augustus Hamilton, who had set himself the task of cataloguing a great deal of material which had never up to then been noted down. His entry for Item 841 states simply: 'Iron Helmet, found in Wellington Harbour'.

The helmet is clearly of European origin. Measuring twenty-nine centimetres high by nineteen centimetres at its widest, it is a dark rust-brown colour with scattered areas of pitting. Some rivets still remain and the former positions of others are represented by holes. One of its most distinctive features is a high metal crest running almost its full length from the brow to the nape of the neck. A large hole in the rear part of this crest might simply be due to corrosion; alternatively it could indicate that the helmet's wearer had suffered a massive blow to the back of his head.

How could such an exotic object have come to be in Wellington Harbour, on the opposite side of the world from Europe? Who could it have belonged to?

Before attempting to answer these questions we need to consider another strange artefact retrieved from the same harbour some twenty or so years later. On 9 June 1926 a dredge belonging to the Wellington harbour authority was removing sand and mud from the seabed not far from the old Maori village at Petone when it scooped up a large iron cannonball. The object at first gave rise to excited speculation, but then it was stowed away in a safe and forgotten. Many years later it was rediscovered and put on display in the Wellington Maritime Museum.

There would seem to be a prima facie case for theorising that the cannonball and the helmet, two early European military items recovered from the same harbour and quite possibly the same part of it, are in some way linked. Strangely, however, New Zealand's museum authorities have gone to considerable lengths not just to deny any suggestion of a connection between the two but even to debunk the very idea that either object could be of historical importance.

The helmet, known unofficially as the 'Spanish helmet', does not even seem to merit permanent display in the new Museum of New Zealand, Te Papa Tongarewa despite the fact that several investigations over a period of many years have all confirmed it to be of sixteenth century origin.

The first attempts at identifying the helmet were made by the Colonial Museum's Augustus Hamilton only a year or two after he first recorded the item. He sent a description and a photo to the *Connoisseur* magazine in London, which published a response in July 1906. The article advised:

> In the photograph we recognise the skull piece of a European helmet and have no difficulty in assigning it to the spacious times of Good Queen Bess [Elizabeth I of England] somewhere about the period 1560–1570...It is quite clear that the skull piece belonged to a 'close helmet' of the sixteenth century.[4]

In 1937 a new director of the museum (it had by then been renamed the 'Dominion Museum') sought an opinion about the helmet from the Hispanic Society of America. In reply, a Mr Granesay, described as being 'of the Metropolitan Museum of Art', replied that the item dated to 'roundly about 1580'. This date was endorsed by the next person consulted, AM Kennard, of the Tower of London Armouries who, in a letter written in 1957, described the helmet as 'a very international type, but may well have been made in North Italy, where so much armour was produced'. Kennard made the further interesting observation that the helmet 'might have been taken on board as part of the equipment of a ship's officer'. The following year a Mr C Blair, of the Victoria and Albert Museum, London, wrote supporting the earlier opinion that it was 'a close helmet of about 1580'.[5]

Despite all this accumulated evidence, however, the Wellington museum authorities seem to have striven in recent times to devalue the helmet's historical significance. A 1983 report compiled by one of its staff gave a 'possible date' for the helmet as seventeenth rather than sixteenth century—though it failed to put forward any evidence in support of this statement. Strangely, the report even questioned whether the helmet had been found in Wellington harbour at all—and this notwithstanding the unequivocal written statement by Augustus Hamilton that it *had*. And most surprisingly of all, the 1983 report also stated that the helmet 'does not appear to have lain in the sea for any length of time'—a claim which flies in the face of a specialist report from the New Zealand Government's Department of Scientific and Industrial Research (DSIR). This report affirmed that:

> It is feasible that the helmet could have lain in the harbour provided that it was in still, stagnant water, or covered in mud, in oxygen-free conditions. That it may have lain in the sea for an unknown period of time can be supported by the size of the pittings observed in the helmet. The type of pitting, wide, flat, large and uneven, is typical of oxygen attack under seawater.[6]

We cannot be sure where exactly the helmet was found, but accounts have suggested this was also in the upper harbour. This area is particularly muddy, the very reason why it was being dredged when the cannonball was retrieved. Here the Hutt River (depicted, as we have seen, on Mendonça's chart) deposits large amounts of silt on the sea floor—as much as six centimetres of black, oxygen-deprived material every year—so it would fit very well the DSIR's requirements for preservation of an iron helmet over a long period of immersion.

We will return later to the question of the helmet. What of the cannonball? It is cast iron and weighs approximately thirteen kilograms. The Wellington Maritime Museum claimed that this was too large to have been fired by the guns that would have been carried by a sixteenth century ship from Spain or Portugal, but this is incorrect. Accurate, detailed records from the time of the Spanish Armada in 1588 show that iron balls of around thirteen kilograms were standard ammunition for just the *medium*-size cannons carried by these countries' ships in the sixteenth century. Their really big guns fired cannonballs weighing close to twenty-three kilograms.[7]

In 1995 the museum attempted to have the cannonball dated by the New Zealand Institute of Geological and Nuclear Sciences. This was done by taking a sample and having the carbon content analysed. The reasoning behind this was that before the mid-eighteenth century, iron was manufactured using charcoal from recently felled trees, so if this was the period when the cannonball was cast, a relatively recent date should be produced by the radiocarbon test. Conversely—or so the museum claimed—an older radiocarbon date would mean the cannonball was made using coal, and therefore it would have to be of eighteenth century or later origin.

The laboratory tests produced a radiocarbon age of 23 000 years. The museum then publicly announced that 'the ball was manufactured using coal and thus dates from the eighteenth or nineteenth century', adding for good measure that 'another myth' [of a Spanish ship] had now been discredited.

In jumping to this conclusion, however, the museum authorities committed a fundamental error. Coal is not formed 'many thousands of years before present' as they mistakenly assumed—it is invariably formed many *millions* of years ago. It has to be because of the very slow natural processes by which it is created. Most of the bituminous coal found in Europe—the coal with which an eighteenth century European cannonball would probably have been manufactured—is about 300 million years old. A leading Australian expert in the scientific study of coal dismisses the museum's conclusion that 'coal 23 000 years old' was used to manufacture the cannonball as 'entirely unbelievable'.[8]

In fact, if the Wellington radiocarbon date is anywhere near correct it would point to the very opposite of the conclusion reached by the museum authorities—that the cannonball must have been manufactured *before* the eighteenth century! This can be said with certainty because the effective limit for radiocarbon dating is about 40 000 years; for samples much older than this the radioactivity has decayed to such an extent that it becomes impossible to measure. As the age of coal is reckoned not in tens of thousands but millions of years, a coal sample should produce no detectable radiocarbon emissions.

But the Wellington researchers reported that they *did* detect such emissions. How can this be explained? Rather than suggesting that the source was coal, as those interpreting the research wrongly concluded, the result might even be taken as prima facie evidence that the cannonball was manufactured using charcoal, with the older than expected radiocarbon date due perhaps to an unexplained error in the laboratory procedure.

There is, though, one other possible explanation which could explain the radiocarbon date. In theory, at least, the cannonball could have been made using peat as a fuel. Peat is the one carboniferous product which could give a radiocarbon date of 23 000 years. On the face of it this explanation seems improbable because peat is an inferior fuel to charcoal. Nevertheless it would have been technically possible for peat to have be used as part of the manufacturing process. And this explanation would also unambiguously point to an origin for the cannonball well prior to the eighteenth century.

Unfortunately, no second test was carried out by the radiocarbon dating laboratory to check the initial, seemingly anomalous, result. If a second test had been performed, and if this also had given a radiocarbon date of around 23 000 years, then the peat theory would have to be taken very seriously.

In summary, then, we are left with an iron helmet identified by all the experts as of probable sixteenth century origin and a cannonball which on the available evidence appears to date from about the same period.

In the case of the helmet, the Wellington museum authorities from various sources over a period of half a century obtained estimated dates for its origin ranging from about 1560 to 1580. But these dates are somewhat arbitrary and there is evidence for believing it could be older. Iron helmets of this general type, known in Europe as a *salett*, were developed as early as the 1400s and became very widely used, taking many different forms. At first they were closed helmets (not 'close', as the *Connoisseur* magazine stated) but from about 1450 onward they could be opened—the front part of the helmet being transformed into a movable visor which could be raised by means of pivots placed on either side of the crown. At about the same time the nape of the neck was given greater protection by a curved plate to which was sometimes attached plate bands which also moved on pivots.

This is precisely the type of helmet found in Wellington harbour. Further, as with the Wellington helmet, the *salett* often had a raised crest running along its centre, while the inside was lined with cloth, leather or padded cotton to make it sit more comfortably and soften the blows of the enemy. The Wellington helmet would have had this type of padding, as evidenced by the holes for the rivets which would have held the padding in position.

As further evidence of the helmet's likely Portuguese origin, helmets of a very similar design can be seen today in the Lisbon Military Museum. In Portuguese, this type of helmet was called a *bacinete* or *celada*, and it was in use prior to and during the sixteenth century. Several examples in the Lisbon museum are notable for having a high ridge almost identical to that of the Wellington helmet. Furthermore, two particularly well-preserved examples indicate that some of the holes visible on the Wellington helmet which seemed to have no particular purpose are in fact intended for the attachment of side plates to protect the wearer's cheeks.

From all this evidence it would seem to be a reasonable conclusion that the Wellington helmet was made at some time between about 1450 and 1580.[9] That being the case, why does the Museum of New Zealand refuse to treat it as an important historical find?

The institution has justified its position by putting forward three alternative theories to explain the artefact's mysterious discovery on the harbour bed—all of them unsubstantiated and, it must be added, unconvincing.

First, its 1983 report claimed that the helmet might have been used as a ship's ballast. But this suggestion overlooks the fact that the helmet's metal is a mere two millimetres thick and that it weighs less than two kilograms, making it entirely unsuitable for such a purpose. And even it we were to assume for the sake of argument that a sixteenth century helmet was for some strange reason part of a nineteenth century vessel's ballast, the museum's theory implies that on arrival in the harbour a sailor, for no apparent reason, descended into a stinking, vermin-infested ship's bilge, fished out this one particular object—and then climbed back up on deck simply in order to toss it overboard!

The second suggestion—that the helmet may have been brought out as a 'souvenir' by some early English immigrant—equally strains credulity. It is well known that space on nineteenth century immigrant ships was at a premium, with many migrants having to content themselves with just a single sea chest for their most essential belongings. Are we to believe that than an immigrant in these circumstances would pack into his precious chest a completely useless antique iron helmet only to apparently take it out and drop it into the sea when he arrived at his destination?

Finally, and now moving in a different direction, the museum's report theorised that the helmet might originally have been given to an unknown Maori chief—who later, it was assumed, must have lost it. The museum drew a parallel with the presentation of suits of armour to the chiefs Hongi Hika and Titore by the British kings George IV and William IV in the early nineteenth century. But not only is this theory lacking in substantiation, it flies in the face of known facts. The two chiefs named, and their descendants, treasured their gifts. In Hongi's case the armour was buried with him, while Titore's descendants presented his armour to the Colonial Museum for safe keeping in 1908. To assume that some other chief would have had so little regard for his gift that he would carelessly 'lose' it in the harbour lacks credibility. And in any case this tenuous theory fails to address the central question of how and when the unknown chief would have come by the helmet in the first place—and why such a gift would not have been documented, as the others were.

Striving to invent such elaborate theories is unnecessary when there is a realistic explanation supported by evidence which can account for the discovery of both the helmet and the cannonball in Wellington harbour. That explanation is the presence in the harbour of a sixteenth century Portuguese vessel—Mendonça's flagship. And we can make an intelligent guess as to *why* the two objects were in the harbour as soon as we look the experiences in New Zealand waters of the European explorers Tasman and Cook. Both of these foreigners were either attacked or threatened with attack (apparently without provocation) by Maoris in war canoes—Cook on at least two occasions; Tasman once, with fatal results.

In Cook's case, off Cape Runaway near the eastern tip of the North Island, he recorded in his Journal on 31 October 1769 that:

> At 9, five Canoes came off to us, in one of which were upwards of 40 Men, all Arm'd with Pikes, etc; from this and other Circumstances it full appear'd that they came with no friendly intention, and I at this Time being very busey, and had no inclination to stay upon deck to watch their Motions, I order'd a Grape shot to be fir'd a little wide of them. This made them pull off a little, and then they got together either to consult what to do or to look about them. Upon this I order'd a round shot to be fir'd over their heads, which frightened

> them to a degree that I believe they did not think themselves safe until they got ashore.

and again off Mercury Bay, a little further north, on 4 November:

> Between 5 and 6 o'Clock in the Morning several Canoes came off to us from all Parts of the Bay; in them were about 130 or 140 People. To all appearances their first design was to attack us, being all completely Arm'd in their way; however, this they never attempted, but after Parading about the Ship near 3 Hours, some times trading with us and at other times tricking of us, they disperse'd; but not before we had fired a few Musquets and one great gun, not with any design to hurt any of them, but to shew them what sort of Weapons we had and that we could revenge any insult they offer'd to us.

Cook's policy of displaying the firepower of European ships' guns to forestall serious trouble was prudent and as circumstances were to show, well justified. There is every reason to believe that Mendonça, if approached by Maori war canoes would have done much the same thing. Opening fire with cannons was the standard Portuguese response to hostile native inhabitants in any number of lands, from Sierra Leone in West Africa to the Malabar coast and Sumatra. However, the discovery in Wellington harbour of both a sixteenth century military helmet of the type likely to have been used by a Portuguese ship's officer, and a cannonball offshore from the Maori village of Petone, would suggest a confrontation between Maori and Portuguese more akin to what befell the Dutchman Abel Tasman than the experiences of the Englishman, Cook.

On 19 December 1642, the Dutch captain had ordered his two ships, *Heemskerck* and *Zeehaen*, to anchor in what is now Golden Bay, near the northern tip of the South Island. Tasman subsequently named it 'Murderers' Bay' and it was to be his first and only New Zealand port of call. Despite noticing that his ships were being surrounded by large numbers of armed Maoris in canoes, Tasman neglected the precautions taken by Cook.

Here is the Dutch skipper's account of the events which followed:

> Saw 7 more craft come from land, one of which (projecting in front high and sharp, manned by 17 men) paddled round behind the *Zeehaen* and a second (in it 13 sturdy men) came before the Ship [the *Heemskerck*] not a half cast from us...when the small boat of the *Zeehaen* set out back from the ship They who lay before us between the two ships began to Paddle So vigorously to it that about a little more than half way to our ship [they] Struck the *Zeehaen's* small boat with their stem on the Side, Dashed over that Same violently, whereupon the foremost in this canoe of rogues pushed the quarterm^r^ Cornelis Ioppen in the neck several times with a long blunt pike so fiercely that he had to fall overboard, whereupon the rest of them set to with short thick pieces of wood...and their paddles, overpowering the small boat, in which violence 3 of the *Zeehaen's* people were killed, and the fourth through the heavy blows was mortally injured. The quarterm^r^ and 2 more sailors swam towards our ship and [we] Sent our sloop for them into which They got alive.
>
> After this monstrous deed and detestable Thing, the murderers let the small boat drift, pulled one of the dead into their canoe, and drowned another. We and those of the *Zeehaen* seeing this shot hard with muskets and cannon, but although [we] did not indeed Hit them, [they] nevertheless hastened back and paddled for land out of shooting range. We fired many shots with our forward upper and bow guns by and about their vessels, but struck none.[10]

The wary Captain Cook fired early warning shots; Tasman also fired—but too late to save some of his men. Nevertheless, his demonstration of the lethal power of his big guns saved his ships from the threat of further attack. And so it would have been with a Portuguese vessel under attack from Maoris in Wellington harbour. The finding of an iron helmet and a cannonball in the harbour, the latter, at least, not far from the site of a Maori stronghold, should not be at all surprising.

We may recall here the opinion of the Tower of London's expert on armoury that the helmet might have been part of the equipment of a ship's officer. As for Petone, near where the cannonball was found, this place was at the time of British settlement in 1839 a stockaded *pa*, or village, of the Ngati-Awa people. The name (more correctly spelt *Pito-one*) means 'end of the beach'. The village was strategically placed on the harbour foreshore at the edge of the Hutt River delta. But the Ngati-Awa would not have been in residence when the Portuguese sailed in; at that time the occupiers were the Ngati-Ira people, later to be driven out by invading Ngati-Awa from the north.[11]

We can now picture a similar event occurring off Petone to that which befell Tasman. A ship's boat is suddenly attacked by Maori war canoes; in the ensuing skirmish a soldier in the Portuguese boat is struck a heavy blow on the back of his helmet by a Maori wielding a hard greenstone *patu* club—or perhaps an iron weapon which he has just seized. The soldier either loses his helmet or is knocked overboard with it still on his head. Mendonça, watching from the carrack's quarterdeck, orders his gunners to open fire with one or more of the ship's cannons. The Maoris turn in fright and paddle for the shore. Helmet and cannonball sink into the mud of the harbour bed—where they remain until rediscovered some 400 years later.

The Maori had no form of writing, and endemic warfare between tribes often disrupted the passing on of oral traditions. There is, however, one surviving local story which sounds very much like an account of the coming of Mendonça's ship. In the early 1900s a New Zealand historian, WJ Elvy, interviewed an elderly Maori in the Marlborough Sounds area—the northernmost part of the South Island and the region to which the surviving Ngati-Ira from Petone had fled after their expulsion by the Ngati-Awa. Elvy was told of a tradition describing the arrival long before of a strange foreign sailing ship. The men on board were like no others the Maori had seen. They were armed with swords and lances, and some wore shiny coats of armour that caused the Maoris' stone weapons to bounce off. According to the tradition the strangers were friendly at first, but trouble arose when they stole Maori women. There was fighting, with men on both sides killed. After a time the ship sailed away again.[12]

The source of this story seems to have been prisoners from a conquered tribe who had been kept as slaves. It is interesting that unlike other Maori traditions describing the arrival of foreigners which mention two ships and almost certainly refer to Abel Tasman, the story told to Elvy records the presence of only a single ship, apparently a large square-rigged vessel.

But the really extraordinary fact is that this ship evidently had a multi-racial crew! The tradition relates that while some of the men on board the foreign vessel were fair-skinned, others were darker than Maoris. This detail would be absolutely correct for a Portuguese ship and probably no other nationality—for as we have seen, because of a chronic shortage of manpower, Portugal's ships were almost invariably crewed by a mix of Portuguese officers and African, Indian or Malay sailors. The tradition's reference to swords and plate armour also clearly points to a sixteenth century vessel, not one from Cook's time or even that of Abel Tasman. And finally, these details are consistent with the discovery in Wellington harbour of the iron helmet and cannonball.

There is yet another intriguing piece of evidence which provides further proof that the Portuguese were the true European discoverers of Aotearoa. On the Dieppe maps which show the *Illa do Magna* (our Island of Mahogany) there is one, and only one, other place name. The Dauphin and Henry II maps spell it in Norman French as *Ye de Saille*—'Sail Island'.

A solitary island that resembled a sail would be precisely the sort of feature one might expect a ship's navigator on a pioneering voyage to note on his chart. And it would surely have to be an obvious landmark, so one might reasonably expect to find an island somewhere off the north-east coast of New Zealand with which it could be firmly identified. But a search for such an island draws a blank. There is no offshore island with the name 'Sail Island', or anything at all similar, and nor is there any prominent, solitary island with another name which looks like a sail. Is the map at least this time in error?

The answer, once again, is that it was the French and not the Portuguese who got it wrong. The Vallard map, as we might by now expect, gives a name for this island closer to the original Portuguese than the other Dieppe charts. At first glance it looks like *Illa do Sail*—another version of 'Sail Island'. But on a closer examination we find that the last word has no 'i'—it is not *Sail* but *Sal*. Does 'sal' have a meaning in Portuguese? It does indeed. It is the word for 'salt'. But 'Island of Salt'? This seems even less probable than 'Island of Sail'.

There is, nonetheless, an explanation—one as remarkable as it is unexpected. And here, once again, we have the ever-observant Captain Cook to thank for the answer. Rising out of the waters of the Bay of Plenty on the North Island's east coast is a 320-metre-high active volcano called White Island. In a state of intermittent eruption, for much of the time it emits a towering plume of steam or fine ash—a phenomenon which would be visible from the deck of any ship long before the island itself came into view. White Island was one of the first landmarks noted by Cook after the *Endeavour* rounded East Cape, some sixty kilometres away. No doubt it would have been an equally significant navigational feature for Mendonça to record on his chart.

But 'Island of *Salt*'? What could that name possibly indicate? The clue lies in the pages of Cook's Journal. Sailing across the Bay of Plenty on 1 November 1769, the English navigator recorded that he was naming this island 'White Island' 'because as such it always appear'd to us'.

And why was it white? The answer is that its active crater constantly produces large amounts of two chemicals. One is sulphur, and the other—the one particularly relevant in this case—is gypsum, a white mineral salt which looks very similar to common salt. Over centuries, the volcano's gentle eruptions had coated the island with a glittering mineral crust. Hence Cook's 'white' island, and hence also the likely reason why Mendonça took it to be an island of salt—for this is what its shimmering white slopes could well have appeared to be from the deck of a passing ship. And this discovery would have important potential for a sound commercial reason: salt in the sixteenth century was a highly prized commodity, essential for the preservation of meat and many other foodstuffs. So valuable was it that in Portugal the trade in salt was a royal monopoly and by some reckonings salt was even more valuable than gold. And here, apparently, was a whole island of it! Little wonder it was recorded on the map.[13]

Since European settlement began in New Zealand in the mid-nineteenth century White Island has erupted with varying intensity no fewer than thirty-five times. These eruptions have caused significant changes in the island's appearance—as has commercial sulphur mining which was carried on for some years in the early twentieth century, despite being temporarily interrupted by a volcanic mud flow which buried the mine buildings and ten workmen. In spite of these changes, however, a large headland on the ocean side of the island is said to still stand out as an area of brilliant white. In the Vallard map the legend *Illa do Sal* appears in the Pacific Ocean off the south-eastern corner of the prominent feature representing Great Barrier Island, but thanks to the map's foreshortened scale it is also directly opposite the bay in which White Island is located.

If yet more evidence be needed that 'Sal' was the name the Portuguese voyagers gave to White Island it can be found at the remote Cape Verde Islands, in the Atlantic Ocean. This formerly uninhabited group, lying 400 nautical miles west of the coast of Senegal, was discovered by a Portuguese expedition sent out by Henry the Navigator in 1455. Most of its ten main islands are named after Christian saints, but there is one notable exception—the Island of Sal. And why was this particular island called Sal? Because, as we might guess, it has extensive salt deposits, supporting an industry which even today provides one of the Cape Verde Islands' few exports.

If Mendonça did indeed sail up the east coast of his 'Island of Mahogany', should there not be Maori traditions telling of this visit? There is, in fact, at least one account highly suggestive of such an event. It was first recorded in written form by a tribal chief at some time in the late 1800s, but it was not until a century later, in 1991, that the chief's account came to the notice of historians.

The very striking—even poetic—narrative begins with some Maoris in a fishing canoe setting out before dawn from a village on the coast south of East Cape, a promontory which we have noted appears in Mendonça's chart. The sea was calm and there was no wind. As dawn broke, the canoe reached the fishing grounds. There it floated and the men let down their anchor and began to fish. The narrative continues:

> When the sun was up a boat was seen paddling towards them, coming from far off. It was a long boat with many paddlers on either side and rows of people in the middle, with a leader standing up and also a man in the stern. It came straight towards them. The crew of the fishing canoe were afraid, thinking that this was a war canoe...As the boat came towards them, however, its paddles were lifted on board...It turned and floated just like the fishing canoe. The local people saw how long it was, and the numbers of people that sat in the middle.
>
> Now the men on either side of the boat took up their fishing lines, baited them and threw them into the water. Soon they were all hauling in fish—four, five or six on the hooks of each line, on both sides of the boat. The people in the canoe no longer felt afraid, they were excited to see these people catching so many fish. Their steersman ordered them to raise their anchor and they paddled toward the place where the fish were biting. As they paddled closer, the people in the boat rolled up their lines. The local canoe drifted and its crew looked at these people—they appeared very strange. The man in the stern of the boat stood up to haul in the anchor; all the people in the middle grabbed the anchor rope and as they hauled it in, they chanted. This is how the words sounded:
>
> Ka whakatakotoria
> Ki te ika te wa o tu
> E ko te tae o tu
> E kore rarii
>
> Once the anchor was on board those people took up their paddles, and as they all moved they spoke in their language. This is what it sounded like:
>
> Pakepakeha, pakepakeha
> Hoihoi hii, hoihoi hii
> Hihi hii, hihi hii
>
> Now they could be plainly seen. They were *turehu*—fairy people—misty-looking, fair, pale and red, like red ochre. That was the way their faces looked. The prow of the boat turned and they

> paddled back the way they had come. The leader and the steersman stood up again. In no time their boat seemed to rise up on the sea: it looked as though they were paddling in mid-air, and finally they were lost in the clouds.
>
> Then the local people knew that these were *turehu*—fairies, evil gods, still-born spirits, whistling spirits. They were sighted many times, before and after that. Their *haka* (chants) are still remembered, and the place where they chanted was called 'Haka of the god'.
>
> The sighting of this *turehu* boat was long before the arrival of Captain Cook's ship. When the old men and women saw Captain Cook's ship they called out 'It is an island, an island floating from afar. Here it is, coming towards us'. When they saw its sails they cried out 'Aha, ha! The sails of this travelling island are like clouds in the sky!'[14]

Who were these strangers who appeared off the coast of Aotearoa long before the coming of Captain Cook? The Maori description of them—'fair, pale and red, like red ochre'—can leave little doubt that they were Europeans. The only other Europeans known to have reached Aotearoa prior to Cook were the men of Tasman's expedition, but Tasman did not even know that New Zealand *had* an east coast; having approached from the west, he assumed that Cook Strait was a large bay on the western edge of a southern continent. So we are left with the Portuguese of Mendonça's expedition—who, as the Vallard map tells us, must indeed have sailed up the east coast, almost certainly in the late southern summer of 1522-23.

The Maori account seems to tell us that what the fishermen in the canoe were gazing at with such wonderment was a ship's longboat. The strangers' prowess in fishing ('four, five or six [fish] on the hooks of each line...') is also strongly suggestive of a Portuguese vessel, fishing being an art at which Portuguese sailors excelled. And what of the powerful, almost mystical account of the *turehu* boat 'rising up on the sea' and being 'lost in the clouds'? This could be a description of the longboat being hauled up on board the mother ship, whose white sails (as with Cook's ship) were identified with the clouds.

Evidently the *turehu* sailors were chanting in a foreign language. But what language? Could the Maori fishermen have been repeating, without understanding them, the words of some old Portuguese sea song or shanty? The Maori had no written language prior to the coming of Europeans, so (unlike English) Maori is today written phonetically. Written Portuguese is also largely phonetic, but with a few anomalies, one of the chief of which is that there is no letter 'k' in Portuguese—that sound being represented by a hard 'c'.

So how much, if anything, can be discerned in the Maori version of the turehu chants? A professional Portuguese translator provides some interesting suggestions. The very first word of the first chant, 'ka', is identical to the Portuguese word 'cá' (pronounced 'ka') meaning 'here'. The next syllable, 'wha', is similar to the Portuguese 'va', meaning 'go', while 'kata' calls to mind 'contar', meaning 'tell'. This suggests that the remaining word, 'kotoria', might be 'história', Portuguese for 'story'.

In the next line, the first two syllables, 'ki te' bring to mind the Portuguese word 'quite' (pronounced 'ki-ti'), meaning 'quit'. 'Ika' could be 'ficar', meaning 'to stay', while 'te' in Portuguese means 'you'. As in the first line 'wa' could be 'va', or 'go', while finally 'o tu' (repeated twice), are the *exact* Portuguese words for 'oh you!'. In the third line, 'e ko te' sounds not unlike 'escute!' (pronounced 'eskoot'), meaning 'listen!', while 'tae' could have an equivalent in 'd'aí', usually meaning 'from there', but used as an interjection with the meaning of 'good!' or 'well done!'. Finally, 'e kore' is not unlike 'escore!' (sounded 'eskor'), a command meaning 'brace!' or 'make fast!'.

The above would give a loose translation something on these lines:

> Here go tell the story
> Quit, stay or go, oh you!
> Listen, well done, oh you!
> Make fast!

The words may not carry a great deal of meaning but then sea shanties were generally not noted for their literary content. They were action songs, a means of keeping time, whether in rowing a boat or hauling on an anchor rope. It is of course possible that the similarities between the Maori sounds and Portuguese may be just coincidence, but at the very least they provide food for thought. As for the word 'pakepakeha' attributed to the *turehu* sailors in the second chant, the Portuguese linguist suggests a similarity between 'pake' and the Portuguese words 'para que' (pronounced quickly as 'prake'), meaning 'what for?' Curiously, 'pakepakeha', subsequently became an alternative Maori term for a fair-skinned, human-like being. The New Zealand historian Ann Salmond sees its use in this *haka* as a likely origin for the present-day Maori word 'pakeha', signifying a European person.

The Maori tradition that this 'vessel of the gods' was sighted many times both before and after the fishing encounter could well be further evidence pointing to this being Mendonça's ship, for the map evidence indicates that his exotic vessel would have been within view of people on land for a large part of its voyage up the coast.

Just before the point where *Illa do Magna* comes to an end on the sixteenth century map there is one very interesting detail. Here the chart depicts a narrow neck of land with a clearly defined harbour on either side. Could this be the isthmus on which Auckland now stands? Auckland, New Zealand's largest city, fronts the Waitemata harbour on its eastern side and has the Manukau harbour, opening to the Tasman Sea, at its back door. The position shown for the isthmus and twin harbours in relation to the Vallard map's Great Barrier Island is exactly right.

This raises an intriguing question: if these really are the Waitemata and Manukau harbours, how did Mendonça know that the two were so close together? It seems out of the question that his ship could earlier have sailed into the Manukau, for the dangerous west coast entrance to this harbour is guarded by a particularly treacherous bar and fierce currents. It would seem to mean only one thing—that a party from Mendonça's expedition must have landed on the shores of the Waitemata and marched inland far enough to sight the twin harbour to the west. It would have entailed a journey of just a few hours. Mendonça and his men would have only needed to climb one of the higher extinct volcanic cones which dot this isthmus to get a panoramic view of the entire region. And if they did so, they could hardly fail to have come into contact with Maori people living there.

There is no known Maori record unequivocally pointing to such an encounter, but this is hardly surprising given the fact that the strategic Auckland isthmus changed hands several times between the sixteenth century and the early nineteenth century. As now, in Maori times this strip of land was the most sought-after piece of real estate in the whole of New Zealand. Its fertile volcanic soil was ideal for cultivating the Maoris' staple crop, the sweet potato or kumara, while its extensive harbours produced a bountiful harvest of fish and shellfish. Archaeological research indicates that by about the sixteenth century the population of the isthmus had probably reached its peak, perhaps numbering several thousands. It seems then to have been the nearest thing to a Maori metropolis; its numerous volcanic cones crowned with elaborate fortresses, or *pa*, providing refuge in case of surprise attack for tribes people working in the fields below.

This *Tamaki-makau-rau*, or 'Tamaki-desired-by-many' as the isthmus was known to the Maori, was, as the name implies, viewed with envy by other powerful tribes. The increasingly ruthless wars that occurred from about the sixteenth century onwards were often genocidal in nature, with the vanquished either enslaved, or killed and eaten. When the first British arrivals appeared on the scene about 1835 they found the entire isthmus almost totally depopulated, its long-abandoned gardens and hilltop forts overgrown with bush and scrub.

An oral tradition has, however, survived telling of the arrival at Tamaki long ago of a vessel manned by *turehu*—apparently the same mysterious people who were observed by the Maori fishing party off the East Cape. According to this fragmentary tradition, these 'spirit people' had sailed to the isthmus 'from beyond the deep'. Also known by the name of *patupaiarehe*, they were said to be like ordinary men in size and general appearance (no women are mentioned), but with reddish hair and pale skins. The tradition says that the strangers had hardly set foot on shore when fighting broke out. But the conflict seems to have been resolved peacefully, because the story goes on to reveal that sexual relations later occurred between some of the newcomers and women of the Maori tribe. The descendents of these *turehu* unions were said to be notable for their fair skins and 'golden' hair.[15]

In this context, some intriguing questions are raised by a discovery made in 1841 by an early British missionary, the Reverend William Colenso. While visiting a Maori village north of Auckland in that year, the missionary found a strange-looking ship's bell made of bronze being used by the women as a cooking pot. Colenso took the bell away with him. Much later, the mysterious lettering inscribed around its lip was identified as antique Tamil script from South India.

The 'Tamil Bell', as it is popularly known, is now in the collection of the Museum of New Zealand, Te Papa Tongarewa. Its exact age has still to be determined, but it is thought possibly to date from the sixteenth century or earlier. How it came to be in the Maori village has never been explained. Several theories have been put forward, including one that it came from a supposed shipwreck of a Chinese junk. While there is no evidence which would link the bell with certainty to Mendonça's expedition, it is exactly the sort of item we might expect to be carried in a Portuguese vessel coming from southern India, either for shipboard use or as an item of trade.

Finally, there is one other little-known fact which supports the belief that it was the Portuguese who discovered New Zealand's North Island. This evidence is in the form of a British chart published in 1817 in which 'Cooks Straits' are also given the name 'Gulf of the Portuguese 1550'. Under the main title 'New Zeeland' the cartographer has written: 'Discovered and named by Tasman, 1642, but whose Eastern Coast was known to the Portuguese, about the year 1550'.

What do we know about the authenticity of this map? Although the chart is English, it is in fact derived from a map of French origin. It carries the legend: 'A chart of the Indian Ocean improved from the chart of M. D'Après de Mannevillette drawn by LS de la Rochette'. An earlier version of this chart appeared in 1803, followed by the improved 1817 edition.[16]

It has been suggested (by some British historians) that de la Rochette must have obtained his information from the 'Dauphin' map, presented to the British Museum by Sir Joseph Banks in 1790. Perhaps—but the Dauphin map contains no reference to any 'Gulf of the Portuguese' and in any case there seems little justification for postulating this roundabout connection to the English aristocrat, Banks. After all, there existed in France the tradition of the great French school of hydrography founded at Dieppe by Pierre Desceliers early in the sixteenth century—so what would be more natural than that de la Rochette might draw on maps or other information he may have discovered in Dieppe or Paris concerning a sixteenth century Portuguese voyage along the east coast of New Zealand? A voyage beginning at the very strait which two and a half centuries later would be rediscovered by the English navigator, Cook.

As for Mendonça's chart of the North Island, in terms of overall accuracy it seems to fall short of the standards of his earlier chart of the coast of Australia. There are several possible reasons for this. We should not overlook the fact that map-making was still in its infancy, and islands also tended to be more difficult to draw than a mainland coast. In some sixteenth century European maps even such well-known places as the British Isles are barely recognisable. A Jean Rotz map of Northern Europe, for instance, shows a channel leading all the way across Britain linking the North Sea to the Irish Sea in the vicinity of the England-Scotland border—and Rotz even had Scottish ancestry!

A second consideration is that Mendonça would have had no reason to devote any more time than was absolutely necessary to charting this remote island in the south seas. It was plainly not the Island of Gold—and it was in Spanish territory anyway. For both these reasons Portuguese ships were unlikely to come here again in the foreseeable future—so a fairly rough map would do. Another possibility to be considered is that sickness, including perhaps scurvy, might by now have started to affect Mendonça's crew.

So after clearing Great Barrier Island, the captain's first priority after so long and arduous voyage would have been to hoist full sail and set a northward course for the far-off Portuguese base at Malacca, carrying with him his precious charts and other information about the new lands his ships had discovered. As the outlying hills of the 'Island of Mahogany' began to sink below the horizon he might also have reflected that there was still no sign of Magellan (who in fact had already traversed the Pacific far to the north). In the southern hemisphere it would probably be early autumn, 1523. By now it would have been about a year since Mendonça had set sail from Malacca.

For the return journey he could have taken one of two routes: through the as yet unnamed Torres Strait with its coral reefs, or the longer but safer course around the north of Papua New Guinea. Probably it would have been the latter. This route would have offered the added inducement of the possibility of a stopover at the Moluccas where he could expect to fill his ship's hold with a profitable cargo of spices—which with any luck would more than recoup the cost of the voyage.

Here we should consider one other possibility—might Mendonça have sailed the northern route by way of the islands of Tonga, as Abel Tasman did 120 years later? The idea is not quite as far-fetched as it might at first seem. Not only did it take Tasman a mere fifteen days to reach Tonga from the northern tip of New Zealand; it seems the Dutch captain had even obtained from some unknown source Portuguese charts covering at least part of this area of the western Pacific. This little-known fact is revealed in a journal from the Dutch voyage. On his way back to Java from Tonga on 15 February 1643, Tasman called a meeting of his seven-man Ships' Council to decide the course that should be taken. The journal records Isaack Gilsemans, merchant and Council member, as commenting:

> Should we consider the Islands sailed to [by Tasman's ships] to be the Salamonis [Solomons] since [we] have found the Same in longitude

> and latitude according to the charting of the portuguese; said Islands cannot have been seen by Schoutens, apparently [it] Could be the Land of Nova Guinea, *according to the charting of the portuguese* (Author's italics).[17]

Gilsemans' reference to 'Schoutens' is to be seen in the context that Tasman knew the position of Tonga because in May 1616 the Dutch ship *Eendracht*, commanded by Jacob Le Maire and Willem Schouten, sighted two of the lesser Tongan islands after crossing the Pacific from the island of Juan Fernandez.

Near the top of the Vallard map, at about the latitude of Cape York's Princess Charlotte Bay, we can see depicted a curious pair of islands. The larger one is called *Illa do Aljofar* and the smaller *Illa Carma*. What could they be? 'Carma' is probably a misspelling of the Portuguese 'calma', meaning 'calm'. 'Aljofar', however, appears to be a Polynesian-derived word and in fact it has a curious similarity to the second part of the name of Tonga's capital, Nuku-alofa. However it would be rash to jump to conclusions, because Tongatapu, the main island of Tonga on which Nuku'alofa is situated, does not look much like the Vallard island Aljofar.

Further east the Vallard map depicts a second puzzling group of islands, two of which have names. The larger of these is called in Portuguese *Illa do Tubaros*, or 'Island of Sharks', and the other very small one is *Illa Ronda* (Round Island). These do not seem to match any islands in the south-western Pacific. Could they have any connection with Mendonça's voyage? A possible alternative explanation is that 'Tubaros' may be intended to depict the island of 'Tiburones' (the Spanish word for 'sharks') discovered and named by Magellan on his trans-Pacific voyage. 'Tiburones', though, is thought to have been Flint Island, an uninhabited island several hundred miles north of Tahiti in eastern Polynesia—and in that case it would lie far beyond the Vallard map's north-eastern border.

As for the remainder of Mendonça's voyage, regardless of which of the two possible routes he took on the return journey from his 'Island of Mahogany', he would have been sailing at this season into the tropics at the right time to take advantage of the strengthening south-east trade winds which with luck would have blown him steadily on his way. After perhaps calling at the Spice Islands or Java he would have set course for Malacca, and then India. He would need to have reached Goa, the major Portuguese naval base for the Far East, no later than September 1523 in order to be sure of avoiding the onset of contrary winds heralding the winter monsoon. But assuming he left *Illa do Magna* (New Zealand) in March of that year, six months should have been more than enough time for him to cover the distance.

What then? Remembering that he was sailing under royal patronage, we can reasonably assume that his next move would have been to join a fleet homeward bound for Lisbon so that he could report his discoveries to the King in person. And thanks to an extraordinary chance discovery, we now know for a fact that this is exactly what he did!

In January 1990 a gardener digging in the grounds of a water sports club at Cape Town's Table Bay unearthed a flat stone which had writing inscribed on it. The stone was twenty-five centimetres long by ten centimetres wide. After it had been cleaned, archaeologists from the South African Museum found that the inscription was in Portuguese. Reconstructed and translated into English it reads:

> The *Vitória*, Captain Cristóvão de Mendonça, arrived here on the 26th of Ma [March or May] 1524. A coin is buried here. All well. God be praised.[18]

The inscription is written in a curious kind of shorthand in which, for instance, the date 1524 is written as just '524' and the month is abbreviated to 'Ma'—hence the doubt over whether this signifies March or May ('Março' or 'Maio' in Portuguese).

South African museum authorities have not the slightest doubt about the stone's authenticity. Several other tablets of this kind, known to Cape Town archaeologists as 'postal stones', have been found near the old shoreline where they had been left—first by the captains of passing Portuguese ships and later by ships of the Dutch, French and English. Letters wrapped in packages of tarred canvas were sometimes concealed beneath these stones—but unfortunately in the case of the Mendonça stone neither a coin nor any package was found.

Portuguese ships seldom stayed long at Table Bay. On the outward voyage from Lisbon their spring departure invariably brought them in sight of Table Mountain's 1000-metre-high summit during the southern hemisphere midwinter, and they were anxious to push on past the Cape to Mozambique in time to catch the winter monsoon which would speed them to India. On the return journey their captains could relax a little once the dreaded Cape was behind them, but now their preoccupation was to complete the last leg home up the Atlantic. A few huts are known to have been built on the shore at Table Bay—and it may well have been on the site of one of these that the Mendonça stone was found.

The *Vitória* (in English, *Victory*) was not, as one might perhaps have supposed, the ship which Mendonça commanded on his outward voyage in 1519. Official Portuguese records show that the *Vitória* sailed from Lisbon for the East with a small fleet in 1522 under the command of a captain named Diogo de Melo, arriving at Goa in August 1523. Mendonça probably reached Goa coming from the opposite direction at about the same time as de Melo. After an arduous voyage lasting in total about two years, much of it in uncharted tropical waters, his own ship would have been in urgent need of a major refit. We know from the official records that he took over from de Melo as captain of the *Vitória*, joining a homeward-bound fleet carrying spices. Perhaps Mendonça's cargo may even have been spices that he himself had brought back to Goa from the Moluccas or Java as he returned from his search for the Island of Gold.

The homeward-bound fleet would have sailed from Goa either in December 1523, or in January or February 1524. As the tablet Mendonça left at Cape Town informs us, he arrived there on the 25th of either March or

May 1524—most probably the latter. He would have sailed again for Lisbon after a short stay for water and revictualling. And we also know for certain that he did reach Portugal, where he was rewarded for his services with a lucrative overseas posting by a grateful King John III.

But did anything untoward befall Mendonça's ship on the final leg of its voyage, from Cape Town to Lisbon? There are some intriguing clues suggesting that this may indeed have been the case. We will look at them in the next chapter.

9

The poet-captain of Dieppe

I often wonder why, for this odd fantasy,
I Europe leave, and why it lies to me
To circle Africa so near around,
Nor can I yet contented be
Until the coasts of Asia shall be found,
To such an effort am I tied;
My head is fired, my spirit has not died.
So, making ready, I am filled with joy,
If questioning still the ends of my employ.

JEAN PARMENTIER ON HIS VOYAGE TO SUMATRA, 1529
(TRANSLATION: DAVID BEERS QUINN)

With Christopher de Mendonça sailing northward through the Atlantic Ocean on the last leg of his long voyage home to Portugal, it is to France and the Dieppe cartographers that we must now turn in order to uncover the final secrets of the charting of the 'Island of Gold'.

On 21 October 1529, two weather-beaten French ships dropped anchor off the south coast of the island of Sumatra. As their emaciated crews stumbled ashore on a small, uninhabited island they fell to their knees in thanksgiving. They were near starvation and sick with scurvy, but there was still cause for rejoicing: they had at last reached their goal—the fabulous Indies. No Frenchmen had ever sailed so far before.

Who were these travellers? Why had they come and what did they hope to find?

The voyage of the brothers Jean and Raoul Parmentier is little known outside France, yet it was an epic of its day, comparable in some ways with the great round-the-world voyage of England's Sir Francis Drake fifty years later. It brought French mariners to within a few days' sail of the north-west coast of Australia.

Even more importantly from an Australian perspective, its story offers vital clues to the puzzle of how the cartographers of Dieppe managed to get their hands on closely guarded Portuguese charts showing the East Indies and a strange southern continent that looked suspiciously like the fabled 'Island of Gold'. How could such a cartographic coup have been accomplished when all Portuguese discoveries were state secrets whose disclosure to any foreign power was a crime punishable by death?

For an answer we must first look at the intense political rivalries that existed in Western Europe in the opening decades of the sixteenth century. At this time France was in a state of undeclared war on the high seas with Spain and Portugal. The cause of this confrontation was the notorious Treaty of Tordesillas. Sanctioned in 1494 by Pope Alexander IV, this treaty purported to divide both the newly discovered and all the yet-to-be-discovered regions of the world exclusively between the two Iberian Catholic powers. Pope Alexander, it should be added, just happened to be a Spaniard himself.

As was soon to be the case with England under Queen Elizabeth I, France, as an emerging maritime power, was outraged by the papal ruling and determined to ignore it. As the French King Francis I was later to sarcastically remark: 'I should very much like to see the passage in Adam's will that divides the New World between my brothers, the Emperor Charles V [of Spain] and the King of Portugal'. French shipowners, for their part, proclaimed their rights to freedom of the seas, continuing to dispatch vessels across the Atlantic to trade on the newly discovered coasts of central America and Brazil. The response of the Spanish and Portuguese was brutal: any French ships that trespassed in their preserves were ordered to be seized and their crews killed.

In retaliation, fast and well-armed French corsairs began intercepting and plundering their enemies' homeward-bound shipping. Particularly lucrative targets for these privateers were the heavily laden Spanish vessels carrying gold and silver back from the Americas, and the big Portuguese carracks homeward-bound from India with valuable cargoes of spices. These French corsair captains should not be thought of as freelance pirates; they operated under a royal warrant authorising them to attack and board Spanish or Portuguese ships and seize their cargoes. The warrant even specified that when the corsairs returned to their home ports the booty was to be divided up by an official committee—one portion going to the corsair's owner, one to the captain and crew, and the third to the king's treasury.

It was what one French historian has called a 'muted war'—essentially a game of tit-for-tat. Rules of engagement followed by both sides were designed to prevent the conflict escalating into all-out warfare. Usually the enemy ships themselves seem not to have been regarded as legitimate prizes. When a corsair vessel belonging to the French shipowner Jean Ango was seized by the Portuguese in 1529, for instance, he was able to successfully seek reparations from the Portuguese Crown.

One of the most important maritime centres of France—and indeed the whole of Europe at that time—was the Channel port of Dieppe. And it is here that many of the threads connecting the earliest maps of the Southern Land and secret Portuguese voyages of discovery in the region of Australia come together.

Dieppe got its name from the Norman 'diep' meaning 'deep inlet'. By the sixteenth century it had developed a proud seafaring tradition which was one legacy of its Viking origins. Dieppe adventurers reputedly sailed to the West African coast of Guinea to set up a short-lived trading post as early as 1365, a century before the Portuguese.

By the beginning of the 1500s, flourishing international trade had produced a new class of wealthy merchant princes in Normandy. Among these nouveau riches, few could rival the wealth of the Dieppe shipowner

Jean Ango. Scion of a wealthy bourgeois family from Rouen, by the 1520s Ango was the owner of a powerful fleet of ships engaged in often lucrative trade to such far off destinations as Brazil, Guinea and the coasts of North America. In addition he held the official posts of Collector of Taxes and *Avitailler* (supplier of weaponry and provisions) to the French fleet, besides being a counsellor and close confidant of King Francis.

Ango had no love for either the Portuguese or the Spanish, neither of whom had shown the slightest compunction about attacking his ships if they were discovered trading to the Americas. On the coast of Brazil, Portuguese policy was not only to attack any French vessels and kill their crews, but at the same time to terrorise any native tribes who dared to trade with the French. An early instance of this policy occurred in 1504 when three French merchant ships trading at Bahia, on the Brazilian coast, were set upon by a squadron of Portuguese vessels. The French ships were either burned or captured, and their crews massacred.

Jean Ango's greatest coup came in 1522 when off the Strait of Gibraltar a squadron of his corsairs under the command of one of his best captains, Jean Fleury, ambushed three Spanish caravels returning from Mexico with the fabulous treasure of Montezuma. After boarding and plundering the main treasure ship, Fleury's flagship the *Salamandre* returned triumphantly to Dieppe laden with a cargo described as 'priceless' and said to have included gold vessels, bracelets, rings, jewels made up as flowers and birds, mosaics of precious stones, 600 of the finest pearls and 'an emerald in the form of a pyramid as big as a hand'.

In other encounters with Spanish and Portuguese ships Fleury seized documents and charts as well as treasure. In 1527, however, his luck finally ran out. His ship was surprised and overpowered by a fleet of Spanish vessels and Fleury himself was taken prisoner. King Charles of Spain (by now the Holy Roman Emperor) rejected a ransom offer of 30 000 ducats. In reprisal for Fleury's depredations against Spain's American treasure fleets, the King ordered that his captive be tortured and executed.

Jean Ango was an archetypal Renaissance leader: well educated, intelligent and a patron of the arts. At the height of his power his fleet totalled twenty-three ships, all said to be well-armed and equipped. In 1517 he arranged a visit to Dieppe by the great Leonardo da Vinci. Planning new and ever more ambitious voyages for his ships, he drew around him a select circle drawn from Dieppe's leading intellectual figures. These *savants* included the poet-captains Jean Parmentier and Pierre Crignon, the famed cartographers Pierre Desceliers and Nicolas Desliens, as well as Nicolas Vallard (sometimes spelt 'Valland'), after whom the Vallard Atlas was later named. Vallard, although described as passionately interested in geography, seems not to have been a cartographer himself—although we cannot be quite certain on this point. French documents describe him as a Dieppe bourgeois: in other words, a prominent businessman. This leaves open the question of who actually drew the Vallard maps—and this also is something we will return to later.

Ango's think tank of intellectuals assembled regularly in his newly built mansion, aptly named *La Pensée* (The Thought) and reputedly one of the finest chateaux in France. Here, returning ships' pilots, geographers, mathematicians, clerics and other learned guests would assemble and beneath an ornate portrait of Francis I, partake of an elegant dinner, during which they would share their thoughts on how the civilising influence of France might be extended to those parts of the world that still remained to be discovered. (The chateau still exists, preserved as a historic place under the name 'Manoir d'Ango').

This was also to be the beginning of what was to become known as the Dieppe school of hydrography, a loose-knit institution which would soon make Dieppe famous throughout Europe. The school's inspirational figure was Pierre Desceliers, abbot of the monastery at the nearby town of Arques. Desceliers was far more than just a simple priest: at a time when the Catholic Church was still the main source of learning in Western Europe, it should come as no surprise that he was also highly respected as a mathematician and scholar. It was he and the pupils of his Dieppe school who compiled several of the beautiful but enigmatic maps of the great southern land which he called *Java la Grande*.

By the second decade of the sixteenth century the Portuguese were firmly established in the Far East and their ships were bringing back to Lisbon cargoes of exotic spices which, on the markets of Europe, brought huge returns for their owners. The cargo of cloves carried back to Spain by the one remaining ship of Magellan's fleet, for instance, is said to have repaid the entire cost of the voyage. As for Portugal, the landed price of a *quintal* (fifty-eight kilograms) of pepper in Lisbon was only a fraction of the sum that had to be paid in Alexandria or Venice for the same amount brought from the Indian Ocean by the laborious and costly traditional route—up the Red Sea in dhows, and then overland by camel caravans. One envious Venetian banker estimated that spices the Portuguese could buy for one ducat on the coast of India would cost sixty to one hundred ducats by the time they reached Venice by the Red Sea route—largely because of the innumerable customs duties and bribes that had to be paid along the way. The Cape route might be longer, but the financial returns were immense.

In Dieppe, Jean Ango eyed the Portuguese spice monopoly with covetous eyes. Why should France's sworn enemies be allowed to enjoy these immense profits unchallenged? And there was another thing: seamen's rumour had it that the Portuguese believed they were on the verge of discovering the fabulous Island of Gold. These same lands were known to have been one of the goals of the renegade Portuguese captain, Ferdinand Magellan, in his voyage in the service of the Spanish king. Magellan had failed to achieve his objective and had perished in the attempt. Perhaps France might yet forestall both her rivals?

Ango discussed a bold scheme with the members of his think tank—Desceliers, Desliens, Vallard and Jean Parmentier. Was it possible that some of his ships might be able to sail to the region of the Spice Islands and obtain valuable cargoes of spices and perhaps an even richer cargo of gold? Ango was aware that in 1444 the Venetian Nicolas de Conti had travelled overland to the Orient reporting on his return that there was 'no country more famous for gold than Sumatra'. Reports had also reached Ango of how in 1520 a Portuguese captain, Diogo Pacheco, had sailed down the coast of Sumatra on a secret voyage whose aim was to find the 'Isles of Gold'. Why should his captains not do the same? And besides, his ships could surely top up their cargoes with many exotic and valuable items from the East such as camphor, benzoin, cloves and rhinoceros horn.

There was one big problem, though: for the project to succeed his ships would have to make their journey without touching at the coast of India and without sailing through the narrow Malacca Strait, since at both these strategic places their Portuguese enemies maintained a formidable military presence in the form of both forts and warships. This would mean pioneering a new route direct from Madagascar to Sumatra. In theory it should be possible—but his captains would need reliable charts and sailing directions. Could these be obtained, and if so, how?

The abbé and cartographer Pierre Desceliers pronounced the plan feasible and set about the task of preparing charts for the voyage. Ango's captains were given orders that in future, whenever they intercepted and boarded a Portuguese or Spanish ship their first priority should be to force entry into the captain's cabin and seize his charts before he had time to destroy them. They were also to interrogate the captain and his officers closely to discover anything they might know about the sailing route to the Orient.

Frequently the abbé travelled from his residence at Arques—the curiously named *La Corne du Cerf*, or 'Stag's Antler'—to the port of Dieppe to talk with returning ships' captains and seamen. During these visits he often stayed at the house of another of the city's famous cartographers, Jean Rotz, a Dieppe pilot born of Scottish parents whose family name was originally Ross. In the meantime Ango's spies were busily gleaning all the maritime secrets they could unearth in the halls and taverns of Lisbon and Seville. This would not have been easy, for as we know, Portuguese charts—especially those pertaining to the Orient—were state secrets. Any Portuguese seafarer rash enough to divulge such secrets to a foreign power faced mandatory execution, often by particularly gruesome methods designed to act as a deterrent.

It is now that the plot thickens and a mysterious character enters the scene. A character who was to play a key role in the French voyage to the East and who may well also have had a crucial influence on important details of the maps and charts which Desceliers was preparing. His name was João (John) Afonso—although that was only the first of at least four different names under which he was known at various times.

Afonso was a skilled Portuguese navigator. It is rumoured that in his youth he had sailed with the great Vasco da Gama on his epoch-making voyage around the Cape of Good Hope to India in 1497. There is no documentary proof of this, but nevertheless by some means or other he had gained expert knowledge of the sea route to the Orient which could be of great value to the French expedition, not least in enabling its ships to avoid the dangers that lay ahead in the form of islands, shoals and reefs. And this was not all. According to some accounts Afonso had also sailed as a pilot with another Portuguese expedition whose purpose had been to discover and map the fabulous Island of Gold. If this was indeed true—and as we shall see there are strong grounds for believing it—then there was every reason for Ango to want to recruit Afonso into his service.

And recruit him he did. Just when or how this happened is not entirely clear. Quite possibly he may have been taken captive—or even induced to defect—by Ango's corsairs while returning from the East in a Portuguese ship. This is supported by an account of the French voyage which states specifically that an [unnamed] 'interpreter of Portuguese and Malay' who joined the crew was a 'renegade' from a ship which had been intercepted by corsairs. At all events Ango clearly possessed the financial resources to have made the Portuguese mariner an offer too good to refuse. And interestingly enough, 'interpreter' is precisely what we find Afonso listed as among the ships' company by the time the French expedition was being readied for sea.[1]

This would be the most audacious French voyage of its time. Ango allocated two of his best corsairs for the expedition. One of them had been given the same name as Ango's chateau, *La Pensée*, while the other was called *Le Sacre* (The Consecration). Both ships were about twenty metres in length. *Le Sacre*, the flagship, measured 200 tons, the smaller *Pensée* 126 tons—both roughly comparable in size with the two bigger ships of Vasco da Gama's fleet on his first voyage to India. In the style of Portuguese vessels of the period the French ships were rigged with square sails on their fore and main masts and lateen sails on the mizzen masts. A map illustration from the period shows what could be one of these two vessels, or a ship very much like them. It is characterised by a small, low forecastle—quite unlike Portuguese carracks with their towering forecastles—and a steeply raked three-tiered poop deck with two mizzen masts, the lower deck bristling with cannon muzzles.

To command the two ships Ango chose his friend and confidant Jean Parmentier and Jean's younger brother Raoul. Both the Parmentiers were respected mariners—Jean had been the first Frenchman to take a ship to Brazil. Yet it was surely a uniquely Gallic touch that they were selected to lead the expedition not just on account of their nautical ability but also because they were acclaimed poets and 'humanists'.

The journals of the voyage are preserved in the Dieppe Municipal Library. When the two little ships put out from the Normandy port on a spring day in April 1529 the mood was one of optimism. As well as sailors, gunners, cartographers and men of science, the ships' complement even included a band of musicians whose task would be to provide entertainment on fifes, trumpets and tambourines.

Le Sacre and *La Pensée* had been carefully provisioned for their long voyage. Drinking water, which tended to go putrid on long voyages, was supplemented by a generous ration of alcoholic beverages: two and a half litres of wine per day for the Breton crewmen and the same quantity of cider for the Normans.

The ships carried charts prepared by Desceliers showing latitude and longitude, known coastlines and ports, islands, reefs and compass directions.

Among them would almost certainly have been one showing the mysterious southern continent which Desceliers called 'Java la Grande'. All of these charts would have been drawn in a flat plane with degrees of longitude wrongly shown as equal to degrees of latitude—for no cartographer had yet devised a method of compensating for the curvature of the earth. But despite this shortcoming Desceliers' charts were among the best of their day and would be vital for the success of the expedition.

Utilising Afonso's experience with similar Portuguese voyages, the two French ships, sailing in close company, avoided the mistake of hugging the West African coast with its contrary winds and currents. Instead, after crossing the equator on 11 May 1529 they sailed south-west almost to the coast of Brazil where they were able to catch favourable winds which would carry them far into the South Atlantic. The two ships sailed on well out of sight of land for over a month until sightings of the sun taken each day through their astrolabes told them they were nearing the latitude of the Cape of Good Hope. Now they confidently turned east, and after weathering a fierce storm, passed south of the notorious Cape in late June without even seeing it.

So far they had made excellent time. Their ships had proved themselves capable of covering as much as 150 nautical miles a day with a following wind. They had passed the southern tip of Africa less than three months after setting out, a swifter passage than the Portuguese were usually able to manage. But now they made their first mistake: by deciding to sail through the Mozambique Channel, the standard Portuguese route between Mozambique and the island of Madagascar, rather than heading further east into the Indian Ocean. On entering the channel they encountered calms and headwinds and it was not until the end of July that they finally sighted the coast of Madagascar.

This would be the first time they had set foot on land since leaving France four months earlier. It was a bungled affair and a foretaste of things to come.

With the ships at anchor in a bay, crewmen rowed two small boats ashore in the hope of obtaining much needed fresh water and provisions from the assembled natives. Four of the sailors, among them a crewman known as

Jacques l'Ecossais (Jacques the Scot), were persuaded to accompany some of the natives to a village a little inland. Rashly, they left their weapons in the boats. Suddenly the rest of the landing party were startled by a loud scream, which they recognised as the voice of Jacques. Then they saw the bosun of the *Sacre* and a third man running for their lives out of the underbrush, only to be speared to death by a hotly pursuing party of natives. Sailors guarding the boats sounded a trumpet to alert their shipmates busy filling water casks. The landing party retreated to the ships, leaving the bodies of the Scot and two French sailors sprawled on the beach.

The grisly scene is recorded in a map of the region by the contemporary Dieppe cartographer Jean Rotz. It may even have been a first-hand impression, for as we shall see there are some grounds for believing that Rotz may actually have been sailing with the expedition.

Jean Parmentier, humanist and man of peace, refused demands by hotheads among his crew for violent reprisals. During a second landing the sailors had managed to replenish their fresh water casks, but they had obtained none of the fresh fruit and vegetables they so sorely needed. Parmentier retired to his cabin to agonise over why *les sauvages* had turned against him after he had gone to great pains to treat them well. After christening the place *Cap de la Trahison* (Treachery Cape) the expedition sailed on in search of a more hospitable landing place.

Dangerous shoals and reefs now further slowed their progress. It was not until the beginning of the second week of August that the two ships passed the island of Mayotte, off the northern tip of Madagascar. The following day they sighted the fertile volcanic island of Grand Comore. Here they might have obtained all the fresh fruit they needed, but they were unable to find a safe landing place. To the dismay of the sailors, Jean Parmentier once more ordered them to sail on—then retired to his cabin to seek inspiration by composing another poem.

By now some of the crew were beginning to display the unmistakable signs of scurvy: swollen hands and gums, arms and bodies covered with pustules, loosened teeth and a general feeling of lethargy. Soon, two deaths from the sickness occurred. They were to be the first of many.

Beset again by calms, it was not until mid-September that the ships came in sight of a group of large, low-lying islands which looked promising for landing. But where were they? At this point the charts the ships carried were confusing and an argument broke out on board. Some sailors even thought they might have arrived at the Moluccas. Afonso settled the argument. He recognised the islands from a previous visit and declared that they had reached the archipelago of the Maldives.

Jean Parmentier went ashore on one of the inhabited coral islands with some of his officers and crew. The local people were Muslims and they seemed suspicious. Fearing another episode of treachery, Parmentier refused to allow any more of his men to go ashore. His sailors managed to replenish their water casks, but the only other provisions they were able to obtain were a few dozen coconuts. The *Sacre* and the *Pensée* weighed anchor and sailed on, bypassing the Portuguese base at Malacca by setting a direct course for Sumatra.

It was another month before the exhausted and near-starving crews of the two ships sighted the Sumatran coast. After enduring more than six months of fearful hardship they had finally reached their goal, the fabled islands of the Orient. The sailors' spirits rose. Now, surely, they would have fresh food in abundance and be able to fill their ships' holds with the spices, gold and other treasures they had come so far to find.

By great good fortune the ships had come upon two small offshore islands which were uninhabited. Here, at last, the emaciated crews were able to stagger ashore in safety. The next six days were spent recuperating, bathing in fresh water, catching fish which abounded in the shallows and eating wild herbs to counter the effects of scurvy. But, for several crewmen who were already on the point of death, the change of diet came too late.

Another day's sail brought the ships off the mountainous south-west coast of Sumatra. They had made a landfall near a port which they called Ticou (recognisable on a modern map as the small coastal town of Tiku). At first the visitors were well received but when the French tried to exchange their trade goods the Sumatrans showed themselves to be hard bargainers, 'worse than the Scots', as Jean Parmentier put it. Deals he made one day were repudiated the next. Exasperated, Parmentier went back on board with his merchandise. It was an unwise move. The Rajah of Ticou, who earlier had entertained the French captains at a banquet, retaliated by taking Raoul Parmentier and another French officer hostage. Afonso bravely offered himself as a hostage in Raoul's place but the French captain refused to allow his Portuguese comrade to sacrifice himself. A tense armed stand-off followed until finally the hostages were released unharmed.

Far worse was to come. Jean Parmentier was suddenly stricken with high fever and acute dysentery. He lapsed into delirium and eight days later was dead. By then some others in the crew, among them Raoul Parmentier, had symptoms of the same illness. Five days after the death of his brother, Raoul, too, succumbed. Bad drinking water on shore is generally held to have been the cause, but the possibility cannot entirely be discounted that the rajah may have had the troublesome French captains poisoned.

Devastated by the sudden loss of both their commanders and disillusioned by their reception at Ticou the French set sail further down the Sumatran coast, anchoring off another small port called Indrapoura, which still exists today. But here, too, they were met with suspicion. They set off yet again, sailing uncertainly for two more days toward the south-east in uncharted waters. But now there were rumblings of discontent among the crew. Where were they? The optimists thought they might soon reach the fabled 'Land of Java', where they would be able to fill their ships with spices and perhaps gold. But others declared that to go on would be madness and could end only in disaster.

The surviving officers called a ships' council to decide what to do. Terra Java with its lure of gold—a land which in their maps also included an ill-defined Australia—was their ultimate goal. How close were they to it? No one was sure. Should they go on?

On 28 December 1529 in true democratic fashion the issue was put to a vote of the entire ships' company. Amazingly enough, despite all the setbacks they had endured, a majority of crewmen were still in favour of pressing on—provided they could take on fresh provisions. This they did, and the two ships sailed further down the Sumatran coast. But trade here proved even more difficult, and finally it was agreed that they should turn back while they still could. First, though, they would return to Indrapoura, where they believed they could at least obtain a good cargo of pepper. But here once again they encountered disappointment. Four crewmen were drowned in a capsize on the harbour bar, and after more haggling all they had managed to obtain in exchange for their mirrors, trinkets and Parisian red cloth was a little gold, rice, flour, chickens, and less than half a ton of pepper. It was a meagre return for so long and costly a voyage.

Time was now pressing. It was nine months since they had left Dieppe, and they must sail without delay if they were to catch the favourable monsoon winds across the Indian Ocean and pass the dreaded Cape of Good Hope before the onset of winter storms. On 22 January 1530 the two little ships unfurled their weather-worn sails and weighed anchor. This time they boldly set a new course across the Indian Ocean direct for the Cape.

The return voyage was a miracle of seamanship. Despite the deteriorating state of both vessels and their depleted crews, it took only two months for them to reach the Cape. A few days out into the South Atlantic they were separated by a fierce storm—but thanks to an incredible stroke of luck, several weeks later they came in sight of each other again off the Cape Verde Islands. In late May or early June 1530—the record is not clear as to the exact date—the *Sacre* and the *Pensée* limped back into their home port of Dieppe. It was not the triumphant homecoming they had hoped for. Both captains and many of the crew were dead and their cargo was pathetic, in stark contrast to the rich spice shipments brought home by the Portuguese—cargoes which often repaid the cost of a voyage eight or ten times over.

The hard lesson for the French was that an expedition of this length without intermediate revictualling ports such as those the Portuguese had established was too ambitious an undertaking. The *Sacre* and the *Pensée* had in fact sailed to within about 1500 miles of the islands fringing the north-west coast of Australia, their hoped-for 'Isle of Gold', although their crews had no way of knowing this. As events transpired, it would be nearly a century before any French ship would come so close to Australia again.

But though French voyaging to the Orient stopped for the time being, there was no setback to the Dieppe tradition of map-making. On the contrary, the decades following the Parmentiers' Sumatra voyage were to be a golden age of French cartography in which all of the famous 'Dieppe' and other Normandy world maps which survive today had their origin. Twenty-two of these maps or atlases have come down to us. Some of the atlases contain numerous individual maps, fifteen in the case of the 1547 Vallard Atlas and no fewer than fifty-six in an atlas compiled in 1555 by Guillaume le Testu, a Huguenot from Le Havre.

So far as is known, the first Dieppe maps were the work of Jean Rotz, who, as we have seen, was a member of Jean Ango's inner circle. Rotz's *Boke of Idrography*, begun in 1537, was originally a huge chart more than four metres in length. Rotz subsequently converted it into book form, adding to his maps written text encompassing 'the whole of hydrography' and marine science. Completed in 1542, the atlas was intended for presentation to Francis I of France in the hope that it would obtain him a position at Court. Rebuffed by the French king, Rotz changed his allegiance to England, moving with his family to London and dedicating his *Boke* to England's Henry VIII. He was rewarded with the post of Royal Hydrographer, which he held until Henry's death in 1547.

One of the notable features of this remarkable atlas is a spherical world map in two hemispheres, drawn according to the author 'as exactly and truly as possible, drawing as much from my own experience as from certain experience of my friends and fellow navigators'. This is probably the first map of those which survive to show an outline of the enigmatic southern continent which Rotz called 'The Londe of Java'.

Rotz's accompanying map of the Eastern Hemisphere contains vivid pictorial scenes from the Parmentiers' voyage to Sumatra. The French expedition is known to have carried an official artist, and there is a distinct possibility that this artist was none other than Rotz himself. There is a reference in the annals of the voyage to *Jean le peintre* (Jean the painter), while Rotz affirms in his 'Boke' that he was at sea on 15 January 1530—the very time when the Parmentier expedition was at Sumatra. Furthermore, the expedition's *peintre* was also expected to serve as a chart-maker, a task for which Rotz was supremely well qualified.

The considered view of Helen Wallis, former curator of maps at the British Library and the person responsible for a recent facsimile edition of the Rotz *Boke*, is that 'there is very good reason to believe that he [Rotz] had sailed with the Parmentiers to Sumatra'.[2] As we saw earlier, in his map of East African coasts Rotz illustrated his chart of Madagascar with a vignette vividly depicting the natives' killing of the three French sailors. His Sumatran illustrations include an elephant, an accurate drawing of a house on stilts and a military procession of the Rajah of Ticou. All of these scenes are drawn with the same verve and conviction of one who has witnessed them at first hand.

Rotz placed his Sumatran illustrations in the area north of the Malay peninsula—for the simple reason that this was the only place on his map with enough room for them. In some later Dieppe maps, notably the Vallard, similar Sumatran scenes are located for convenience in the interior of Terra Java, alias Australia.

We have to make the qualification that the Rotz atlas was 'probably' the first of the Dieppe series because there are two other possible contenders for this title. One is the famous Harleian planisphere—also known as the 'Dauphin chart' because it was dedicated to Henry the Dauphin who became Henry II of France in 1547. This world map, like the Rotz atlas in the collection of the British Library, is the one on which most of the arguments and counter-claims concerning a perceived outline of the coasts of Australia have been based. The Dauphin is anonymous and undated but it is generally held to have been drawn some time between 1537 and 1543. Its interior of 'Jave la Grande' pictures huts and camels, as well as animals that look like deer.

The other very early map is believed to have been drawn by none other than the Portuguese renegade, João Afonso. Having formally changed his allegiance to France and assumed the new French-sounding name Jean Alfonse, he sailed to Canada with the French explorer Jacques Cartier in 1534. Alfonse (as we will now call him) is credited by some with the discovery of Davis Strait, between Canada's Baffin Island and Greenland. He made another voyage to Canada in 1542 as chief pilot for Cartier and the Count Jean de Roberval, who was attempting to found a French colony at Quebec. Alfonse returned to France the following year, and this seems to have been his last major voyage. We learn that he settled down with a French wife at the village of Saint Onge, near Cognac. He also seems to have lived for a time under the name Jean Fonteneau—possibly his wife's family name.

At about the time of his retirement from the sea, or perhaps in between his voyages to Canada—known then as 'New France'—Alfonse devoted his energies to writing a *ruttier* (in English, 'rutter' or 'sea book') which went under the popular name of *Voyages Avantureux*. Its full title was *Les Voyages Avantureux du Capitaine Jan Alfonce Saintongeois*. Among many other things, this work is believed to have originally contained a small world map in two hemispheres, including an outline of the mysterious southern continent, which was subsequently plagiarised and published under the name of a Norman poet and cosmographer, Jean Maillard. Alfonse's book was not printed until 1559, several years after his death. When it was finally published, at Poitiers, it was an instant bestseller. Between 1559 and 1605 it was reprinted no fewer than ten times, meeting the public hunger for knowledge of the exotic lands and peoples being discovered by Europe's maritime adventurers.

Voyages Avantureux was arguably the most remarkable navigational treatise of its time: nothing less than a grand tour of all the major sea routes of the world known to voyagers by the mid-sixteenth century, including coasts both newly discovered and 'undiscovered'. It is also at times a rather confusing mixture of descriptions of exotic lands. Some of them Alfonse clearly has seen with his own eyes, while others he is evidently recounting from sailors' tales and hearsay. In the eyewitness category are his accurate descriptions of the east coast of Canada:

> Passing the land of Labrador, the people all wear clothes of skin and have their houses underground. The low land here is the coldest in all the world, covered with pine forests and nothing else. This coast is dangerous because of the ice floes in this sea which resemble islands.[3]

But speaking of some other regions he is considerably less reliable. This is certainly the case with his description of the lands which he claims lie southwest of the Strait of Magellan. Of these he wrote:

> Past the Strait of Magaillant the coast turns to the southwest more than two hundred leagues, then turns west-southwest and then west. This coast is not all discovered, except that a ship has sailed by it more than two hundred leagues, which then was forced to run northwest and west-northwest, when it came upon a coast running north and south. In this eastern land one sees many pine trees and many good rivers. The people who have been seen here are white and of good size. They have three months of night and three months of day in their great night and great day.
>
> From this cape the coast veers west for a good hundred and fifty leagues and has many islands and banks, a low land with hardly any mountains. From here the coast turns northwest to the island of Horphie, which is twenty or thirty leagues from the land of Jave. Horphie is at seven degrees from the Equator on the Antarctic side to the southwest of the island of Molucca. To the northeast of the island of Horphie is the island of Cellan, where one finds mace. One then comes to the island of Molucca and several others.[4]

These passages have been taken by a few commentators, foremost among them Roger Hervé, former curator of maps at the *Bibliothèque Nationale* in

Paris, as evidence that the 'lost' Spanish caravel *San Lesmes* must have sailed this way heading west from the Strait of Magellan in 1526, eventually coming across the north-south coast of New Zealand. But this theory ignores some vital facts. As we now know, there is *no* land at all south-west from the Strait of Magellan's Pacific exit apart from the Antarctic, and it would have been entirely out of the question for any sixteenth century ship to have sailed westward even remotely close to the ice-bound coast of Antarctica in the teeth of constant south-westerly gales. Alfonse's description of a ship having sailed past a coastline running south-west from the Strait of Magellan for 200 leagues (at least 600 nautical miles) must therefore be dismissed as pure fiction and from that it follows that so too must the alleged discoveries on the same fictitious voyage of the coasts running north-south and then west almost to the Moluccas.

What Alfonse actually seems to be trying to describe here is none other than the fabled 'Great South Land' or 'Terra Australis Incognita'. This mythical continent was still being shown on some Dutch maps a full century later linking Tierra del Fuego to the north coast of New Guinea and it was only finally consigned to the realms of fiction by Captain Cook in 1775.

Some critics have cast doubt on Alfonse's reliability as an author because of his descriptions of polar regions. In a second book, *La Cosmographie*, he states:

> However, I have been in a place where the day has lasted three months, counted by the revolution of the sun, and I did not want to wait longer for fear the night would overtake me...[5]

This has been taken by his detractors as a false claim that he had sailed almost to the Antarctic—but this is a mistaken view. The context makes it clear that Alfonse is talking in general terms about both the Antarctic and the Arctic and his knowledge of sub-polar regions would clearly have been acquired during the voyages which we know for a fact he made to Canada with Jacques Cartier—during which the expedition wintered over—rather than to any first-hand knowledge of Antarctic regions.

Turning to the Far East, Alfonse describes mysterious coasts below the equator which can hardly be anything other than those of Australia. Proceeding west from the vicinity of the Moluccas, he advises:

> ...the coast turns west and west-southwest to the Trop-bonne [Sumatra]. This coast is called the coast of Jave: it has mighty straits of which no one has yet seen the end. It forms a land with many islands and many dangerous small islets, where the people are like those of Brazil and worship the Sun and the Moon. Between Trop-bonne and the continent of Jave the sea is so dangerous that ships do not pass.[6]

And again:

> Past Trop-bonne the coast veers south-west and south-south-west more than a hundred leagues, and there are some mountains, many islands and great rivers, with harbours of which no one has yet seen the end. And hence veers the coast due west fifty or sixty leagues, and here makes a cape. This cape is fifteen degrees beyond the line of the coast of the Antarctic and to the west. From this cape one sees an island, which is more than two hundred leagues long and a hundred wide, but what people live there no one knows. The coast of this cape turns south for forty leagues and then turns straight to the southeast. No one has yet gone beyond to make discoveries in that distant sea, but it is believed it goes to the south-east coast of the land of Magellan.[7]

Alfonse's 'coast of Jave' is clearly not the island of Java that we know today. Java is a long, narrow island with nothing that could fit the description of 'mighty straits of which no one has seen the end'. Furthermore, Alfonse's description of the inhabitants of 'Jave' as 'people like those of Brazil, worshipping the Sun and the Moon' is particularly telling. What he seems to be saying here is that these people are not Muslims but heathen savages, similar to the natives of Brazil, who are 'all Caniballes and eat their enemies'. And this reference has an echo in Desceliers' maps and several other Dieppe charts, where a prominent bay in the north-west of *Terre de Java la Grande* has been given the name *Baye Brasille*.

Alfonse's use of the term 'firm land of Jave' also gives us a clear indication that this is not an island but a continent. Later in his book he adds the detail that 'in the land of Jave, in certain places, is found gold...' His description of the sea between Sumatra and 'the continent of Jave' as being 'so dangerous that ships do not pass' has similarities to the warning given to the Portuguese adventurer Diogo Pacheco by Malay captains when he questioned them about their sea route from southern Sumatra to the 'Island of Gold'. The coast of this land, he was told, was very dangerous because of its many shoals and reefs.

Apart from his own claim of having been there, Alfonse's books provide further evidence for believing that before he switched his allegiance to France, he did actually sail with a Portuguese expedition to explore the coast of *Jave* (Island of Gold)—possibly with Mendonça. His detailed description of the coastline 'south of Trop-bonne' certainly seems to support this view. The coast which he describes as running south-west 'more than a hundred leagues', with some mountains, many islands, and great rivers and harbours 'of which no one has yet seen the end' fits the coastline of north-west Australia, and nowhere else in the region. Alfonse's undiscovered island of 200 leagues must be classed as a fiction and may well be a reference to the mythical *Isle des Geans* (Island of Giants) shown in this part of the Indian Ocean in the Dieppe maps. But apart from that, his description of a long mainland coast running first south and then far to the south-east toward the Strait of Magellan in seas where 'no one has yet gone' sounds very much like a general description of the south-west coast of Australia as depicted in the Vallard Atlas.

Turning to the Moluccas, Alfonse gives remarkably detailed sailing directions for this area of sea, even to the extent of advising on depth and location of the best shipping channels. This, together with the advice he proffers on how to go about trading for spices and the best trade goods to offer, seems convincing evidence that he had sailed to the Spice Islands himself. As for the largely unexplored seas to the east of the Moluccas, Alfonse ventures the opinion that in this region are 'a good five thousand islands, none of which have been discovered'.

In more specific vein he continues:

> To the south-east of the Moluccas a good hundred and fifty leagues...one comes to the Island of White Men, where one crosses the diametre line which is named Estual. On that route, which is more than eight hundred leagues, are the islands Magellan found when he passed the strait to come to Molucca, and one of these islands has more than sixty leagues of coastline, good harbours, and is inhabited by white people. The land is good, neither too high nor too low and shows itself as well populated. The islands of Horphie and Sophie are the richest islands in all the world, so much gold, silver and precious stones. From these islands was carried the gold and silver for the Temple of Solomon in Hierusalem.[8]

Alfonse makes several other references to the legendary 'gold-rich' island of *Horphie*. In his *Cosmographie* he calls it 'Dorfy' and says it lies to the east of Java, some 900 or 1000 leagues from Peru. Variants of the name appear in Dieppe maps of the region. In an atlas by Le Testu it is called *Terre de Offir* and located on the northern coast of *Terre Australle* (Australia) opposite Timor.

Alfonse's other 'sea book', *La Cosmographie*, was published in 1544, some fifteen years before *Voyages Avantureux*, though it is generally believed to have been written after it. It appeared under the name of 'Jean Fonteneau, known as Alfonse de Saintonge, Captain-pilot of [King] Francis I'. In the book's introduction his publisher, Georges Musset, calls him 'the most expert navigator in France', while Alfonse himself proudly claims to have been a navigator for forty-eight years on all the seas of the world, and in all that time to have never lost a ship.

La Cosmographie contains no fewer than seventy sketch maps of various parts of the world. His map of the Far East is a fairly crude affair in comparison with the more polished Dieppe atlases, but the interesting thing about it is that it shows a deeply indented north coast of 'Java Major' which resembles far more the coast of Australia's Arnhem Land than the coast of Java. At its eastern extremity this coast ends in a gulf which looks not unlike the western side of the Gulf of Carpentaria. Within this gulf is a large island with the name 'Java Minor'—identifiable with the Indonesian island of Flores.

In the text accompanying these charts, Alfonse asserts:

> This Jave is a land which extends to beneath the antarctic pole; on the western side it joins la terre Australle [the Southern Land], and its eastern coast joins the land of the Strait of Magellan. Some say that it is made up of islands. *But from what I have seen of it, it is a continent*...The land called Jave Mynore is an island. But la Grand Jave is a continent.[9] (Author's italics).

A little later on he expands on this point:

> Grand Jave is a continent. And around it are numerous islands. This Jave is attached in the west to the Strait of Magellan and in the east to la terre Australle in accordance with the roundness of the earth. And from what I understand [it] goes across the antarctic pole, and between the one and the other it is believed there are many inlets unknown to us, and we do not know whether they separate the said land into islands. And there has not been discovered so far more about Jave because of the great cold that surrounds the antarctic pole.[10]

'From what I have seen of it, it is a continent...' In this sentence Alfonse leaves us in no doubt that he has actually sailed to the coast of 'Java Major' himself. And assuming this claim to be true, he must surely have sailed there with Mendonça's expedition—since this is the only Portuguese expedition to Australia at this time of which we have any verifiable record.

This brings us to what, from an Australian perspective, must be counted the most astonishing statement in all of Alfonse's works. It is a revelation whose significance seems to have been overlooked even by those historians who have taken the trouble to read his books for themselves. It comes after Alfonse returns in his narrative to the Strait of Magellan and begins to describe the coast of South America running to the north-west. Two hundred and twenty leagues up the coast from the strait, he says, one comes to the *Cap de la Tormente*, 'at forty-six and a half degrees above the Antarctic pole'. Alfonse's *Cap de la Tormente* (Cape of the Torment) has been identified by French historians as Cape Tres Montes, one of the most prominent headlands

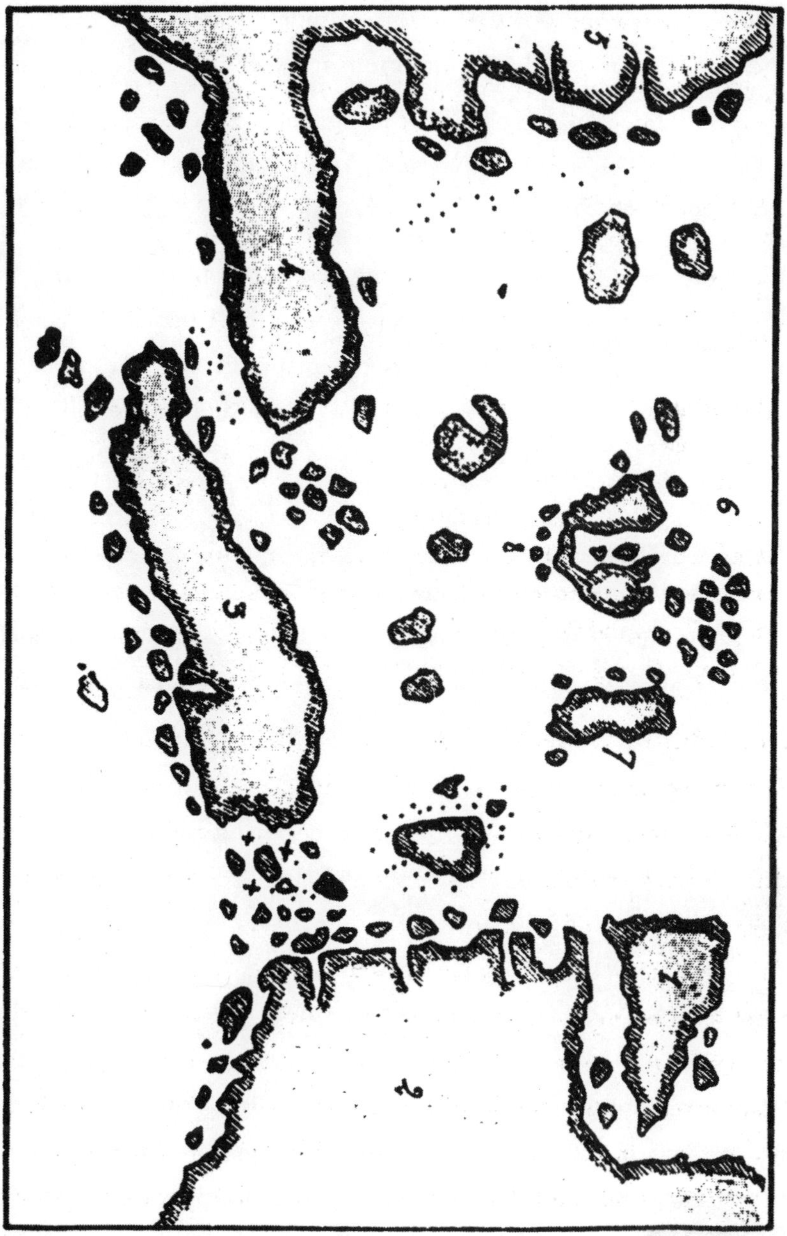

Jean Alfonse's sketch map of the East, published in his book *La Cosmographie* in 1544. He writes that his numerals represent: (1) *Java minor*, (2) *Java major* (Australia), (3) *Terre de Samatrez* (Sumatra), (4) *Melacque* (Malacca/Malaya), (5) *La Chine* (China), (6) *Molucque* (Moluccas), (7) *Ceslam de Muscade* (Seram), and (8) *Isles du Clou* (Islands of Cloves).

in southern Chile, lying at precisely this latitude. From here Alfonse takes a gigantic leap right across the Pacific. From the *Cap de la Tormente*, he says, a line of direction can be taken to:

> Le Cap Formose which is called "Beau" [Beautiful] which is on la terre australle's west coast in the Pacific sea... [11]

Le Cap Formose—where have we come across this name before? It is, of course, none other than the title given to the tip of that huge, enigmatic promontory which appears in all the Dieppe maps of eastern Australia, going also under the names *Cap de Fremose*, *Cap de Freinose* and (in the Vallard map) *Cap Frimosa*—all meaning 'Cape Beautiful'—a cape which we have identified as Wilsons Promontory.

From the *Cap de la Tormente* to *Le Beau Cap*, says Alfonse, the direction is north-west and the distance 600 leagues.

Alfonse now gives a second bearing. Again on the coast of South America, he notes the position of 'the isle of Fernandez de Magellan [the island of Juan Fernandez]... at twenty-seven degrees[12] above the antarctic pole'. He continues:

> Le Beau Cap is at thirty-nine and a half degrees above the antarctic pole. The direction from the isle of Fernandez de Magellan [to] Le Beau Cap [is] south-west...and the route is 500 leagues. And everything here is in the Pacific sea.[13]

Alfonse clearly had never sailed across the Pacific himself. To be quite certain of this one needs only to compare the longer of his two estimated distances—600 leagues (perhaps 2400 nautical miles) from *Cap de la Tormente* to *Cap Formose*—with the actual distance from Cape Tres Montes to Wilsons Promontory, which is over 6000 nautical miles. Alfonse evidently was either not aware of, or did not believe, the distance of 4000 leagues which the Italian adventurer Pigafetta, who sailed across the Pacific to the Phillippines with Magellan, claimed (with only slight exaggeration) for the length of that epic voyage.

Alfonse was by no means alone in his error. Even long after Magellan's epic crossing others had still not grasped the true immensity of the Pacific. For instance his contemporary, the great Jean Rotz, in his atlas of 1542 has a map depicting Central America and the Caribbean. This optimistically shows the Spice Islands located in the Pacific at little more than a stone's throw from the west coast of Mexico—or to be more precise, about as far away as Cuba is from Panama! Next to his Spice Islands the cartographer has written: 'The Indis of orient [which] the portingals doeth occupy'. Rotz's version of Australia (The Londe of Java) is not shown on this particular page, but its eastern coast does appear in a companion map of his atlas—and it is about as close to the Americas as his Spice Islands!

As with Rotz, Alfonse's estimates of distances across the Pacific may be hopelessly too short, but now comes the crucial test: what of the chart location which he confidently gives for *Le Beau Cap* (*Cap Formose*)—'thirty-nine-and-a-half degrees south'? How does this compare with the coordinates for the modern equivalent of *Cap Frimosa* on our realigned Vallard map? Alfonse has given us a very precise latitude, even confidently narrowing it down to one half of a degree. But surely we cannot reasonably expect him to be as accurate as that?

The answer is stunningly conclusive. Thirty-nine and a half degrees south is the *actual* latitude for Wilsons Promontory *to within one quarter of a degree*! The promontory's southernmost tip lies at 39¼ degrees, and the 39th parallel bisects it. And this calculation was made by a ship's navigator in the early sixteenth century! Even a global positioning satellite of the twenty-first century could hardly have done a much better job.[14]

The implications of this are far-reaching. Alfonse cannot possibly be referring to the position of *Cap Fremose* as depicted on the original Vallard and other Dieppe maps, jutting eastwards far out into the Pacific Ocean. He states firmly that while his *Cap Formose* faces the Pacific, it is on the *west* coast of *Terre Australle*. This at first seems to contain a hopeless contradiction. How could this cape possibly be on the west coast, when the west coast of Australia faces not the Pacific but the Indian Ocean? It is also a puzzling statement in the context of Alfonse's own concept of a *Terre Australle* extending from near the polar regions of the South Pacific all the way to Java—for this coast would also by definition have to be east-facing.

Alfonse does not appear to have thought about this contradiction too deeply, but a closer examination of his text makes it apparent what he really means. What he surely wants to convey is that while his *Cap Formose* borders the Pacific, it also has a *western* coast. And this is also an accurate description of Wilsons Promontory. This southernmost cape of mainland Australia does in fact have one coast facing east toward the South Pacific and another coast facing west toward the Great Australian Bight and ultimately the Indian Ocean.

Alfonse could not have been basing his description of *Cap Formose* on something he had seen in a French chart because all the Dieppe maps depicting Terra Java without exception have *Cap Formose* (or its namesakes) protruding far out into the Pacific Ocean with a north coast and a south coast but nothing that could possibly be called a west coast. Alfonse has evidently written down in his book important facts he has obtained in person, either from another Portuguese mariner—or, quite possibly, because he may even have sailed to this cape himself with Mendonça. He knows the name and latitude of this important promontory and he is able to pass on two different bearings for it—one north-west from *Cap de la Tormente* and the other south-west from Juan Fernandez. And both of them are correct. Finally, he knows the seemingly contradictory fact that *Cap Formose* has a west-facing coast.

These are the plain facts he records in his book. And in doing so he confirms for us, beyond all reasonable doubt, that the correct identification for the Vallard map's mysterious *Cap Frimosa* is indeed the headland now known as Wilsons Promontory.

Could Alfonse possibly even be the mystery author of the Vallard map of Terra Java? It seems unlikely. For one thing, the fact that the Vallard cartographer has failed to understand the meaning of some (but not all) of the Portuguese place names on the sea charts from which his map must have been derived is strong evidence that whoever drew this map was not a native Portuguese speaker but rather a Frenchman with a rough but not fluent knowledge of Portuguese. Furthermore, Alfonse was a pilot, not a cartographer, and the maps in his *Cosmographie* are so crude in comparison with the Vallard Atlas that there would seem to be no real possibility of the cartographer in each case being one and the same person.

This is not to say that Alfonse may not have played some role in the compilation of some of the Dieppe maps, perhaps as a kind of consultant. Roger Hervé, the former curator of maps at the Bibliothèque Nationale in Paris, has theorised that it may even have been Alfonse who brought to Dieppe copies of the Portuguese charts which Hervé, like most other historians, is convinced were the basis for all the subsequent Dieppe maps of the East.

Hervé could well be right. It has generally been assumed that these charts were copies obtained surreptitiously from originals held under lock and key in the Portuguese state archives in Lisbon. But what if they had not come from Lisbon at all, but instead been seized from a Portuguese ship on its way back from India—in other words, from Mendonça's carrack *Vitória*, homeward-bound in 1524 after his momentous voyage to discover the Island of Gold? We know for a fact that at this very time the captains of Jean Ango's corsairs were under orders to confiscate as a top priority the charts of any returning Portuguese and Spanish ships which they boarded. So an origin of this kind for the Dieppe cartographers' charts of Terra Java might actually seem the more likely possibility.

So far as any conclusive answer to this question is concerned, what records there are of corsair voyages provide no real evidence one way or the other. Hardly any names of intercepted ships seem to have been recorded—or if they were originally noted, the records have been lost. One interesting fact, though, emerges from a letter which King John of Portugal sent to his ambassador in France in 1530. This complained that since the beginning of the century more than 300 Portuguese ships had been intercepted and plundered by French corsairs.[15] This would be an average of ten ships a year, a figure which in some years would have been the equivalent in theoretical terms of almost the entire Portuguese fleet returning from the Indies (though in reality there would have been other Portuguese ships coming back from destinations such as Brazil and West Africa).

And now, one other possibility: what if the French had managed to get possession, not just of copies of the Portuguese charts of Terra Java but the *originals*? If so, it is not impossible that these might even have been the only such charts in existence. We need to remember that in the sixteenth century all copying had to be done by hand and in the case of maps this was a difficult undertaking requiring the services of highly skilled professionals. Specialists of this sort would have been few and far between in the often chaotic Portuguese empire of the East, so the task of copying the charts of the newly discovered land of Terra Java might conceivably have been put off until the originals could be delivered to Lisbon. And if Mendonça's ship was indeed intercepted, this could have provided the opportunity for the Portuguese mariner João Afonso (soon to become Alfonse) to change sides. And this could also have been the moment when he 'stole away' with the vitally important Terra Java portolans, as Roger Hervé has suggested.[16]

All this might sound speculative except for one intriguing fact that fits the scenario. Within little more than a year or two of Mendonça's return from the East, the eminent Dieppe cartographer Pierre Desceliers is known to have been working with a collection of valuable portolans seized from an enemy ship. The ship in question was evidently Portuguese because the portolans themselves were said to contain Portuguese-language place names. Desceliers used these and other charts to make a planisphere—a world map in four sections described as being, more than seven *pieds* (feet) long and four-and-a-half *pieds* wide.[17]

The theory that Desceliers and the other Dieppe map-makers (including of course the cartographer of the Vallard maps) were working from Portuguese originals, rather than copies, would also explain another puzzling fact—why no maps of Terra Java have ever been found in the Portuguese archives. It has usually been assumed that the Dieppe maps were based on copies clandestinely taken from original charts held in the Torre do Tombo, the national archives in Lisbon, and that those originals must have all perished in the great Lisbon earthquake and fire of 1755. But if Jean Ango's corsairs had managed to seize the *original* charts of Mendonça's epic voyage of discovery and if there were no other copies in existence, that would put an entirely different complexion on the matter. And it would also provide a further good reason why no features even remotely suggesting the coasts of Australia are shown on any of the numerous officially sanctioned Portuguese world maps of the sixteenth century.

True, no original Portuguese charts have been found in Dieppe either. But in the final analysis, the precise means by which the Dieppe cartographers came into possession of their source material is not of vital importance. The magnificent series of Dieppe maps of Terra Java are proof enough that in one way or another the French map-makers acquired the vital Portuguese charts they needed. And in Dieppe, as in Lisbon, there was later to be a catastrophic loss of much irreplaceable archival material, this time not by the forces of nature but at the hands of an English fleet which in 1694, during one of the lesser European wars of the seventeenth century, bombarded and destroyed most of the town, including the buildings housing its archives and graphic records.

This brings us finally to the vexed question of why all of the Dieppe cartographers without exception made the fundamental mistake of assuming that the north coast of Australia—their Terra Java, or Jave la Grande was one and the same as the north coast of the *island* of Java.

The Portuguese themselves could have had no doubt about the true nature of Java as an island. As early as 1512, as we have seen, when Portuguese ships made their first voyage between Malacca and the Spice Islands, the pilot Francisco Rodrigues produced a set of portolan charts correctly depicting Java as a long, narrow island with a clearly defined south coast. Another

Portuguese captain, Francisco de Sá, who sailed along the Java coasts in 1526, reported that:

> Java resembles the shape of a pig crouched on its forefeet, before whose snout passes the channel [Sunda Strait] which is much used by our ships...*the south coast* is not much used by us because it does not have ports or harbours.[18] (Author's italics).

Perhaps the French error of confusing the island of Java with Australia was just a straightforward mistake by Jean Rotz, or whoever was the first Dieppe cartographer to compile a composite map of the eastern hemisphere from the individual Portuguese sea charts which the French had purloined—and then with every subsequent repetition this first mistake became an entrenched fact.

But there is another possibility: could the error have been the result of deliberate Portuguese misinformation? Spreading abroad the belief that the only way of reaching the region of the Spice Islands was by sailing north of Java—where interlopers could be intercepted by the Portuguese naval forces based at Malacca—would have perfectly suited Portugal's political aim of excluding other nations from its lucrative spice trade.

And there was something else to be gained, too. The assumption that Terra Java (Australia) was a southward extension of the island of Java led inevitably to a huge westward displacement of Australia on the Dieppe maps. The north-west corner of the Australian continent now had to correspond in location to the western end of the *real* Java—a westerly displacement of some twenty-five degrees of longitude or approximately 1500 nautical miles. This westward shift would have suited Portugal very well indeed, for it would demonstrate to any Spaniard who might get hold of a Dieppe map that the lion's share of this vast new land (which might yet prove to be the fabled Island of Gold) lay within the Portuguese hemisphere and not that of Spain.

And who could have been better placed to plant such an idea than the renegade João Afonso? Could Afonso (Alfonse) have been a double agent? We have no hard evidence for this but two things could give rise to a suspicion. The first is that North American historians studying Cartier's voyages to Canada, on which Alfonse sailed, recently discovered in the archives a secret report by 'a Spanish spy' (unnamed) detailing preparations for the French expeditions.

And secondly there is the very curious fact that in 1532, a decade or so before his Canadian voyages, Alfonse was offered a pardon by Portugal's King John III. We know this for certain because a letter to the King from his ambassador in France, Gaspar Vaz, has been discovered in the Portuguese state archives, the Torre do Tombo. In this letter, Vaz reported that he had given instructions for:

> João Afonso, the Portuguese, who is and lives at La Rochelle [to be told that] he should return to Portugal and I should endeavour that Your Highness would forgive his wrongs and grant him some benefit—and this man [I sent] pressed him so much that they agree what you will see from the letter he wrote to Your Highness and I am sending by this courier. If you agree that this should be done you should send the pardon very soon.

A draft document was found with the ambassador's letter in which the king pardons Afonso and gives him a safe-conduct for his return to Portugal. This pardon is not dated but was probably written in 1533.

Alfonse's 'wrongs' were doubtless his defection to the French and the use of his valuable navigation skills in their service. One early Portuguese account describes him as 'an intelligent and daring pilot' who served France 'out of spite' because his many services to Portugal were not rewarded.

What emerges from the cloak-and-dagger episode referred to in the Portuguese ambassador's letter is that Alfonse apparently agreed to some secret proposition, though he evidently declined the offer of a safe return to his homeland, because he continued to serve the French as pilot or captain of several ships, a career culminating in his voyage as chief pilot with de Roberval's 1542 expedition to Canada.

Spying for a foreign power seems to have been nothing unusual in the sixteenth century. The noted Dieppe cartographer Jean Rotz, as we saw earlier, moved with his family to England in 1542 in order to present his great *Boke of Idrography* to Henry VIII and, so he hoped, gain fame and advancement at Henry's court. Rotz, however, soon became disillusioned by the lukewarm reception he was accorded in London. Documents which have recently come to light, reveal that he agreed to become a secret agent for the French in order to secure his return to France following Henry VIII's death in 1547.

So did Alfonse become a double agent? It is hard to shake off the suspicion that the renegade master mariner must have agreed to perform some important tasks in the service of Portugal to warrant King John extending the unusual favour not just of a pardon but also (in the ambassador's words) 'some benefit'. Providing intelligence about French voyages could have been one such service. Persuading the Dieppe cartographers that their Grand Jave, alias Terra Java, was attached to the island of Java and that there was no seaway south of it might have been another very good reason for reward.

In saying this we may be doing Alfonse a grave injustice. As far as the Dieppe cartographers are concerned, however, the fact remains that their fundamental mistake of confusing the north coast of Australia with the coast of Java was to be the source of much subsequent difficulty and error. And as we shall see in the next chapter, this was to prove particularly the case when they attempted to compile a map of the west coast of this vast island continent.

10

To a distant sea and 'a cape which no one has yet gone beyond'

A century has passed away since any European vessel appeared in these seas, and we shall soon be able to judge that it was not without reason that voyagers had so long abandoned so dangerous a navigation.

François Péron, on the perils of Australia's west coast waters, in *A Voyage of Discovery to the Southern Lands*, 1809

We saw earlier how the Portuguese adventurer Diogo Pacheco landed on Australia's north-western Kimberley coast and was killed there by Aborigines during his ill-fated search for gold in 1519. But what of the other Portuguese voyagers who undoubtedly went on to explore and chart almost the entire west coast of Australia? Their identity may be uncertain, but in investigating their achievements some strange and unexpected discoveries will come to light—including that of a remarkable coral reef second only to the Great Barrier Reef and a maritime tribe of black people who built stone huts and spoke a language that resembled Latin.

The French naturalist François Péron, who sailed with Nicolas Baudin's expedition of 1801–03, was by no means the only voyager to remark on the hazards facing any ship's captain attempting to navigate and chart the coasts of western Australia. Off the rugged north-western part of this coast in particular, mariners had to contend with fierce currents and a dangerous cordon of countless reefs, shoals, rocks and islands.

Matthew Flinders gave the north-west coast a wide berth on the grounds that it had already been charted by the Dutch, but in reality the Dutch charts were often inadequate and unreliable. Abel Tasman, the discoverer of Tasmania, who had been sent by the Dutch East India Company in 1644 to investigate and chart this same coast of 'The Southland', found it so daunting that for the most part he sailed cautiously well outside the outermost islands. As a consequence the charts he produced contained little useful coastal detail. Even now, modern charts warn that large areas of this coast are 'inadequately surveyed'—or sometimes not even surveyed at all.

Some fifty years after Tasman's expedition another Dutch captain, Willem de Vlamingh, did a more thorough job in charting the southern section of 'New Holland', as it was by then called. In his report, de Vlamingh dismissed the country as 'dry, sandy and but little adapted for the habitation of animals, still less of men'. Offering few if any secure harbours, this coast was also a dangerous lee shore, the graveyard of a succession of Dutch East Indiamen bound for the Javanese port of Batavia.

All this explains some of the difficulties the Portuguese must have met in attempting to chart the west coast of Australia. Yet, as we shall see, they were undoubtedly there—a century or more before Tasman and the other Dutch captains who have invariably been credited with the discovery and first European mapping of the coast of western Australia.

Who made these Portuguese voyages and when? On these questions the official Portuguese chroniclers are resolutely silent. They record in great detail many other stirring events at this time involving ships and commanders and battles in the East, but nothing concerning the exploration of unknown lands. The reason, as we now know, is that voyages of this kind were classified as secret, and it was the task of the State censors to ensure that discoveries of any potential value to rival nations were not revealed.

The proof that such voyages were indeed made, however, has always been there waiting to be recognised. It is to be found in the same series of detailed yet little-known sixteenth century French charts which we have already examined, most of them showing almost the entire west coast—and all of them bestowed with descriptive place names of undeniable Portuguese origin. As with the later Dutch voyages, the Portuguese mapping of Australia's west coast seems to have been difficult, fragmentary and disjointed. Whether the portolan charts it produced were the outcome of several voyages or just a single one cannot be judged with any certainty.

Was Christopher de Mendonça involved? It would have been out of the question for the explorer of Australia's east and south coasts to have made a major expedition down the west coast in the brief period that would have been available to him prior to his departure from Goa, probably in January 1524, on his voyage back to Portugal. But what about the year *before* his east coast expedition? If the Portuguese chroniclers are to be believed, he had actually set sail from Malacca on his mission 'for the discovery of gold' with a fleet of three ships toward the end of 1521, only to be diverted at the request of the Portuguese commandant, at Malacca to help complete the building a fort on the north coast of Sumatra.

This story does not entirely ring true. Mendonça's expedition to find the Island of Gold, it will be remembered, was on the direct orders of the supreme authority, Portugal's King Manuel, so one can doubt whether Mendonça would have been under any obligation to take contrary instructions from the local Portuguese commandant, particularly when they concerned such a menial job as helping to build a fort. The accounts in the chronicles of what Mendonça allegedly did during the several months he was supposed to have been at Sumatra are vague, to say the least.

Could the reality be that this was another piece of deliberate Portuguese misinformation—a cover story for a secret voyage of exploration down the west coast of the supposed Island of Gold? That would be entirely in character. All we can say with any assurance is that given the difficulties that would have had to be faced in charting this coast in the early sixteenth century, the attention to detail and the scientific knowledge that parts of the Vallard map in particular reveal are fully consistent with professionalism shown in Mendonça's charting of the east coast.

Several maps of the French Dieppe school attempt to depict the west coast of Australia. Jean Rotz's World Map of 1542 shows only the north-western part of the coast. Pierre Desceliers produced more extensive maps in 1550 and 1553, while Guillaume le Testu published several imaginatively decorated versions a few years later. But as we shall see, despite the often Gallicised place names there can be no real doubt that all these maps are derived ultimately from Portuguese originals.

All of these maps have one huge mistake in common: the Dieppe cartographers were of one mind in deciding that their continent of 'Terra Java' or 'Java la Grande', as it is variously called, was a southward extension of the *actual* island of Java. This monumental error was to create huge problems when they tried to piece together the 2500 nautical mile-long outline of this coast from the collection of Portuguese charts that had fallen into their hands.

The map in the Vallard Atlas of 1547 is once again the best source, although the maps of Desceliers offer significant additional details. A common feature is that all of these maps depict a long coastline facing west toward what is quite clearly the Indian Ocean—and which Desceliers actually calls 'La Mer des Indes Orientalles'. The Indian subcontinent is shown at the northern end of this ocean, with Madagascar and the coast of Africa at its western edge. The Indian Ocean in the Vallard map is decorated with drawings of bizarre sea monsters. On land, the illustrations of people and habitations are once again (as with the eastern Terra Java map) evidently derived from Sumatra and the Parmentier brothers' epic voyage to that island in 1529–30.

All of these Dieppe maps appear to have been compiled from at least four separate portolans. As far as the north coast of Terra Java is concerned, as with the eastern Terra Java chart it is clear enough that this strange-looking region is in reality the north coast of the island of Java. Not only is the coastline depicted on the map recognisable, however roughly, as that of present-day Java; many of the place names on it are readily identifiable—for example *Japarra* (modern-day Japara), *Tubany* (Tuban), *Curibaia* (the exact Portuguese spelling for Surabaya) and *Canal do Sunda* (Sunda Strait), to name just a few.

This coast has probably been superimposed on what would originally have been a chart showing the coast of Australia's Arnhem Land. There are some superficial similarities between these two coastlines, and with the north coast of Java having already been thoroughly mapped, the Dieppe cartographers might well have decided that their Arnhem Land chart was merely an inferior version of the same coast and hence to be discarded.

At the beginning of the north-western coast of Terra Java the first two names in the Vallard map also clearly relate to the *real* island of Java. They are *Terra Plata* (Flat Land) and *Rio Tigri* (Tiger River)—the latter evidently a reference to Javanese or Sumatran tigers, since there were no such animals to be found in Australia. The same name, this time spelt *Rio Tigory*, also appears on the southern shore of the wide river depicted at this part of the map. The 'river' itself can best be seen as a confused attempt to reconcile on the map the existence of the actual sea gap between the south coast of Java and the Australian mainland.

From this starting point we can attempt to retrace the historic voyage made by Mendonça—or whoever it was who first sailed through these treacherous waters of north-western Australia.

The first harbour shown on the Vallard map south of the 'Tiger River' corresponds in shape and location to Anson Bay at the mouth of the Daly River, south-west of the present-day port city of Darwin. And now we come to a sure sign that this coast is no longer the green and fertile land of Java. The name appearing on a prominent headland here is *C Vermello*, in correct Portuguese 'Cabo Vermelho', meaning 'Red Cape'. Few names could be more evocative of the parched, sun-scorched earth of Australia. And surprising as it may seem, we can tell exactly what the sixteenth century Portuguese voyagers

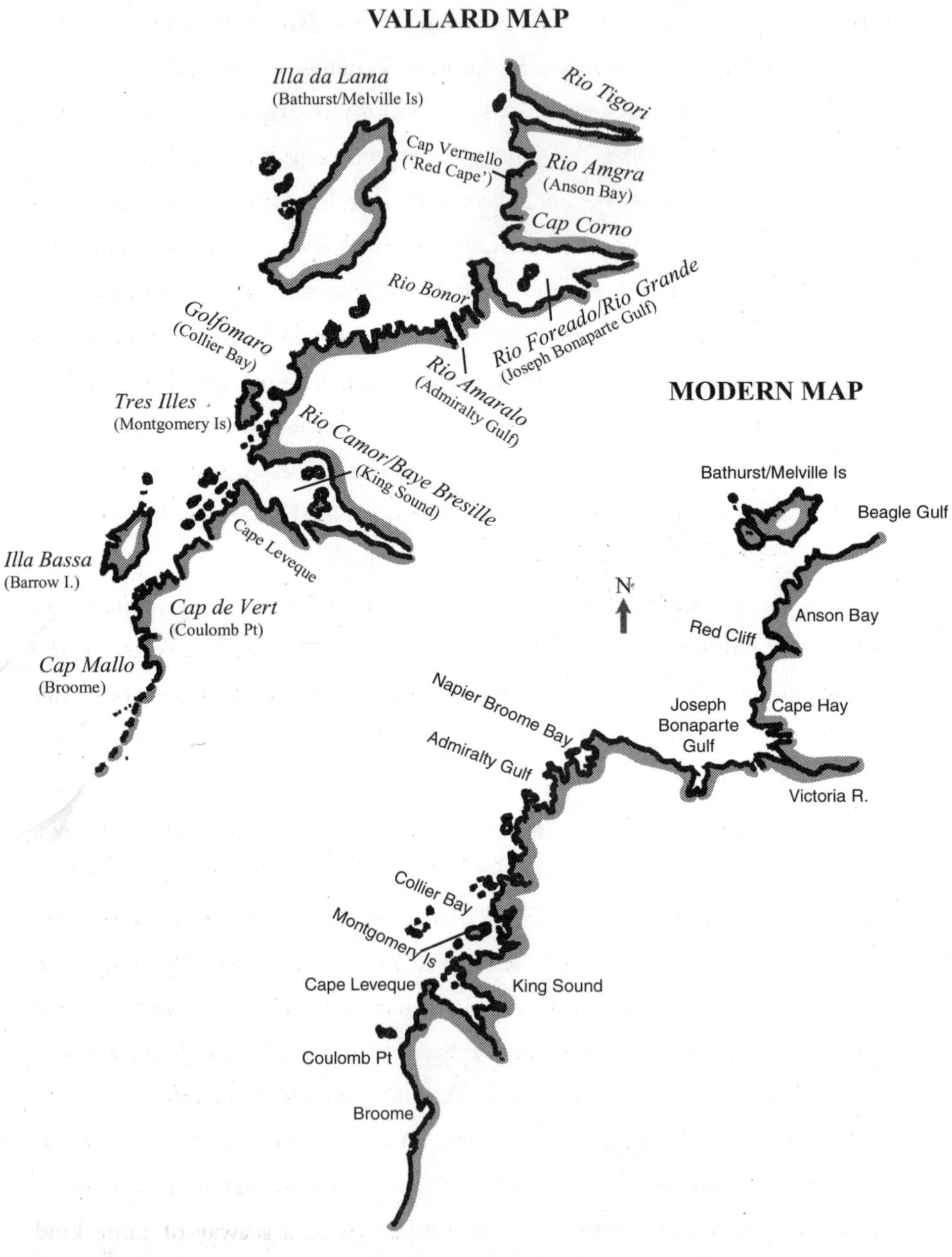

Australia's north-west coast as depicted in the Vallard map (left) and a modern map. The Dieppe cartographer who drew the Vallard map assumed Australia was a southern extension of the island of Java – hence the displacement of the Bathurst/Melville islands, whose correct position had ben usurped by Javanese islands.

were looking at: it can surely have been none other than the headland south of Anson Bay which today has an almost identical name—Red Cliff.

South of their aptly named Red Cape the Portuguese noted a smaller bay which appears on the Vallard map as a neat semicircle. It has not been given a name, but it can be identified as Hyland Bay, which has exactly this shape. Further on again we can see depicted a deep, narrower estuary with the apparent name *Rio Patoa* (perhaps, as we have seen on the east coast, a garbled version of *Rio Patos*, meaning 'River of Ducks'). This is equally clearly the harbour today known as Port Keats. Immediately south again from Rio Patoa the Portuguese ships sailed past a fairly narrow, pointed headland to which they gave the very distinctive name of *Cap Corno*, or 'Cape Horn'. This description fits the horn-shaped Cape Hay, near which the coastline abruptly turns away toward the south-east.

Passing their 'Cape Horn' the explorers found themselves sailing into a broad waterway which, as depicted on their chart, clearly represents the mighty Joseph Bonaparte Gulf. Nearly 200 kilometres wide at this point, this gulf extends far inland. In the Vallard map it has an indistinct name which appears to be *Rio Foreado*; a later Dieppe map generally attributed to the cartographer Pasterot calls it *Fourgado*. In Portuguese it was probably 'Rio Folgado', meaning 'Wide River'. In the Vallard map this name is written alongside a feature on the southern side of the Joseph Bonaparte Gulf corresponding to the mouth of the Ord River and the Cambridge Gulf. The Ord, in recent times dammed to form a vast irrigation lake, is one of the two biggest rivers flowing into the Bonaparte Gulf. The other, the Victoria River, is correctly shown in the Vallard map extending well into the interior.

In most other Dieppe maps of this region, including those of Rotz, Desceliers and Desliens, this is the site of a very peculiar cartographic feature. The French map-makers knew that there had to be a seaway of some kind south of their 'Little Java' (the real Java)—for after all, the Spaniard Juan del Cano had sailed this way in command of the last surviving ship of Magellan's fleet, the *Victoria*, on his way back to Spain in 1522—and this obviously could not have occurred if 'Little Java' and 'Great Java' were joined. The awkward problem was resolved by deciding that the Bonaparte Gulf—'Rio Grande' as it is called on most of the other maps—was the entrance to a navigable channel leading all the way to the Bali and Lombok straits at the eastern end of Java (or, in an Australian context, to the Roper River on the Gulf of Carpentaria).

The cartographer Jean Rotz depicts this fictional waterway in his 1542 *Chart of South-East Asia*. It also appears in the Dauphin map, while Pierre Desceliers, in his *Java la Grande* map of 1550, shows two labourers apparently in the act of widening the channel or, alternatively, perhaps filling it in! In the Vallard map, the place where the channel would be is obscured by a large picture of two men holding spears, so we cannot be sure whether our cartographer believed in the existence of this waterway or not. Perhaps he was hedging his bets.

As in the other Dieppe maps, the Vallard chart depicts at this point, a very large island offshore shaped rather like the outline of a glove or mitten. But there is no such large island anywhere near the entrance to the Joseph Bonaparte Gulf. Is it fictional or does it belong somewhere else?

Its name, *Illa de Lama*, gives an important clue. 'Ilha de Lama', which would have been the original wording, is Portuguese for 'Island of Mud'. Where might we expect to find a large 'island of mud'?

If we look further north on a modern map we can see that the shape of this peculiar island in the Vallard map in fact resembles a composite drawing of the Bathurst and Melville islands, just north of Darwin. These two islands are low-lying and separated only by a very narrow channel which even the much later Dutch voyagers failed to discover. The southern end of this channel we can see depicted on *Illa de Lama* as a large bay. Most significantly of all, Bathurst-Melville island is surrounded by wide, mangrove-fringed mud banks which lie exposed to the tropical sun at low tide. An 'Island of Mud' indeed!

But why would this island have been depicted so far from its correct position? The answer is not hard to discover. Once the coastline of Java had been superimposed on Arnhem Land there was no vacant spot remaining among the many real islands in the sea off Java's north coast where this large intrusive newcomer could be accommodated. So the Dieppe cartographers had to assume that it had been drawn in the wrong place on the portolan from which they were working. The simplest way to rectify this 'mistake' would have been to move the island to the nearest vacant place on their maps—off the north-west coast of Terra Java.

There is evidence that the Portuguese themselves not only knew the correct location of Melville Island but in later times were even in sporadic communication with its native inhabitants. These indigenous people were the Tiwi, who had lived here in isolation for thousands of years and had developed a language and culture quite distinct from that of the Aborigines on the mainland. A Dutch naval lieutenant, DH Kolff, who surveyed some of the lesser-known islands of the region in the 1820s, wrote that:

> When Captain [Phillip] King first visited Melville Island...the natives appeared on the beach and called out to our voyager, '*Ven aca*', the Portuguese term for 'Come here!'[1]

Returning to the Joseph Bonaparte Gulf, off which Bathurst-Melville island has found a new home in the Dieppe maps, at the gulf's south-western entrance the Portuguese voyagers came to and depicted on their chart a very prominent peninsula. It has no name, but its shape, resembling that of a pointing finger, corresponds closely to Cape Londonderry. Behind this cape lies Napier Broome Bay, where Diogo Pacheco made his fateful landing in 1519. In this bay also is the island where the two Portuguese cannons described in Chapter 2 were discovered at an Aboriginal sacred site. The bay itself has been given the name *Rio Bonor*, or 'Good River', suggesting that the Portuguese expedition charting the coast may well have put in here.

The next named feature in the Vallard map is *Rio Amaralo*, probably a garbled version of the Portuguese 'Rio Amarelo', or 'Yellow River'. It is shown at approximately the position of Admiralty Gulf. The cryptic name

Amconio, shown next, is probably a corruption of Cabo Conico, or 'Conical Cape'. This would correspond in both position and appearance to the distinctive Cape Voltaire described by some coastal sailors as towering up from the sea 'like a black, abandoned castle'. West of this 'Conical Cape', Bigge Island seems to be depicted, with *Terra Alta* (High Land) indicated toward the interior. The next wide inlet shown, *Golfomaro* (Golfo Mare or Tidal Gulf), can be identified with Collier Bay which, as the Portuguese name implies, is a place of fierce tidal currents.

Now we are coming to another very distinctive feature on the chart: once again a huge gulf fed by two big rivers. This corresponds both in size and approximate outline to King Sound a gulf some fifty kilometres wide at its entrance and extending 100 kilometres inland. In the Vallard map, King Sound has the name *Rio Camor*[*o*]. Probably the nearest we can come to this in Portuguese is 'Rio Comodo' or 'Commodious River'—an apt enough description for this broad waterway.

In most of the other Dieppe maps, though, King Sound has been given a quite different and rather strange name—*Baie Bresille*, meaning 'Brazil Bay'. Not, it must be added, because any Portuguese voyager imagined that he had suddenly reached South America. It is in all probability a reference to Aborigines, who we can find described elsewhere (for instance in the words of the Portuguese-French navigator Jean Alfonse) as 'people like those of Brazil'—heathens who worshipped the Sun and the Moon.

And curiously enough, three people in the act of worshipping a golden disc of the sun are actually depicted here in Pierre Desceliers' map of 1550. The devotees are clad in knee-length tunics, so they can hardly be interpreted as a realistic attempt to portray Australian Aborigines. The trio are kneeling before a satyr-like idol, which on the map has been placed just inland from *Baye Bresille*.

Around the approaches to King Sound and nearby Collier Bay in both the Vallard and Desceliers maps are shown scattered groups of islands corresponding approximately to the Buccaneer Archipelago. In this area, some

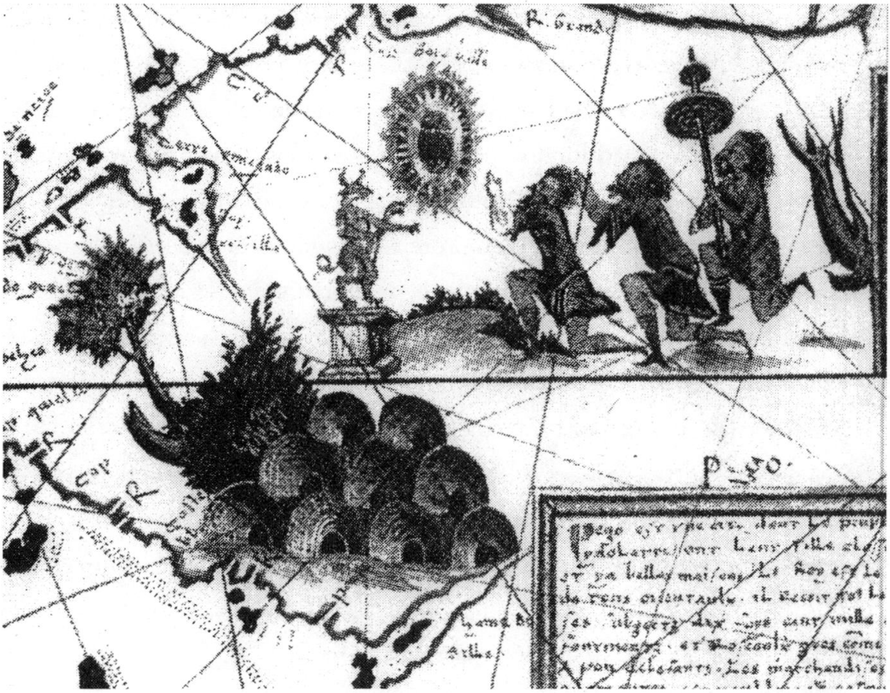

Sun-worshippers and beehive huts: detail from the Kimberley region as pictured in Pierre Desceliers' world map of 1550. Western Australia's Ningaloo Reef can be seen delineated alongside the cluster of huts. (*British Library*)

strange and little-known stories raise intriguing questions about how much contact the Portuguese may actually have had with the indigenous 'people like those of Brazil'.

In the early 1930s, Bernard C Ryder, a Fellow of the Royal Geographical Society, chartered a seventeen-ton pearl lugger for an expedition to King Sound. Even at that comparatively recent time the area was little-known and Ryder described it in a subsequent report as 'wild and remote'. Day after day, smoke columns from distant Aboriginal fires signalled the white men's approach. Languages were often completely different from one tribe to the next—and all of them proved incomprehensible to the two Aborigines who had been taken on board as interpreters at the pearling port of Broome. Communication was only possible through sign language.

It was in the waters of the Buccaneer Archipelago that something extraordinary occurred. Sailing among the islands of this archipelago the expeditioners found themselves among what Ryder called 'a strange tribe of blacks' who were paddling dugout canoes—quite unlike the Aborigines on the nearby mainland, who possessed no craft of this kind. Ryder was also surprised to find that the islanders were 'of magnificent physique, most of them over six feet in height, and broad in proportion'.

What Ryder found most astonishing of all, though, was that when one of the blacks was coaxed into paddling his canoe close to the lugger, he dipped his hand over the side into the water and said 'agua!'. Ryder, who was knowledgeable in the classics, was taken aback. 'Agua' sounded remarkably like 'aqua', the Latin word for 'water'. Encouraging his black visitor to identify more objects in his own language, Ryder came to an extraordinary conclusion: this tall islander was speaking 'a dialect with a strong resemblance to Latin'. Recalling the affair later, he wrote of the islander:

> He was brought aboard and given water. As he squatted down I pointed to the water and repeated his word 'agua'. I pointed to his dog. 'Caningo' he repeated laconically. It seemed strange, for one could literally converse with this wild man. The word 'caningo' bore an extraordinary resemblance to canine and dingo. Other words I obtained in the brief time were 'apita' for his head, and 'oombooroo' for arm. A dead fish was 'mat' in his dialect, and in truth it was very 'morte'. I pointed to the mast of the vessel, and he grunted 'oobra'. I could not follow that until I realised with a shock that he pointed to the *shadow* of the mast.
>
> Before he left the vessel we gave him a pannikin of tea; but he would not drink it until he saw us do the same, for probably he had never seen coloured water before. After drinking the tea he calmly scooped up the tea-leaves, which he chewed with evident relish.[2]

In fact, the list of 'Latin' words which Ryder wrote down phonetically shows that what the man in the dugout canoe was *really* speaking was a dialect of Portuguese. Ryder evidently was not familiar with this European language, which is closely related to Latin. The Latin term for water, 'aqua', is very close to what the islander said but 'agua' is the *exact* Portuguese word.

What of the rest of Ryder's vocabulary? 'Canino', a Portuguese familiar term meaning literally 'doggie' is quite obviously the equivalent of 'caningo', while the Portuguese word 'cabeça' (head) can readily be seen to be related to 'apita' when it is written phonetically in English, as 'cabisa'. From his knowledge of Latin, Ryder was able to see that in using the word 'mat' his visitor was referring not to the fish itself but to the fact that it was dead (morte): the Portuguese word is virtually the same—'morto'. As for the islander's word for the shadow of the mast, 'oobra', this in modern Portuguese would be 'sombra'.

'Oombooroo', which Ryder wrote down as the word for 'arm', at first seems to be an exception. It bears no obvious resemblance to Latin, let alone to the Portuguese term for an arm—'braço'. Could this be a purely Aboriginal word from a different linguistic source? As written, it certainly looks Aboriginal. In fact, however, there is a quite different explanation—and one which seems to have eluded even Ryder. 'Oombooroo', particularly if rewritten as 'omboro', is clearly the Portuguese word for 'shoulder'—'ombro'. What we have here, in other words, was a simple misunderstanding: the islander thought Ryder was pointing to an adjacent part of his anatomy—and responded in near-perfect Portuguese!

Ryder seems to have made no attempt to follow up this astonishing encounter. He describes the natives' dugout canoes as 'quaint'—without explaining what has led him to this odd choice of adjective. Were the canoes decorated in some unusual way perhaps? As the black man paddles off into the distance, though, Ryder is sufficiently puzzled to ask himself whether there could be 'a mixture of some Spanish blood in the native's make-up, from some early vessel.' Unfortunately, he leaves this question up in the air. In his defence, it must be said that it seems the Englishman's main objective was not so much to study indigenes as to retrace the coastal voyage of his seventeenth century compatriot, the buccaneer William Dampier (after whom the Buccaneer Archipelago was given its name). Latin-speaking blacks were just a passing curiosity.

So what are we to make of this extraordinary discovery? One fact of major significance which Ryder does record is that his 'strange tribe of blacks' lived on what Europeans had named the Montgomery Islands. So this would be the obvious place to look for clues.

The Montgomery Islands lie at the north-eastern edge of the Buccaneer Archipelago, surrounded by an extensive coral reef. Most of them are low-lying and sandy but there is one exception, High Cliffy Island lying nearest to the mainland. Though only about a kilometre long, this island is remarkable for being the location of hundreds of mysterious stone structures. Some of these are little more than windbreaks, but many others appear to be the lower walls of circular dwellings up to three metres in diameter. These walls have been made from flat blocks of stone laid one on top of another. With just one narrow entrance the structures resemble the lower part of a traditional beehive—or perhaps an Eskimo igloo. The walls of the best-preserved examples are fully a metre high.

Sue O'Connor, an archaeologist, visited the island in 1985. In her subsequent report she commented with surprise that although two other investigators had mentioned these structures, 'nothing I had read or been told prepared me for the extent of them'. O'Connor thought that the round stone walls might originally have been roofed using wooden uprights covered with a mix of spinifex leaves and clay. Some of the structures even had stone floors carefully paved with slabs of sandstone and two of the round houses contained grindstones.

How had the island-based people of the Buccaneer Archipelago come by these exceptional house building skills? Excavating the floor of one of the structures O'Connor found it littered with large numbers of stone artefacts. On this basis she theorised that the round houses were probably used in the wet season for the manufacture of stone tools, possibly for trade. Significantly, she also noted that the people using High Cliffy Island, the

Jaudibaia, had their own language and were of a different physical appearance to the Aborigines on the mainland (as were the islanders whom Ryder encountered). Unusually, they were also an exclusively island-based clan, owning no mainland territory. Until as recently as the 1920s, clan knowledge of how to exploit the abundant marine resources of the archipelago was still being meticulously passed down from one generation to the next.[3]

This leads us to yet another thought provoking 'coincidence'. In Pierre Desceliers' world map of 1550, just below the picture of the three sun worshippers, there appears a drawing of a cluster of nine beehive-shaped huts. This picture has usually been assumed to be no more than a flight of imagination by the cartographer and indeed similar illustrations are to be seen lower down the Australian coast near the map's equivalent of Perth as well as in the middle of Africa. But the now proven existence of a substantial settlement of beehive-shaped houses on High Cliffy Island raises the question of whether Desceliers' illustration could be something more than fiction after all. Could it be based on something the cartographer had been told? Unlikely, perhaps, yet not totally impossible.

A decade after O'Connor's visit to High Cliffy Island a West Australian linguist, Dr Carl von Brandenstein, in 1994, put forward an intriguing theory that the Jaudibaia (whom he called the Yawujibaia) might originally have been black slaves from Africa taken to the island by their Portuguese masters to mine the semi-precious stone chalcedony (a form of milky quartz) found there. Von Brandenstein, regarded as an authority on Aboriginal languages, has also remarked on what he believes is a resemblance between the High Cliffy Island round houses and African kraals. His theory of a deliberately planned slave settlement raises interesting questions, but it is not as yet supported by other accepted historical or archaeological evidence.[4]

What cannot by any means be disputed, though, is that 'baia', the last part of the name 'Jaudibaia', is the exact Portuguese word for 'bay'. As for the first part of name, 'Jaudi', this has a very close similarity to the Portuguese word for 'people'—'gente' (pronounced 'jenti'). In other words, what we seem to have here is the name 'Bay People'. And what description could be more fitting for this maritime, island-based clan?

Perhaps more plausible than von Brandenstein's 'slave settlement' theory is that the Jaudibaia could have been descendants of survivors from a Portuguese ship, wrecked somewhere on the coast centuries ago—bearing in mind, as we have already seen, that Portuguese vessels were generally crewed by a mixture of races, including black Africans. Assuming the survivors had taken Aboriginal wives it would help to explain the clan's seafaring skills, their unique 'Latin'-influenced language and also the fact that they possessed no mainland territory. A possible shipwreck connection was indeed the first thing that occurred to Bernard Ryder when another man from one of the Buccaneer islands drew for him in the sand the outline of an old cannon, at the same time waving his hand toward the north.

Could the apparent Portuguese legacy in the Buccaneer Archipelago be in some way related to the voyage or voyages which resulted in the charting of the western Australian coast? We shall probably never know the answer to this question because within just a few years of Ryder's visit the Western Australian state government forcibly removed the Montgomery Islanders from their homes.

It was a callous and brutal act but we should not be surprised by it, for it was in keeping with white attitudes at that time. In a paper on his travels which he published in the French journal *La Géographie* in 1935, Bernard Ryder records how in Broome he saw sandalwood-laden camel trains driven by turbaned Afghans and then witnessed:

> A group of Aboriginal prisoners working on a road, guarded by white policemen armed with rifles. The prisoners were manacled together with chains and bars, like natives of the Congo.[5]

As for the unfortunate Montgomery Islanders, they had no say in the matter of their deportation, for in the 1920s (and indeed even up to the 1960s) the original inhabitants of Australia were denied citizenship rights or any voice in the running of their own country. Bernard Ryder records that the state government had recently taken it upon itself to 'rent' the Montgomery Islands to a white man for use as a fishing base. So these free island people, who had committed no offence, were herded together and transported against their will to 'native reserves' and mission stations on the mainland, in the territory of alien tribes whose language they could not even understand. Demoralised, regimented, deprived of their marine livelihood and forced to rely on the white man's charity, their numbers rapidly fell into decline. According to official reports, by the mid-1970s their clan had died out.

Returning now to the sixteenth century maps of this north-west region of the Australian coast, Jean Rotz and some other Dieppe cartographers at this point show a curious island named variously *Isle Nege*, *Isle Negie* or *Ye de Neige*. 'Neige' is French for 'snow', so the name would translate as 'Island of Snow', or 'Snowy Island'. But in this tropical location it is clearly not a reference to snow at all, but rather a figurative term for white sand or something similar. We have come across a similar allusion to white sand in the Vallard map's name for Botany Bay—*Baia Neve*, 'Bay of Snowfields'.

There is one seemingly obvious candidate for 'Snowy Island' on a modern map—a small island called White Island. This isolated speck, lying far out from the mainland, was given its distinctive name many years ago because roosting sea birds had covered it in white guano. It would probably be sighted by any ship sailing at a safe distance from this dangerous stretch of coast. White Island, however, lacks one vital detail, a large, elliptical-shaped reef which Rotz has drawn around his *Isle Nege*. This feature, on the other hand, bears a striking similarity to the extensive coral reef which we noted earlier encircling High Cliffy Island. A similar large but unnamed island or reef which could perhaps represent the Montgomery Islands is also shown in the Vallard map in approximately the same area. Opposite it is the legend *Tres Illes*—'Three Islands'.

Could *Isle Nege* possibly be High Cliffy Island, site of the mysterious stone 'beehive' structures? As a landmark, High Cliffy Island is neither snowy nor even particularly white, but there is nevertheless one thing suggesting a possible linkage. That clue is to be found in the fact that chalcedony, the semi-precious stone which the island produces, is also known as white agate. Could sixteenth century Portuguese mariners have learned of the existence of this snowy, quartz-like gemstone on High Cliffy Island? Could they even have attempted to trade for it? And could this in turn be linked to the purpose of the island's mysterious stone round houses? At present we can only pose the questions; further archaeological research may perhaps one day reveal answers.

If we turn now to the Vallard map's depiction of the entrance to King Sound we can readily identify the headland of Cape Leveque, where the buccaneer William Dampier, the first Englishman to set foot on Australian soil, made his landfall on the coast of New Holland in 1688. Just inside this headland (unnamed in the Vallard map) lies one of the main objectives of Bernard Ryder's expedition—the bay where Dampier's pirate ship, the *Cygnet*, was careened for repairs.

From Cape Leveque onward the coastline is shown in the Vallard map continuing, correctly, toward the south-west. Two bays are named. The Portuguese voyagers evidently decided they were river estuaries, because on the map they are called *Rio Palano* and *Rio Meriqna*. The first name should probably be 'Rio Plano', meaning 'Smooth River'. The second is more difficult to decipher; the closest modern Portuguese word is probably 'meritorio', giving the meaning of 'Meritorious River'. Their present-day equivalents would be Pender Bay and Beagle Bay. A short distance further on we come to a headland called *Cap de Vert* (Cabo Verde or Green Cape). This cape is shown in approximately the position of the headland of Coulomb Point. Interestingly enough, Coulomb Point is still green today—its bushland protected as a nature reserve.

From *Cap de Vert* onward the Vallard coastline correctly turns almost due south in the direction of Roebuck Bay (named after Dampier's second ship) and the port of Broome, noted historically for its pearling industry.

Gantheaume Point, at Broome, seems to be the feature indicated on the map as *Cap Mallo*—'Bad Cape'. Why, we might wonder, was it given this rather ominous name? Did something go wrong here? Did the ship, or ships, charting the coast run into difficulties attempting to clear this headland, perhaps because of adverse winds or currents? It would not be at all surprising, for Roebuck Bay is a place of fierce, dangerous currents. We can only speculate on the reason—but the fact is that about here the Vallard map, along with all the other Dieppe maps, takes a very puzzling turn.

Off *Cap de Vert* in the Vallard map we can see a fairly large island with the name *Illa Bassa*—'Low Island'. The very distinctive shape of this island—in particular the prominent pair of horns depicted at its southern end—leaves not the slightest doubt that this is supposed to represent Barrow Island. Some thirty kilometres long, Barrow Island is indeed low-lying, just as the Portuguese name indicates. Today it is a major location for offshore oil drilling. A much smaller island shown to the north of it is in the right place to be one of the Montebello Islands, where the British government conducted controversial nuclear weapons tests in the 1950s. The big problem, though, is that Barrow Island and the Montebellos are in reality to be found not off Coulomb Point (*Cap de Vert*) but about 700 kilometres to the south-west, off the coast of the Pilbara region. What are they doing here, so far from their correct position?

The answer seems to lie in the fact that an entire section of coast roughly equivalent in length to the displacement of Barrow Island has been omitted from the Vallard and other Dieppe maps. This missing coastline, including what is today known as the Eighty Mile Beach, is a low-lying, featureless ribbon of coast on the western edge of the Great Sandy Desert which stretches in a great curve for nearly 300 kilometres from Lagrange Bay, south of Broome, almost to the present-day town of Point Headland. Offering no shelter or safe anchorage it would have been a particularly dangerous coast to approach, and any prudent navigator would have steered well clear of it. Quite probably the Portuguese did not chart this coastline at all—or if they did try to draw it on a portolan, perhaps this chart was so devoid of any features that the Dieppe cartographers discarded it.

Whatever the explanation, a further complicating factor is that beyond *Cap Mallo* (Broome) and Roebuck Bay the coast ought to continue south-west for about 1000 kilometres until it reaches North West Cape. Instead, what is depicted of it in the Dieppe maps abruptly swings off in a completely different direction, south-*east*. What can the reason be?

Judging from the way we know these maps were compiled, the likely explanation is that this is where two separate portolan charts were joined. In other words, the error came about in a similar way to the mistake which led to the creation of the fictitious eastern promontory in the Dieppe maps of Australia's east coast. Perhaps misled by the fact that the northern shore of Roebuck Bay trends south-east for some distance, the original Dieppe cartographer (possibly Jean Rotz) who put together the first composite map of the west coast from individual portolans seems to have made a drastically wrong assumption about how the next portolan should be joined at this point—and all the later French map-makers followed suit.

And it is not difficult to see an important factor which could well be behind this error. Thanks to their blunder in assuming that the north coast of Australia was one and the same as the north coast of Java, the Dieppe cartographers had shifted the northern end of Australia's west coast a huge twenty-five degrees west of its true position. If the section of coast shown in the portolan we are now about to consider had been drawn in its correct alignment, running south-west, this would have vastly compounded the original error. On the Dieppe maps it would have pushed the westernmost point of Terra Java (Australia's North West Cape) far out to a spot near the middle of the Indian Ocean directly under the southern tip of India.

For the cartographers this would have created a very awkward situation. How could it be that a continental coastline lying in this vicinity had not been run into, or even sighted, by any of the numerous European ships traversing the Indian Ocean, including the two French vessels *Sacre* and *Pensée* on their homeward voyage direct from Sumatra to the Cape of Good Hope? Clearly something was wrong. The answer had to be that at about its mid-point the coast of Terra Java must turn east!

How the jigsaw of the west coast portolans *should* have been put together (ignoring the confusion with the Java coast) is shown in the accompanying sketch map. Making allowance for the missing section of coast beyond *Cap Mallo* (Broome), the eastward trending section is pivoted through about ninety degrees so that it now faces in the correct direction, south-west. The adjoining and final southern section of coast, assumed to be a separate portolan beginning at a wide estuary named *Rio Bono*, moves westward with it, but without changing its orientation. Once this adjustment to the alignment of the portolans has been made, virtually all of the remainder of the map (along with its Portuguese-origin placenames) now accurately matches the coastline of south-western Australia.

Now that a prominent, finger-like headland in the Vallard map called *Cap Capar* points north instead of south-west it becomes evident that this is in fact North West Cape. The meaning of the Vallard name is ambiguous. The Portuguese word *capa* does mean 'cape' but in the sense of 'cloak'. As a descriptive name for a headland this is not very convincing. There is, however, an alternative and it relates to the Tropic of Capricorn, a reliable reference point in all the Vallard maps. The meridian of Capricorn is shown on the sixteenth century map passing just north of *Cap Capar*. In reality Capricorn crosses the coast a little below North West Cape, but once the alignment of this section of the Vallard coast has been corrected the line of the tropic comes very close to its correct position. This suggests that 'capar' in this context does not mean 'cloak' at all, but is in reality a garbled abbreviation of the Portuguese word for Capricorn—'Capricórnio'.

It now also becomes clear that Exmouth Gulf, a large inlet on the eastern side of North West Cape, is shown in the sixteenth century chart with name *Gouffre S Fransois*—'Gulf of St Francis'. The importance of this Saint's name in the Portuguese Christian calendar suggests that the ship charting this section

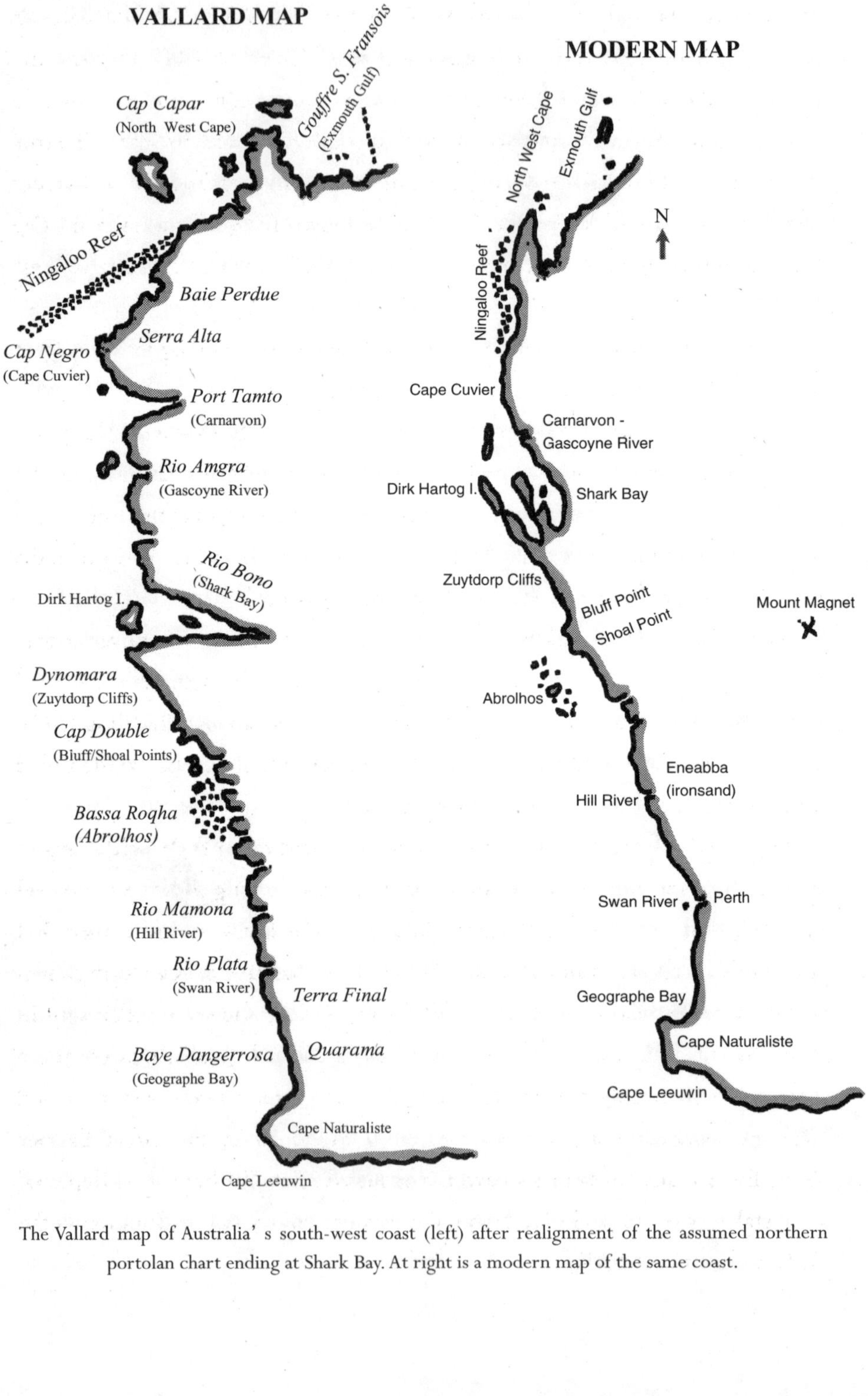

The Vallard map of Australia' s south-west coast (left) after realignment of the assumed northern portolan chart ending at Shark Bay. At right is a modern map of the same coast.

of the coast may have been here on 4 October, the feast day of St Francis. It is in this vicinity, off the north-western end of Exmouth Gulf, that Barrow Island (*Illa Bassa*) should have been depicted. But taking into account the omission of the entire coast between Exmouth Gulf and Broome, Barrow Island as depicted in the Vallard map is not really all that far from its correct place. An island with the name *Illa da Java*, shown lying off the Vallard's *Cap Capar* is in the right spot to correspond to the Muiron Islands, off North West Cape.

Heading south from their 'Gulf of St Francis' the Portuguese voyagers found themselves once more in potentially hazardous waters. The danger is indicated by a strange and very distinctive feature which appears at this point in the Vallard map. It consists of a series of long dotted lines running parallel with the coast for a considerable distance. This feature appears not only in the Vallard map, but also in the Dauphin chart, the Desceliers map of 1550 and a map of 1587 attributed to the cartographer Pastoret.

What can it be? It seems to represent a reef. If so, this should be an important test of the maps' veracity. Is there any reef matching the one depicted in the sixteenth century maps on this part of the west Australian coast? The answer is yes, there is indeed! The strange object in the map can be identified with certainty as the Ningaloo Coral Reef.

Some 260 kilometres long, the Ningaloo is said to be the world's largest west coast coral reef, as well as one of the last surviving pristine coral reef systems. Today a protected marine park, it is the home of more than 500 species of tropical fish and 220 species of coral. Annual visitors to its warm waters include whale sharks, the world's largest fish, known to reach lengths of up to eighteen metres and weigh up to 15 000 kilograms. Despite these impressive statistics, though, the Ningaloo has always been overshadowed by the world-wide fame of its east coast big brother, the Great Barrier Reef. Even many Australians have never heard of it. Yet here it is, depicted unmistakably on maps derived from Portuguese voyages of exploration in the early sixteenth century!

In the Vallard map, the reef is shown running in more or less a straight line instead of following the uneven contours of the coast. It is also depicted extending somewhat further south than is actually the case, suggesting that the Portuguese navigators were content to stay well offshore and make a cautious guess as to where the reef ended, rather than risk approaching dangerously close to it. The Dauphin and Desceliers maps both provide a little more detail, better depicting the reef's curving outline. In the Dauphin map the mainland here has been given the name *Coste Blanche*—the 'White Coast'. Desceliers, in his 1550 map of western Australia, also calls it *Coste Blanche*, though in his revised map of 1553 he changes the name to *Coste Darenes*—'Coast of Sands'. Either description would be absolutely correct for a coastline noted above all for its brilliant white coral sand.

Examining the Vallard map again, we find that a coastal indentation behind the Ningaloo Reef has been given the enigmatic name *Baie Perdue*—a name which we came across on the coast of the Great Barrier Reef and which, as we saw there, could mean either 'Lost Bay' or 'Shipwreck Bay'. It suggests that a vessel encountered difficulties of some kind in this area, or may even have been wrecked here. Was the problem bad weather, the coral reef—or perhaps both? Unfortunately there are no further clues. A prominent headland near the southern end of the reef with the name in the Vallard map of *Cap Negro* (Black Cape) appears to be Cape Cuvier, although present-day maps show just north of Cape Cuvier a headland of a different colour—Red Bluff. A short distance behind this barren stretch of coast, but invisible from the sea, lies a vast salt lake

Some 100 kilometres south of Cape Cuvier is the town and fishing port of Carnarvon, at the mouth of the Gascoyne River. The 865 kilometre-long Gascoyne is the longest river in western Australia, so we might reasonably expect to find its equivalent in the Vallard map. And sure enough, a large river is indeed shown emptying into the sea on this stretch of coast. It has the name, highlighted in red ink, of *Rio Amgra*, meaning 'River Bay'. Not far away is another name, also in red, which appears to be *Port Tamto*—possibly a misreading of 'Porto Tamanho', meaning a 'large' or 'impressive' port. This would be a reasonable description for Carnarvon harbour, which is the only port worthy of the name on this stretch of coast. In this vicinity also is the legend *Serra Alta*, or 'High Range'. This seems most probably a reference to the Brown Range, some twelve kilometres inland from Carnarvon, although intriguingly enough, further inland lie several other taller mountain ranges, one of which has, in English, the very same name as that bestowed by the Portuguese: High Range!

Sailing further south again the Portuguese voyagers soon found themselves at the entrance to a large gulf to which they gave the name *Rio Bono*, or 'Good River'. There is one, and only one, feature on Australia's west coast with which this can possibly correspond, the wide expanse of Shark Bay. In the Vallard and other Dieppe maps with their original orientation, the identity of this 'river' is particularly baffling because it is shown running toward the sea in entirely the wrong direction for Shark Bay, south-west instead of north-west. Once it has been reoriented with the rest of this section of the map, however, it suddenly begins to make sense.

Shark Bay was given its name by William Dampier, who observed when he sailed into it on his second visit to what was then known as 'New Holland' in August 1699 that its waters were seething with shark fins. In the Vallard map of this prominent coastal feature, a large island shown lying across the mouth of the bay is almost certainly Dirk Hartog Island, western Australia's biggest island. The island depicted in the Vallard map is considerably broader than Dirk Hartog Island, but a reasonable explanation for this would be that it is incorrectly shown as including the higher parts of Peron Peninsula, lying immediately behind it when viewed from out at sea.

Dirk Hartog Island was named after the Dutch sea captain who landed here in October 1616. Dirk Hartog is the first Dutchman known to have set foot on Australian soil, and to commemorate his visit he erected a wooden post to which he nailed an inscribed pewter plate. Hartog's plate was removed in 1697 by his countryman, Willem de Vlamingh, who replaced it with a plate of his own (now in the Western Australian Maritime Museum at Perth). Dirk Hartog may have been the first Dutchman to land on this island but the evidence of the Vallard and other sixteenth century maps indicates he was almost certainly not the first European to sight it. Pierre Desceliers' 1550 map also depicts what appears to be Dirk Hartog Island, lying a little to the north of an unnamed cape, probably what is now called Zuytdorp Point.

For the Portuguese voyagers, Shark Bay (their 'Good River') would have provided not only a sheltered anchorage but also a welcome and varied supply of seafood for reprovisioning. In addition to Dampier's sharks, the 150-kilometre-long bay teems with dolphins, turtles, whales and over thirty species of edible fish. The Portuguese may perhaps also have sighted some of the bay's colonies of dugongs (gentle marine mammals which live entirely on a vegetarian diet of seagrass). It was the fondness of these creatures for basking on reefs—though on the opposite side of the Indian Ocean off the shores of East Africa—that in an earlier century had given rise to Arab sailors' legends of mermaids.

By the time William Dampier made his second visit to New Holland, the ex-buccaneer had assumed a mantle of respectability and been given command of a Royal Navy warship. He found the coast in the region of Shark Bay barren and waterless. It was after landing further north that he made the often-quoted comment dismissing its naked black inhabitants as 'the miserablest People in the world'. But Dampier had arrived at the height of the dry season when conditions were at their most inhospitable.[6]

Shark Bay is a complex geographical feature and the original chart drawn by the Portuguese may well have depicted something very different to the simple, V-shaped estuary shown in the Dieppe maps. Its correct shape, if indicated on the portolan, would probably have given the French cartographers a major headache as they attempted to fit it into their wrongly oriented map, particularly as this seems also to be the break point where the final southern portolan begins.

Australia's north-west coast as far south as Shark Bay was a particularly difficult coastline to chart, as all the early navigators agree, and the outcome as shown in the Dieppe maps is in parts confusing. Some sections even appear to have been drawn on a larger scale than others. But the southern section of the west Australian coast which we are now coming to runs in one direction only—almost due south. Its shape is relatively straightforward and much easier to depict. In the Vallard map, even at a glance this stretch of coastline almost 1000 kilometres long is instantly recognisable.

On this southern stretch of coast, moreover, an unusual feature which appears in all the Dieppe maps can be identified with absolute certainty. This is because in the sixteenth century maps it has exactly the same name that we still find on modern maps of this part of the Australian coast—Abrolhos.

The Houtman Abrolhos, to give it the full name by which it is usually known, is a particularly dangerous 100-kilometre-long stretch of low islands and reefs lying off the coast near Geraldton, a town some 420 kilometres north of Western Australia's capital city, Perth. Houtman was the name of a Dutch captain, Frederik de Houtman, who commanded a fleet of eleven ships which sailed from Holland for the East Indies at the close of the year 1618. After crossing the southern reaches of the Indian Ocean, on 29 July 1619 de Houtman's flagship, the *Dordrecht*, sighted at latitude twenty-eight degrees twenty-six minutes a chain of low islands surrounded by coral reefs.

De Houtman cautioned in his report to the East India Company that the area 'lies most treacherously for ships that want to call in at this land'. Unfortunately his warning did not prevent another East Indiaman, the *Batavia*, from being wrecked on one of the Abrolhos reefs just ten years later—an event which led to mayhem, rape and murder. After the *Batavia's* captain, Francisco Pelsaert, had sailed to *Batavia* in the ship's longboat to get help, a religious fanatic by the name of Hieronymus Cornelisz proclaimed himself 'governor' and began a reign of terror among the 250 crew and passengers left on the Abrolhos. By the time Pelsaert returned with a rescue ship three months later, Cornelisz and his henchmen had massacred 125 of their fellow-survivors, including twelve women and seven children. Taken prisoner by soldiers from the relief vessel, Cornelisz and seven other leading mutineers were hanged at the scene of their crimes after having their hands cut off.

A second Dutch ship, the *Zeewijk*, came to grief on the same reefs in 1727. This shipwreck ended more happily, with the survivors managing to reach *Batavia* a year later in a makeshift vessel which they built from the *Zeewijk's* timbers.

But notwithstanding these early Dutch connections, the place name Abrolhos is pure Portuguese. The word has no connection with the Dutch language. 'Abrolhos' is a compounding of two Portuguese words: *abr-*, from the verb *abrir*, meaning 'to open', and *olhos*, meaning 'eyes'. By the sixteenth century *abrolhos* had become a common phrase in the language of Portuguese navigators. Its English equivalent would be the injunction 'keep your eyes open!' In nautical terms it was a warning of dangerous reefs and shoals.

Another significant but little-known fact is that it was Portuguese captains, not the Dutch, who actually pioneered the southern route from the Cape of Good Hope to the East Indies. Portuguese carracks began sailing this faster route in the mid-1500s, half a century before the ships from Holland. The carrack *São Paulo* (referred to in Chapter 3), for instance, sailed to within about 900 nautical miles of the Australian coast in December 1560 before turning north. How close some other Portuguese ships may have come to Australia is a matter for conjecture, but given the difficulty of accurately determining longitude it would not be surprising if a few of them had inadvertently sailed to within sight of Australia's western coast and its treacherous reefs. William Dampier, more than a century later, recorded in his journal that he was very careful to keep *his* eyes peeled as he neared the treacherous 'Abrolho shoal' on his 1699 voyage to New Holland.

As for the Portuguese name for these reefs, if we look at other areas of the Vallard map which includes Western Australia, we find that it depicts and names another 'Abrolho' (the singular form of 'Abrolhos') off the south-east coast of the Malay peninsula, proving once and for all (if further proof be needed) that the name 'Abrolhos' confirms a Portuguese, not a Dutch, discovery. It is also an interesting fact that modern world atlases list two Abrolhos. One is the Australian Abrolhos, while the other is the Arquipélago dos Abrolhos, located off the east coast of Portugal's former colony, Brazil. The Arquipélago dos Abrolhos is a line of low islands lying dangerously in wait for ships sailing south from Portugal toward the port of Rio de Janeiro, just as the Australian Abrolhos lay treacherously across the path of wayward East Indiamen following the southern route from the Cape to Java. Sixteenth-century Portuguese-origin maps of the Indian Ocean also depict an 'Abrolho' lying between Madagascar and the Maldives.

It is also significant that 'Abrolho', the singular form which Dampier used, appears with this identical spelling in the last of Desceliers' maps of *Java la Grande*, the 1553 edition, in exactly the right place—roughly three-quarters of the way down the western Australian coast and next to a group of offshore islands.

The equivalent location in the Vallard map does not have the name 'Abrolho(s)' but the evidence here is hardly less convincing. At this spot the map depicts not only islands but also a large, dotted area—a symbol which in Portuguese charts usually denoted reefs or shoals. Close to this feature is the name *Bassa Roqha*. 'Bassa' (in modern Portuguese written 'baixa' and pronounced 'baisha'), means either 'low' or, in the nautical context 'shoal', while the old Portuguese word 'roqua' corresponds to the modern 'rocha' or 'rock'. Hence the name in the Vallard chart, clearly derived from Portuguese, has the meaning 'rocky shoal', exactly the right description for the Abrolhos.

What is evidently a Gallicised version of this same name, *Basse Roque*, appears at the same spot in two 1587 charts attributed to the Dieppe map-maker Pasterot. At the northern end of its equivalent of the Abrolhos islands the Pasterot maps also have the name *Coste bracque*. In modern French spelling, 'braque les yeux' means 'fix your eyes on'. So this would seem to be the cartographer's own translation of 'abrolhos'. In the original Portuguese chart the name may perhaps even have been a composite one: *bassa roqha—abrolhos* (rocky shoal—keep your eyes open!).[7]

What all this adds up to is clear evidence that Portuguese navigators charted and named the Abrolhos at least seventy years before the Dutch set eyes on it. As usual the Portuguese tried to keep their discoveries secret, but one way or another word of them got out. Frederik de Houtman's father, Cornelius, was thrown into prison in Lisbon as a spy in 1594 after informers reported that he had been overheard questioning seamen about matters concerning navigation to the East Indies. He was released only when the merchants of Amsterdam agreed to pay a huge fine. Houtman repaid his countrymen by leading the first Dutch squadron of four ships to the Indies the following year, with the help of Portuguese charts showing the sea routes to the Indies.[8] As for Frederik de Houtman, he is said to have sailed in Portuguese ships prior to his appointment as a captain with the Dutch East India Company, so it seems likely that this was how he first came to know about the southern route and the Abrolhos. When he sailed that way later under the Dutch flag he, too, may well have been in possession of a purloined Portuguese chart.

If any further confirmation of the Portuguese discovery of this coast should be needed, it can be found just north of the Abrolhos, where the Vallard map delineates another very distinctive coastal feature—a prominent cape with twin headlands and the name *Cap Double* (Double Cape). The same feature can be seen in the Desceliers and Pasterot maps. Does such a cape exist north of the Abrolhos? The answer is again in the affirmative. 'Double Cape' corresponds to the twin headlands of Bluff Point and Shoal Point, some twenty-five kilometres apart and prominent landmarks on an otherwise somewhat featureless coast.

Immediately north again of *Cap Double* the Vallard map indicates a long, unbroken, slightly curving coastline with the name *Dynomara*. This is unmistakably the region of the Zuytdorp Cliffs, a forbidding stretch of coast whose name commemorates another Dutch shipwreck, that of the East

Indiaman *Zuytdorp* in 1712. Bound from Cape Town to Batavia, the *Zuytdorp* is thought to have sailed full tilt into these cliffs at night when its crew, depleted by scurvy, were too exhausted to keep a proper lookout. The name on the Vallard chart is evidently a garbled version of the Portuguese words *duna [do] mar* or 'sea dune', an apt description for this undulating line of cliffs which, seen from out at sea, resemble high sand dunes glowing a vivid yellow in the late afternoon sun.

The *Zuytdorp's* wreckage at this remote spot lay undiscovered for two centuries until a roving stockman came across it in 1927. Australian archaeologists found the wreck site carpeted with thousands of Dutch silver coins. They also found signs that some of the crew had survived the shipwreck and climbed to the top of the cliffs. The evidence comprised glass bottles, clay pipes, barrel staves and the remains of nautical instruments. How long the castaways waited vainly for rescue, as well as their eventual fate, still remains a mystery.

With this section of the Portuguese-derived charts firmly anchored on modern maps we can proceed with some confidence to identify most of the remaining major coastal features further south. There are more surprises in store.

At the southern end of the Abrolhos region the Vallard map names another *Terra Alta*, or 'high land'—probably the range of hills behind Geraldton, rising to a height of some 200 metres. This brings us to an intriguing place name which provides yet more compelling evidence of a remarkable Portuguese voyage of discovery. The feature depicted is a river. Its name could possibly be *Mamora*, but the final consonant looks more like an 'n', making it *Rio Mamona*. In Portuguese there is no word 'mamora' but there is a noun 'mamona', an unusual word which can have two quite different meanings. The first and most common one is 'castor-oil plant', but this would seem to have no possible relevance for a ship sailing off this barren coast of Western Australia. The second meaning as defined in Portuguese dictionaries is a rarely used and very specialised one: 'magnetite dust' or in common language, ironsand.

Could this be the 'Magnetite River'? At first the idea might seem to strain credulity. But wait. Examination of a present-day map of this region demonstrates the fact that truth can at times almost be stranger than fiction. For inland from this possible 'Magnetite River', if we look at a modern map one place name instantly stands out—Mount Magnet!

Situated near the Great Northern Highway linking Perth and the vast iron mining region of Pilbara, Mount Magnet is the centre of a huge conglomeration of magnetic iron ore. Rivers flowing south of the town of Geraldton from the direction of this magnetite country carry black iron grains toward the sea and magnetite ironsand occurs in many places along the coast between Geraldton and Perth. *Rio Mamona* appears to be the Hill River which flows into the sea just south of the small fishing port of Jurien. Given the demonstrated accuracy of the Vallard map elsewhere it should come as no surprise that the Hill River rises close to an area called Enneabba which has the greatest concentration of magnetic ironsand in this entire region.[9]

The Portuguese voyagers could have given the name *Mamona* to this river mouth for one of two reasons. The first possibility might be that they landed here and found ironsand on the beach but this seems unlikely, given the forbidding nature of this coast. The other possible explanation and almost certainly the correct one, is that they found black magnetic iron particles sticking to the grease which they would have put on the bottom of their sounding lead. Determining the nature of the seabed in this way had long been an important navigational technique.

And here we might pause to consider whether in this meticulous recording of a mineral discovery we might detect the hand of Christopher de Mendonça, for the circumstances would seem to have an intriguing similarity to the Portuguese captain's encounter off the coast of Queensland with small pieces of floating pumice, which he recorded in the place name *Pomezita*—'Small pumice'.

The Mount Magnet region was in the earlier years of European settlement also a thriving gold-mining area, a fact which could make us wonder about another river shown on the chart further up the coast opposite the southern end of the Abrolhos. This is possibly the Greenough River, which originates about 100 kilometres west of Mount Magnet and flows into the sea just south of Geraldton. On the Vallard map the river's name somewhat indistinct name appears to be either *Rio Andrio* or *Audrio*, neither of which has any immediately obvious meaning. But if any traces of alluvial gold had been found in the vicinity the name might conceivably be a garbled version of *Rio Aureo*—'Gold River'.

Whether or not the Portuguese voyagers had come across any gold, there can be no doubt about the significant name they gave to the next and final river depicted on this coast. Its name in the Vallard map is *Rio Plata*—the same name given to the river dividing Argentina and Uruguay. In English, the River Plate.

The Argentine river's full name in Spanish is *Rio de la Plata*, or 'River of Silver'. It was given this name in 1512 because the captain who discovered it, John de Solis, a Portuguese in the service of Spain, believed he saw signs there of silver.[10] Interestingly, the Vallard Atlas features on another of its pages a map of the south-west coast of South America and here the Argentine river's name has the shortened form identical to that of its west Australian counterpart—*Rio Plata*.

Where on a modern map should we look for this hitherto unknown Australian River Plate? Does it even exist? The answer is that it is none other than the Swan River, site of Western Australia's capital city, Perth! The Vallard map puts *Rio Plata* in exactly the right place relative to other identifiable west Australian coastal features. And again correctly, it is also shown as the last major river on this coast. We can be quite certain that this is not some absurd mistake because the same name in Gallicised form can be found in the two later Dieppe maps attributed to Pasterot. Here, the name given to the river is *Riviere Plato*.

In the Vallard (as with the Pasterot maps) there is no sign of the low-lying Rottnest Island—named by Dutch mariners who mistook a species of pygmy wallaby for rats—which lies off the entrance to the Swan River. However, in the generally cruder 1550 and 1553 Desceliers maps of the same coast, a final, mysterious coastal feature *does* depict in this region, and in considerable detail, a gulf which has a large river and several islands, labelled *Baye Dangerrosa* (Dangerous Bay).

The name is significant because the French expedition of Nicolas Baudin in 1801 also found this bay to be dangerous. After Baudin's two ships had anchored here to explore the coast, a sudden storm caught them perilously exposed on a lee shore. A longboat was wrecked on the beach and it was only with the greatest difficulty that the ships' officers managed to extricate their vessels and avoid a disastrous, premature ending to their voyage.

Although only in the form of a shorthand sketch, *Baye Dangerrosa* in the Desceliers' maps is tolerably correct in its main elements representing an open bay which extends southward from Perth to Cockburn Sound. The one important feature which at first does not seem right is that *Baye Dangerrosa*, in addition to a scattering of small islets or rocks, is shown having three large islands. In reality there should be only two—Rottnest and Garden Island.

An explanation for this anomaly becomes apparent, however, when we compare the bay shown in the Desceliers chart with a chart of the Perth coast begun in 1801 and completed in 1803 by navigators with Baudin's expedition. Surprisingly this French chart, based in part on an earlier Dutch map, also depicts *three* larger islands—Rottnest, Garden Island (*Ile Buache*) and a third, unnamed island lying near the coast north of the Swan River (*Rivière des Cygnes*). Why the extra, non-existent island? Significantly, it is shown lying close to the coast. The explanation would seem to be that from out at sea the French, like the Dutch and Portuguese before them, mistook the prominent Mullaloo Point, north of Perth, for an island encircled by reefs.

Did Portuguese explorers sail into the Swan River? The shallow entrance would have been difficult to navigate, as Baudin discovered, but the very fact that the Portuguese voyagers named this waterway 'Rio Plata' suggests that as with de Solis in South America, they must have ventured into the estuary in order to have seen what they thought were signs of silver.

According to Australian geologists no evidence of silver is to be found in the vicinity of Perth. There is, nevertheless, one curious fact worth considering. The hills in which the Swan River has its source are the site of one of Western Australia's earliest mines, not of silver, but lead. And it is a geological fact that where there is lead there is also the potential to find silver.

South of Perth the Vallard map offers only two more names. The first is *Terra Final* (Last Land)—a clear indication that the Portuguese expeditioners were aware that they were sailing past the most southerly shores of this vast island continent. The final name on the Vallard map is a single word, *Quarama*. It appears at about the position of the present-day port of Bunbury, at the northern end of Geographe Bay. The modern Portuguese equivalent of this word is *Quaresma*, meaning Lent. Depending on the calendar, the season of Lent begins in either February or March. So this was apparently the time when the Portuguese voyagers were nearing the southernmost point of western Australia.

Unmistakable in the Vallard map is the broad headland depicted at the south-western corner of Australia, a headland which in modern maps somewhat resembles a boot, with Cape Naturaliste at its north-facing toe and Cape Leeuwin its heel. Cape Naturaliste and Geographe Bay were both named by Baudin to commemorate his expedition's ships, but as we now know, this was nearly 300 years after the Portuguese had first sighted them. Rhumb lines radiate in all directions from the Vallard's equivalent of Cape Naturaliste.

Beyond the high land of Cape Leeuwin[11] the coast now begins to turn east toward the Great Australian Bight, fading away behind an artist's sketch of a forest. The immensity of the unknown Southern Ocean must have been an awesome sight. This was surely the place the Portuguese defector Jean Alfonse had in mind when he wrote of that mysterious southern region of his 'continent of Jave':

> The coast of this cape turns south for forty leagues and then turns straight to the south-east. No one has yet gone beyond to make

> discoveries in that distant sea, but it is believed it goes to the south-east coast of the land of Magellan.[12]

Alfonse's surmise was correct. The stormy ocean south of Cape Leeuwin does indeed offer a sea route to the far-off tip of South America. In the Vallard Atlas, it must be added, there is another very faint line running to the bottom of the page which suggests that the map's cartographer may have been reluctant to let go entirely of the popular belief that Terra Java must be attached to a great Southern Land linking both sides of the Strait of Magellan. But in contrast to the *real* coast of west Australia, this tentative extra length of 'coastline' has the appearance of nothing more than a vague assumption, sketched in unconvincingly and lacking either any detail or a single place name. As with his fudging of the fictional 'Rio Grande' south of Java, once again our cartographer may well have been having a bet each way.

What seems clear enough from the Vallard map is that the Portuguese mariners sailed as far as Cape Leeuwin and then turned back. Was there anything in particular that led to this decision? On this question the map could have something dramatic to tell us.

Out in the Indian Ocean, off the map's equivalent of Cape Naturaliste, we can see amid the figures of fantastic sea monsters a lifelike illustration of a ship under sail. Its appearance is that of a small carrack. At first glance it looks just like the drawings of numerous other ships, large and small, which adorn the pages of the Vallard Atlas. But a closer look reveals something unexpected about this particular vessel—its mainmast has snapped off! The jagged stump of the mast can be seen still in place amidships, and what is evidently the main yard (the long spar on which the mainsail would have been hoisted) has been lashed to the side of the vessel.

Is this just a whim on the part of the illustrator or is it meant to tell a story? Another picture of a dismasted vessel can be found in the Vallard Atlas off the coast of Brazil. It is not impossible that the Dieppe cartographer could have had access to accounts of voyages, either orally or from mariners' log books. Could this particular illustration on the map of Terra Java be telling

Dismasted carrack pictured off the south-west coast of Australia. (*Detail from Vallard Atlas*)

us that a ship exploring the southern reaches of this land was hit by a storm which broke its mainmast? Could this even be one of Mendonça's ships?

That, of course, can only be conjecture. But the illustration clearly shows the vessel limping along bravely under a jury rig of a square foresail and lateen mizzen. The ship is heading northward, away from the perilous expanse of the great Southern Ocean, its captain steering perhaps for the relative safety of Timor or the Portuguese base at far-off Malacca.

A scene which suitably brings to an end this chapter in the discovery of Australia.

11

The ignominious voyage of Jeronimo de Figueiredo

We owe to the Portugals the discovery of such parts hereof, as they fell on in their way to the Eastern Indies; of which we have little but the names, as Psittacorum Terra, so-called from the abundance of parrats which they saw on the Coasts.

PETER HEYLYN, *COSMOGRAPHIE*, 1656

On his return to Portugal in 1524 after his epic voyage to discover the Island of Gold, Christopher de Mendonça was handsomely rewarded for his services to the Portuguese Crown. He had come home to find that King Manuel I, the sovereign who had bestowed on him the honour of leading this voyage, had died three years earlier, at the very time when Mendonça's ships were setting out from Malacca at the start of the momentous journey which would take them to the coasts of Australia. But Manuel's successor, John III, was equally committed to maintaining Portugal's power and wealth.

The recognition bestowed on Mendonça by King John took the form of a lucrative posting as commander of the fort and city of Hormuz, at the entrance to the Persian Gulf. Hormuz was not just any old Portuguese fortress. Chosen for its strategic location, it was also a thriving international emporium at the hub of the great sea routes from India and China, where an astute commander could expect to amass a considerable private fortune.

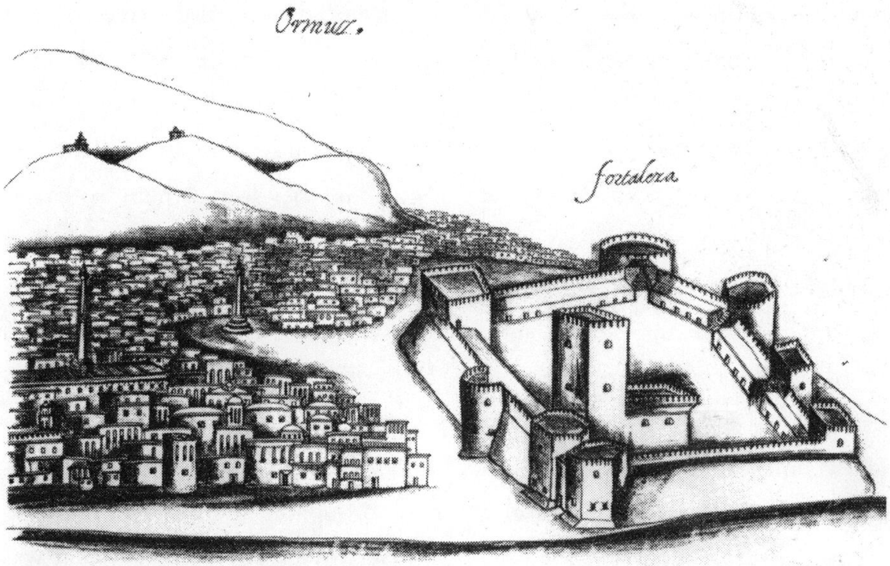

The island city of Hormuz, at the entrance to the Persian Gulf, where Christopher de Mendonça spent his final years. He was given the lucrative post of commandant there as a reward for his Island of Gold expedition. A merchant of Venice who visited 'Ormuz' in 1563 remarked on its wealth and described how 'standing neere unto the waters side [is] a very faire castell, in which the Captaine of the king of Portugull is always resident'. (*From Lendas da India, by Gaspar Correa. Australian National Library*)

Hormuz (formerly 'Ormuz' the name by which the Portuguese knew it) was originally a Persian city. Situated on an island some nineteen kilometres off the southern coast of the Persian (now Iranian) mainland, it was in a position where it could dominate the entrance to the Persian Gulf. Its island location made it secure from land-based attack, but it suffered from the severe disadvantage of having no fresh water. Its water, fresh food and other day-to-day necessities had to be brought across from the mainland and nearby islands.

Added to this disadvantage was the fact that this general region of Persia had a reputation for becoming intolerably hot in summer. Marco Polo, who visited Hormuz during his journey from Venice to China in about 1272 related that 'during the summer season the inhabitants do not remain in the city, on account of the excessive heat, which renders the air unwholesome, but retire to their gardens along the shore or on the banks of the river...' Polo also relates that some years earlier an entire army of more than 2000 men marching across the mainland to attack Hormuz died from heat and thirst after being assailed by the burning desert wind. When the people of Hormuz came to bury the carcases of the dead, he adds, they found them 'so baked by the intenseness of the heat, that the limbs, upon being handled, separated from the trunks...'.

But the problems of heat and lack of water notwithstanding, in Polo's words, Hormuz was a port 'frequented by traders from all parts of India, who bring spices and drugs, precious stones, pearls, gold tissues, elephants' teeth and various other articles of merchandize. These they dispose of to a different set of traders, by whom they are dispersed throughout the world.'

Hormuz might have little to fear from land attack but from the sea it was a different matter. A Portuguese fleet commanded by Afonso de Albuquerque captured the city with little resistance in 1515. Under the Portuguese, in the words of the eighteenth century Edinburgh historian Dr William Robertson, Hormuz became 'the great mart from which the Persian empire, and all the provinces of Asia to the west of it, were supplied with the productions of India; and a city which they built on that barren island, destitute of water, was rendered one of the chief seats of opulence, splendour and luxury in the Eastern World'.[1] It was the richest source of customs revenue in the whole of the Portuguese East.

A vivid eyewitness account of Hormuz under Portuguese rule comes from a merchant of Venice, Caesar Fredericke, who visited the island city in 1563. Translated from Italian for contemporary publication in England, his account tells how 'in this city there is very great trade for all sorts of spices, drugges, silk, cloth of silke, brocardo, and divers other sorts of marchandise.' As for the Portuguese presence, Fredericke observes that 'there is standing neere unto the waters side a very faire castell, in which the Captaine of the king of Portugall is always resident with a good band of Portugalles, and before this castell is a very fair prospect...' The island of 'Ormus' still had a Persian king, the Venetian notes, but he was a king in name only. '[He] is created and made king by the Captaine of the castell, in the name of the king of Portugall.'[2]

At first things went well for Mendonça at Hormuz. Shortly after his arrival there in 1528 he made a voyage to Basra, the southern port of present-day Iraq, on the Euphrates River. From there he sent a courier overland to Portugal with urgent dispatches for King John. The chroniclers tell us that this was the first time this route through hostile Muslim territory had been attempted. It was a hazardous enterprise: the courier had to travel incognito with a camel caravan across the Syrian desert to Damascus, then on to Tripoli (Lebanon), Cyprus and Italy. Had his identity been discovered he almost certainly would have been tortured and executed. But as luck would have it the dispatches reached Lisbon safely in under three months—less than half the time they would have taken by sea around the Cape of Good Hope. King John is said to have warmly commended the enterprise.[3]

Ultimately, however, Hormuz was to prove an unlucky appointment for Mendonça. The next thing we learn about him, in 1532, is that he has just died from disease only four years after his arrival. The exact cause of his death is not stated, but in Portuguese outposts such as this, sickness often took a far greater toll than war. In addition to the oppressive summer heat of Hormuz, two major killers were dysentery and malaria, whose causes were at that time unknown.

As for the significance of Mendonça's 'Island of Gold' voyages, Peter Heylyn, the noted seventeenth century English geographer, entertained no doubt that it was the 'Portugals' who were the true European discoverers of the land we now call Australia. His views on this question appear in an appendix to his monumental, four-volume work, *Cosmographie*, printed in London in 1656. The appendix is entitled *Endeavouring a Discovery of the Unknown Parts of the World, especially of Terra Australis Incognita or the Southern Continent*. In it Heylyn records that the Portuguese had discovered and named a southern land notable for its abundance of parrots, 'Psittacorum Terra'. He suggests that they may also have found the fabled land of Beach, 'a Region yielding Gold and possessed by Idolaters, with the two Kingdoms of Lucach and Maletur'.

Heylyn's list of what he refers to as 'Portuguese' names in the region of his *Terra Australis Incognita* contains a number of names we have come across already in the context of maps derived from Mendonça's voyage. He describes:

> A promontorie called *Cape Hermoso*, in the East parts hereof near the Islands of *Solomon*. Another in the Western part, but as near the Aequinox called Point Primiro. A third in the first bendings of the Coast towards the South called *Cape de Buena Deseada*, or the *Cape of Good Desires*. *Rio de Valcanes*, *Rio de Lorenzo*, Rio de S. *Augustino* on the East of Cape *Formoso* towards the Streights; the Rivers of *S. Peter*, *S. Paul, S. Andrew* and S. *James*—betwixt Cape *Formoso* and *Primiro*'.[4]

Heylyn's 'Primiro' is an incorrect spelling for the Portuguese word 'primeiro', meaning 'first'. As we saw earlier in Chapter 3, a 'Rio Primero' (First River) appears in the Vallard map in about the position of the Daintree River in north Queensland. There is no 'Point Primero' (or Primiro), but on the other hand the very prominent Cape Grafton, just south of the Daintree, has not been given a name. It seems possible that in the original portolan chart the name 'Primero' might have indicated the cape but the author of the Vallard map wrongly assumed it to be the river. The possibility that this cape might have particular importance in the original chart is suggested by the fact that it is a focal point from which rhumb lines radiate.

As for Heylyn's other 'Portuguese' names: his 'Cape Formoso' we have already identified as Wilson's Promontory, the southernmost point of the Australian mainland. 'Hermoso' is the Spanish equivalent of 'Formoso', and since this is said to lie 'near the Islands of Solomon', which were discovered by a Spanish expedition in 1568, we would not expect to find it in the Vallard map—unless Heylyn is confusing 'Hermoso' and 'Formoso'. Rio de Valcanes, Rio de Lorenzo and Rio de S. Augustino do not seem to have equivalents in the Vallard, but this is not surprising since Heylyn says they are 'East of

Cape Formoso towards the Streights [of Magellan]'—which would put them somewhere well out in Spanish territory.

On the other hand, Heylyn's '*Cape de Buena Deseada* or the *Cape of Good Desires'*—located at 'the first bendings of the Coast towards the South' is surely the *Cap Bon Espoir* (Cape of Good Hope) shown in the Vallard map at the location of Point Lookout, the northern headland of Stradbroke Island, off present-day Brisbane. Not only are the two names strikingly similar (bearing in mind that the Vallard name is a French translation), Point Lookout does indeed lie at the 'first bendings' where the east coast of Australia begins to turn definitively toward the south, just as Heylyn describes. And finally, as we have seen, Heylyn's 'Rivers of S. Peter, S. Paul, S. Andrew and S. James' seem also to have recognisable Vallard map equivalents.

One reason why Mendonça's discoveries had to be concealed could well relate to the fact that in 1529 King John paid Spain's cash-strapped King Charles V the considerable sum of 350 000 ducats (in today's terms, possibly in the vicinity of A$350 million) under the Treaty of Saragossa to move the boundary line between the Spanish and Portuguese spheres of influence in the Orient seventeen degrees further east. The ostensible reason for the Portuguese payment is that it eliminated any further Spanish claim to the Spice Islands. And this it certainly did, but unknown to Spain, moving the boundary that distance eastward also gave to Portugal most if not all of the secretly discovered island continent of Australia. And who could tell what riches this vast land might be found to contain in the future?

If we consider the matter in more detail we can see that the existing Treaty of Tordesillas had already provided Portugal with exclusive and hugely valuable trading rights to the entire coast of Vietnam and China, as well as Formosa (Taiwan), the Philippines, and at least the south-western tip of an only vaguely known Japanese archipelago. To the north-east of the Spice Islands, the new line of demarcation agreed under the Treaty of Saragossa gave Portugal only a derisory reward—just a huge arc of open ocean containing nothing more than a scattering of insignificant coral atolls. Immediately to the east was Papua, the 'Land of the Blacks', but this island, although large, was already known to be of little value for trade purposes.

But the Island of Gold—that was another matter! In an Australian context, we can now say that as defined on a modern map, the new demarcation line would have passed just east of Cape York, giving Portugal control of the strategically important Torres Strait. At the time the Treaty of Saragossa was signed, though, the precise longitude of the east coast of Terra Java (Australia) remained undetermined. Mendonça's discoveries on the east and south coasts of Australia were potentially of immense value. And by propagating a fiction that the continent of Terra Java lay directly under the island of Java, Portugal could lay claim not just to most of Australia, but to the *whole* of it!

The Portuguese chronicles tell of only one other attempted voyage to the Island of Gold, some twenty years after that of Mendonça, but it was an ill-planned, botched affair which ended in farce.

In about the year 1541 (the record is not clear as to the exact date) the Portuguese commander at Malacca, Pedro de Faria, came into possession of what he believed was vital new information about the 'Land of Gold'. There are two main sources for this story: one the chronicler Manuel de Faria e Sousa, the other a sixteenth century freelance Portuguese adventurer by the name of Fernando Mendes Pinto. Pinto was an extraordinary character who over a period of twenty-one years travelled to the furthest corners of the Orient, Africa and the Middle East and got into all sorts of scrapes and predicaments. He seems to have led a charmed life. In a book which he subsequently wrote and which quickly became a bestseller, he claimed that he was, 'five times shipwrecked, sixteen times sold and thirteen times made a slave'. Sousa, the serious historian, seems to have drawn on Mendes Pinto's account for much of his own information about the Island of Gold expedition.

Both men say Pedro de Faria had learned that the island was located in 'the ocean of the river of Calandor', five degrees toward the south and 150 or 160 leagues from the tip of Sumatra, surrounded by numerous reefs and fierce currents. The 'river of Calandor' seems to be a reference to the Sunda Strait, between Sumatra and Java. In the sixteenth century this strait also went under the names of the River of Chenano, or Chiama, while in the Vallard map of South-East Asia there is yet another variation: an indistinctly written name which appears to be 'Condor'—a name with an evident similarity to 'Calandor'. This 'Condor' river flows from west Java into the southern entrance of the Sunda Strait. Interestingly, a distance five degrees south from here brings us close to the latitude the northern tip of Arnhem Land. Sailing east at that latitude for 150 or 160 leagues, as the chroniclers also specify, would take a ship toward the north-western coast of Australia.

Pedro de Faria apparently obtained most of his information about the Island of Gold from the ruler of the kingdom of Batas, in south Sumatra, who was seeking urgent military aid from the Portuguese in an ongoing war with the northern Sumatran sultanate of Aceh. The Batas ruler, whose kingdom bordered the ocean in which the Island of Gold reputedly lay, assured the Portuguese that if they came to his aid they would have 'gold in abundance'.[5] Pinto claims in his book that he himself acted as the go-between in these negotiations. So convinced was de Faria about the veracity and importance of his information that he forwarded a written dispatch to King John III in Lisbon urging the speedy exploitation of this great opportunity. The king was apparently impressed, for in due course a royal warrant arrived back from Portugal not only authorising an expedition for 'the discovery and conquest' of the Island of Gold but also decreeing that the honour of leading it should go to a knight of the royal household, Francisco de Almeida, 'a nobleman of experience and valour'.

This Almeida is not to be confused with another more famous Dom Francisco de Almeida who served as Viceroy of India from 1505 to 1509 and was killed in a clash with African tribesmen near Cape Town on his voyage back to Portugal. According to Pinto, the later Almeida (perhaps related to the former viceroy) was granted leadership of the Island of Gold expedition as a reward for many valiant services on the king's behalf in the Spice Islands.

Almeida assembled his expedition and set out from Goa, only to die of fever after his ship reached the Nicobar Islands, north of Sumatra. After this setback the expedition must have been put on hold, for Pinto says that after the news of Almeida's death had reached him in Lisbon, King John ordered that command of the expedition be given to Diogo Cabral, a captain who hailed from the Portuguese island of Madeira. Cabral, however, had enemies in high places and they whispered to the new Governor of India, Martim Afonso de Sousa, that Cabral had personally insulted him. The Governor revoked Cabral's command and gave it to another nobleman, Jerónimo de Figueiredo, of the House of the Duke of Bragança, the same noble house to which Governor Martim Afonso also just happened to belong.

Figueiredo's expedition consisted of a caravel, which would have been his flagship, and two support ships called fustas. The latter were among the smaller Portuguese ocean-going ships, although contemporary descriptions of them vary greatly as to both their dimensions and rig. In general they were long, narrow vessels of shallow draught with a small poop deck, usually covered by an awning. Some are said to have had only one mast, but others had three, with triangular lateen sails on the main and mizzen masts and a large square-rigged sail on the foremast. Fustas also had banks of oars which enabled them to be rowed if there was no wind.

With eighty officers, soldiers and sailors under his command, Figueiredo set sail from the Portuguese naval base at Goa for the 'Island of Gold' in 1542. But 'carried away by his ambition', as Sousa puts it, he forgot his real purpose and detoured to the coast of Burma where he fell in with some 'ships of Mecca'—Arab vessels—which he captured and plundered. He then arrogantly refused to divide the booty among his soldiers ('as they had every right to expect,' adds Mendes Pinto). In retaliation the soldiers mutinied, tied their commander hand and foot, and 'after many other things happened which are better left unsaid' (mutters Pinto) they put him ashore on a beach near Galle, at the southern tip of the island of Ceylon (Sri Lanka), sailing away with the plunder he had denied them.

'And through the guilt of all, but principally the captain,' observes Sousa, 'our king was deprived of the discovery of this island so greatly desired.' Mendes Pinto adds a second pensive footnote. 'From that time on,' he writes, 'no effort has been made to discover this island, which it seems will be very beneficial for the welfare of this Empire if our Lord should some day see fit to let it be discovered'.

Figueiredo's ignominious demise is not surprising if we are to believe the chronicler Gaspar Correia, who describes him as a mischief-maker. It was he, apparently, who informed Governor Martim Afonso that Cabral had cursed him, whereupon Cabral was not only relieved of his command but put in chains.

There is, however, something very odd about the chronology for Figueiredo's voyage. His expedition is said to have set out in 1542. Yet when the mutineers returned to Goa after dumping him at Ceylon, Pinto says they were pardoned by a new Governor of India, Dom João de Castro, in return for agreeing to join an armada being assembled to attack Muslim forces besieging the Portuguese fort at Diu, further up the west coast of India.

What is puzzling about this story is that Castro only took over as Governor in 1545, and the (successful) Portuguese naval attack at Diu did not take place until November 1546. This seems to indicate that at least two, and possibly even as many as three years elapsed between the Figueiredo expedition's departure and the return of the mutineers. Sousa and Pinto are both clear enough about the final course of events: first the trussed Figueiredo was put ashore at Ceylon; then his ships sailed on to Goa. That being so, what happened in the earlier, blank period? Two years would have been ample time for a voyage to the Island of Gold, and for an exploration of its coasts as well. Could this have been what happened? It certainly would make more sense of the piracy episode: returning empty-handed from a long and arduous voyage, it would have been tempting for Figueiredo to try and retrieve some of the expedition's fortunes by seizing the cargoes of Arab merchantmen. And by the same token his long-suffering soldiers and crewmen would have been justifiably incensed at an attempt to deny them their share of the plunder.

In other ways, too, Figueiredo's expedition raises more questions than it answers. King John could hardly have forgotten that the coasts of Terra Java had already been discovered and charted by Mendonça, and perhaps other Portuguese captains. But that would have been a closely guarded state secret to which few if any outside the king's inner circle would have been privy. Was the king simply sending Figueiredo on a fool's errand? Or, since we can be sure that Terra Java would not have produced the hoped-for golden bonanza, had King John been led to believe that Pedro de Faria had some other destination in mind?

Whatever the truth of the matter, Portugal's Spanish rivals were by now discussing the possibility of embarking on a quest of their own to find the elusive Land of Gold. Not content with the huge amounts of treasure already seized in the Americas (much of which had been squandered), by the middle of the sixteenth century Spanish adventurers were turning their thoughts to the Pacific, where they believed that King Solomon's mines—said to lie in Ophir the fabled Land of Gold—still awaited discovery. Their hopes rested on an often-quoted passage in the Old Testament of the Bible which reads:

> And King Solomon made a navy of ships...on the shore of the Red Sea, in the land of Edom. And Hiram sent in the navy his servants, shipmen that had knowledge of the sea, with the servants of Solomon. And they came to Ophir, and fetched thence gold, four hundred and twenty talents, and brought it to King Solomon. [6]

Nowadays the idea seems completely absurd, but in the sixteenth century many Spaniards firmly believed that by sailing westward across the Pacific they would come to King Solomon's mines. The first Spanish expedition of two ships set out from Peru to find Ophir under the command of Álvaro de Mendaña in November 1567. After a voyage of nearly three months they made landfall at a group of islands south of the equator which Mendaña optimistically named the Islands of Solomon. But the adventurers were unable to find the gold of Ophir and after a few months of sagging morale they struggled back to Peru empty-handed.

Mendaña nevertheless remained convinced he was on the verge of a brilliant success, and in 1595, after much lobbying, he finally managed to persuade the authorities to provide him with a fleet of four ships to colonise the Solomon Islands and find the gold he was sure was there. In addition to nearly 300 soldiers and sailors, the Spanish fleet this time carried would-be settlers—378 men, women and children, most notable among whom was Mendaña's wife, the haughty Donna Isabella, whose announced ambition was to find the Queen of Sheba and carry her back in triumph to Peru.

Of course it all came to nothing. The old problem of having no means to calculate longitude correctly led the ships to make landfall not at the Solomon Islands but in the Santa Cruz group, some 200 nautical miles further east. Mendaña firmly believed he had rediscovered the Solomons, but again there was no sign of gold—and no Queen of Sheba. An attempt to establish a colony quickly ended in failure. Many of the settlers died from malaria and other tropical diseases, with Mendaña himself succumbing after naming Donna Isabella as 'governor'. After only two months the survivors gave up and sailed for Manila.

One last cartographic mystery remains to be solved. What could have been the origin of a world map apparently depicting almost the entire coast of western Australia which was published by Heinrich Bunting at Magdeburg, Germany, in 1581, some thirty-five years before the first known Dutch encounter with Australia's west coast? This map, printed from a wood engraving, was purchased by the National Library of Australia in 1985.

Bunting was noted as a theologian rather than a cartographer, and his little map appeared in a popular work he had written, *Itinerarium Sacrae Scripturae*, concerning the supposed travels of the Biblical patriarchs and prophets. Although crudely drawn in comparison with most of the Dieppe maps (it was after all a woodcut rather than an original work of art), its outline of the apparent Western Australia is, in the words of former National Library curator Michael Richards, 'uncannily accurate and in roughly the right place'.

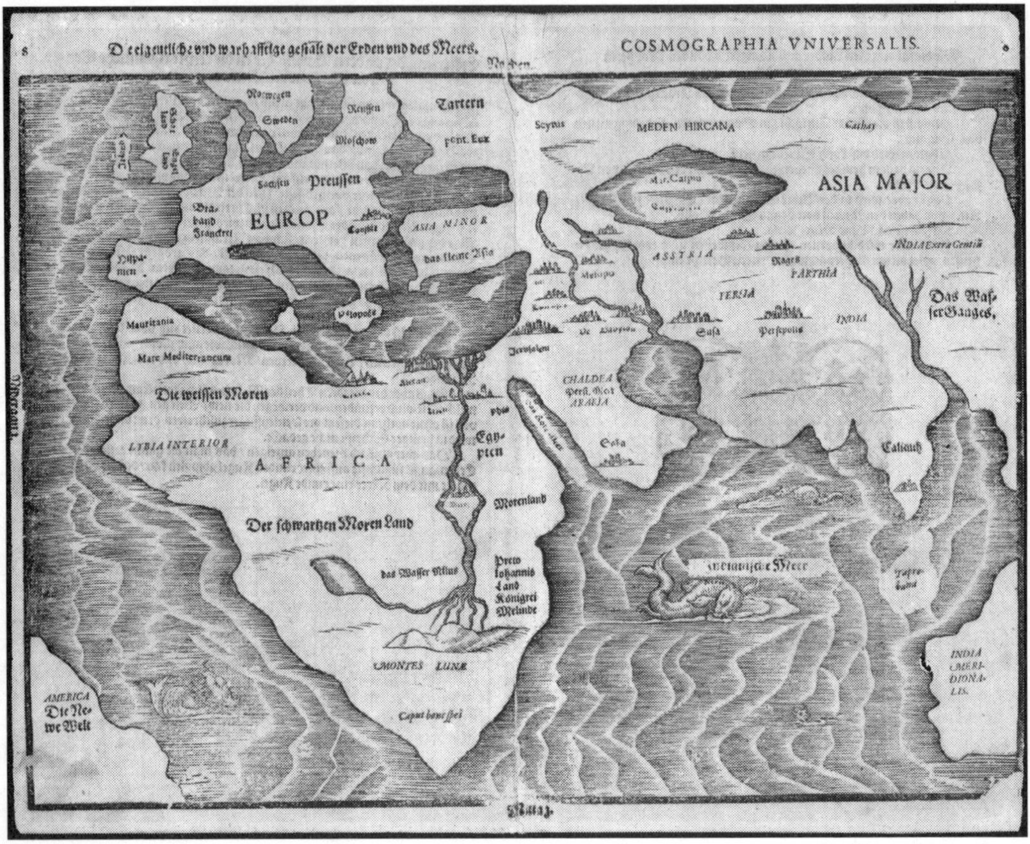

The Heinrich Bunting world map, published in Germany in 1581. Its outline of the coast of western Australia is astonishingly accurate, the more so when compared with the distorted outlines of many other regions in his map, including Europe. Where did Bunting obtain his information – at least thirty-five years before the first Dutch encounter with Australia's west coast? (*National Library of Australia rm2430*)

The point about the map's accuracy is that it does not contain the major errors perpetrated in the Vallard and other Dieppe maps of the same coastline—errors arising from the fact that all of these maps depict western Australia as a southward extension of the western coast of the island of Java. In the Bunting map, the entire west coast of Australia stretching from Cape Londonderry and the Kimberleys in the north to Cape Leeuwin in the far south is instantly recognisable—so much so that no objective viewer could reasonably dismiss it as the product of chance. The prominent North West Cape and the deep indentation of Shark Bay further south can both be seen.

Overall, this land mass which Bunting calls *India Meridionalis*, meaning 'India of the Southern Hemisphere', is far more recognisable than, for instance, the British Isles, which appear in the same map. And his chart goes further even than the Vallard map in depicting also a section of what is evidently the south coast of Australia from Cape Leeuwin extending approximately as far as the Recherche Archipelago, near the western end of the Great Australian Bight.

Bunting clearly was not drawing on the Dieppe maps for his information—so where did he get it from? In terms of style his map is not unlike the rough charts drawn by Jean Alfonse, the Portuguese master mariner who defected to France. Could some unpublished work by Alfonse perhaps have been his source? Curiously enough, although described as a world map, the Bunting map shows only half the world, and that half happens to be the hemisphere allocated by papal decree to the Portuguese. In the Americas the map depicts only the Portuguese territory of Brazil, while in the East it ends near to the present-day border of the state of Western Australia, which in turn lies close to the meridian of longitude originally demarcating the eastern boundary of the Portuguese hemisphere.

The mystery of how Bunting had learned the correct shape of Australia's west coast remains unsolved but what this map confirms once again is that some European navigator had charted the west coast of Australia long before the Dutch first appeared in these waters. And that 'someone' could only have been Portuguese.

As for Portuguese maps of this period, from the 1520s right up to the end of the sixteenth century these do not give even so much as a glimpse of any great continent lying to the south-east of the island of Java. It is not so much that the maps falsify this region to the south, rather that in the main they simply ignore it.

One particularly interesting world map of this period was produced in

1529 by Diogo Ribeiro, a Portuguese cartographer in the service of Spain. This has a blank space where the south coast of Java (the real Java) should be but it also shows a large ship under full sail proceeding through the ocean just off this missing shoreline. Other later Portuguese maps also indicate only open sea to the south. Further along the Indonesian chain, Sumbawa, Flores and Timor are all drawn correctly as islands, confirming once again that the Portuguese knew perfectly well that none of these territories were headlands of a southern continent, as the Dieppe cartographers had supposed.

Why did Portugal fail to follow up its momentous discovery of a vast, unknown continent in the southern seas?

There are a number of reasons, some of them known, others conjectural. Whether it was Mendonça's one and only set of charts which the French had managed to seize, or whether another set reached Lisbon only to vanish in the state archives, we cannot be sure. The fact is, however, that Portugal by this time had its hands full just to maintain its immensely long trading route stretching from Lisbon all the way down the coast of Africa, across the Indian Ocean, around the tip of South-East Asia and out across the tropical seas to the Spice Islands. There were numerous fortresses to be manned on the coasts of Africa, India and countries of the East Indies. Maintaining control over Brazil was a further preoccupation. The spice trade might be hugely profitable, but Portugal was a small country of fewer than two million people and its limited resources were stretched to the utmost simply trying to maintain its far-flung seaborne empire. Fending off hostile local rulers and would-be commercial rivals was a constant distraction in terms of ships and manpower.

The reports Mendonça must have given about the limited trading potential of the almost empty land of Terra Java would not have been immediately encouraging. Perhaps this was not the Island of Gold after all? And in any case the regions to the south of Java were distant from Portugal's vital Spice Islands trade routes. Exploitation of this vast but forbidding new territory could wait until times were more propitious.

Unfortunately for Portugal, that time never arrived. In August 1578, John

III's successor, the fanatical, headstrong young King Sebastião (Sebastian), led a huge but ramshackle crusading army across the Straits of Gibraltar to invade Morocco. Sebastian saw himself in quixotic terms as a 'Captain of Christ'—but what was to have been the triumphant conquest of the infidel Moors turned instead into a rout, with the Portuguese army annihilated. According to popular belief King Sebastian's body was never found, and a Messianic legend grew up that he would one day return. According to other less romantic accounts, however, the king was in fact among the thousands of dead, his naked and bloody corpse having been discovered by a page on the battlefield next day.

With no heir to the Portuguese throne, it was only a matter of time before political manoeuvring brought Portugal under the sway of its bigger neighbour, Spain. Officially it was a union of the two Crowns, but in reality it was a Spanish takeover. The merger brought further misfortunes for Portugal, which found itself embroiled in King Philip II's religious wars against the Protestant Dutch and English. Most of Portugal's best ships were lost taking part in Philip's ill-fated Armada against England in 1588.

A legacy of the war against Holland was the implacable enmity this aroused in the Dutch, who from the beginning of the seventeenth century began to forcefully challenge the Portuguese monopoly in the East. Unlike the French, the Dutch were not to be fooled into believing that the seaway south of Java was blocked by a continental landmass. They sailed boldly from the Cape of Good Hope straight across the Indian Ocean almost as far as Australia, bypassing the Portuguese fortress at Malacca and arriving at their own stronghold of Batavia by approaching Java through the back door—from the south. In taking this route they must have had access to some Portuguese maps, if only because, as we have seen, the Portuguese name 'Abrolhos' appears on their own early charts indicating a particularly dangerous area of reefs and shoals off Australia's western coast.

The Portuguese empire was now entering a period of terminal decline as one military catastrophe followed another. The fort at Tidore in the Moluccas was lost to the Dutch in 1605, and with it went Portuguese control of the Spice Islands. In 1622 the great international trading centre of Hormuz, once governed by Christopher de Mendonça, was retaken by the Persians with the aid of an English fleet.

In 1640 the Portuguese rebelled against their resented Spanish masters, and the Portuguese Duke of Bragança was proclaimed King John IV. But now Portugal found itself having to fight on two fronts: against the Spanish at home, as well as against the Dutch overseas. As for the Dutch, they were not slow to take advantage of this new Portuguese weakness, seizing the strategic port of Malacca in 1641. By 1658 they had conquered Portugal's trading settlements on the coast of Ceylon. And just before the close of the seventeenth century there came the disastrous loss of the great Portuguese bastion of Fort Jesus on Mombasa Island, lynchpin of the East African coast. Fort Jesus fell on 13 December 1698 after a three-year siege by an Arab fleet from Oman, on the Persian Gulf. In one of those great ironies of history, a Portuguese squadron sent from Goa to relieve the beleaguered fort arrived one day too late.

In the East the Portuguese now retained just Mozambique, Goa and their Chinese enclave of Macao, plus a few places of minor importance including half the island of Timor. In 1961 the Indian Army marched into and reclaimed Goa, meeting virtually no resistance. Macao was handed back to the Chinese in 1999.

All that remains of Portuguese power in the East today are the crumbling remains of a few forts on islands to the north of Australia, and the curious anomaly of East Timor, which, after freeing itself first from Portuguese colonialism and then (with Australian help) from Indonesian hegemony, has chosen to retain many of its linguistic and cultural links with Portugal.

East Timor is in many ways symbolic of the Portuguese decline—a tiny, half-island remnant of a once-proud seaborne empire which at the height of its power spanned half the globe. An empire whose intrepid sailors were the first Europeans to discover and chart the unknown coasts of the great southern continent of Australia.

Epilogue

Marco Polo was right, of course. Australia did indeed turn out to be the Land of Gold—though not quite in the way the sixteenth century Portuguese adventurers envisaged. Australia today ranks as the world's third-largest gold producer, with an output of the precious metal amounting to more than US$3 billion a year. Since gold was first discovered in the mid-nineteenth century, total production from Australian gold mines has been worth close to US$150 billion at year 2006 values.

But Land of Gold or not, it is an unfortunate fact that almost without exception orthodox Australian historians and maritime archaeologists, bent on defending at all costs their view that there never was any Portuguese discovery of Australia, have either dismissed all evidence of such a discovery as fiction—or simply ignored it, perhaps in the hope that it will somehow go away.

How else is one to explain, for instance, the total lack of interest shown by mainstream academics in the extraordinary discovery, detailed in this book, of a sixteenth century Portuguese lead sinker in the sands of Fraser Island? It was pure chance that Dr Bill Ward and his scientific team unearthed this 500-year-old artefact. A few centimetres difference either way and their drill hole would have missed it. The fishing net to which the sinker must have been attached would long ago have disintegrated, but it is highly unlikely this net would have been weighted by only a single sinker. That being the case, there may well be other similar sinkers awaiting to be unearthed in the sands at Hook Point.

We can at least be thankful that the only lead sinker so far discovered is today safely stored in the Anthropology Museum at the University of Queensland. But it is nevertheless somewhat ironic that this same university still refuses to acknowledge the fact that this very artefact, held in its own museum, is proof positive that a Portuguese voyage of discovery along the coast of Queensland really *did* take place. For that matter, the artefact itself surely deserves a better fate than to be locked away obscurely in an academic museum with restricted opening hours.

And what about the equally remarkable trawling up of at least one, and most probably two, early sixteenth century Portuguese pottery jars from the sea bed off the coast of south-eastern Australia? A date of around 1500 AD obtained through recognised laboratory techniques for the first pot by a reputable university scientist has been either ignored or dismissed as 'impossible' by the defenders of orthodoxy. But if the date is wrong, why have none of these critics attempted to make their own scientific tests and thus prove their point?

And why have Australian maritime archaeologists expressed not a grain of interest in investigating the exciting prospect that the pots could even be evidence of a sixteenth century Portuguese shipwreck lying on the ocean floor? This is by no means a certainty, but it is at least a possibility. If there did prove to be a sixteenth century Portuguese shipwreck lying in these waters, it would be guaranteed to create an international sensation. The question has to be asked: are Australia's coasts so teeming with 500-year-old shipwrecks that this prospect is not worth even a preliminary survey?

Fortunately, the two pots recovered almost intact from the possible wreck site have not been lost. The first is in the possession of a Sydney oceanographer, while the other is in the keeping of the Eden Killer Whale Museum.

As for the historic sixteenth century Portuguese cannons found by the Royal Australian Navy on Carronade Island in Napier Broome Bay, these are

How it would have looked: this composite map compiled by the author shows how 'Terra Java' appears when the two corrected eastern and western Vallard maps are joined. East-west foreshortening is due to the cartographer having wrongly assumed that the north coast of Australia was the same as the north coast of Java and its nearby islands. But despite this error, the resemblance to Australia is unmistakable.

arguably the oldest European artefacts ever found anywhere in Australia. Yet they are unavailable for public viewing, having for many years past been incarcerated in a naval storage facility on Spectacle Island in Sydney harbour. Not surprisingly, they are said by the Navy to have suffered deterioration during this time, and to be in need of restoration.

It could reasonably be argued that, by removing them from a sacred Aboriginal site with whatever honest intentions but clearly without Aboriginal permission, the Navy has no legal or moral right to these artefacts. A proper course might be for them to be returned to their traditional owners at Napier Broome Bay or—even better—with the consent of the Aboriginal people put on public display in a major national museum, perhaps either the National Maritime Museum or the National Museum of Australia.

The enigmatic stone ruins at Bittangabee (Pettungerbe) also surely deserve archaeological investigation. The official claim that this site was 'never inhabited' and is therefore not worthy of investigation is misleading and incorrect. The indefatigable traveller George Augustus Robinson reported unambiguously that on his visit in 1844 the place most definitely was inhabited—by Aboriginal whaleboat crews—and that there were even gardens and fruit trees. Was this originally the site of a stone stockade erected by Portuguese sailors? Thorough archaeological investigation of the stone ruins would be a major task given the neglected state of much of the site—but it might, nevertheless, produce some fascinating discoveries.

One other intriguing question remains to be answered. Was it Christopher de Mendonça who also mapped the *west* coast of Australia? It is the belief of this writer that he almost certainly did, and that the claim by the official chronicler Barros that his fleet was diverted to Sumatra to help build a fort was no more than a cover story concocted to disguise the fact of the voyage. The departure of Mendonça's fleet from Malacca and its subsequent return there in January 1522 could not be hidden from prying eyes—but where it had been in the intervening period could be concealed without too much difficulty. This hypothesis would also account for the mysterious 'other' deliberate voyage to *Terra Australis* referred to by the Flemish cartographer Cornelius Wytfliet.

The skill with which the dangerous coastlines of Western Australia were charted certainly looks very much like the work of Mendonça, as does the careful recording of mineral discoveries in the form of magnetite and (possibly) silver. On this basis we might reasonably theorise that Mendonça sailed as far as the south-western tip of Australia, saw that a great seaway opened up toward the east, and then turned back to Malacca—or perhaps even sailed direct to India to report his momentous discovery to Governor Lopes de Sequeira in person.

The corollary would be that on his second expedition, when his ships came to the beginning of the Great Australian Bight, Mendonça could have felt reasonably sure that this was the further end of the seaway he had sighted on his previous voyage. Terra Java, in other words, had been shown to be not an extension of the polar regions, as most sixteenth century cartographers mistakenly believed. It was an island continent in its own right.

Chapter notes

Chapter 1: The golden land of Marco Polo

1. See *The Travels of Marco Polo*, Book III, Ch. VIII, p. 335 in Heron Books' Geneva edition.
2. ibid., p. 335-6. Other references to the island of Java and surrounding lands are to be found in Chapter VII and Chapters IX-XVII.
3. RH Major, *The Discovery of Australia by the Portuguese*, London, 1861, p. 9.
4. George Collingridge's monumental work, *The Discovery of Australia*, was printed in Sydney in 1895. A facsimile edition was published by Golden Press Ltd, Sydney, in 1983.
5. KG McIntyre, *The Secret Discovery of Australia*, Sydney, 1977.
6. RH Major, *Early Voyages to Terra Australis, now called Australia*, London, Hakluyt Society, 1859, p. xxvii. One of the 'facsimiles' to which Major refers came to light last century in the library of the Parliament of the Commonwealth of Australia, from whence it was transferred to the National Library of Australia. Interestingly enough, in the National Library's catalogue it is also listed as 'First Map of Australia'.
7. A comprehensive account Portugal's emergence as a world power is given in CR Boxer's *The Portuguese Seaborne Empire 1415–1825*, published in London in 1969. More revealing details on Prince Henry's role in the slave trade are provided in Richard Hall's *Empires of the Monsoon*, London, 1996, pp. 115-119.

8. An intriguing theory about secret voyages is propounded by Richard Hall, op. cit., p.156.
9. Quoted in John Villiers's *East of Malacca*, Bangkok, 1985, p.17.
10. Useful, up-to-date sources of information on sixteenth century European navigation and seamanship include, in particular, JH Parry's *The Age of Reconnaissance*, Los Angeles, 1981, pp. 69-99, and *The Explorers of the Pacific*, Kenthurst, NSW, 1996, by Geoffrey Badger, pp. 23-28.
11. Stevens's account was published by the English geographer Richard Hakluyt in his anthology of travel, *The Principal Navigations, Voyages, Traffiques and Discoveries of the English Nation, 1598–1600,* Everyman edition, Vol. 4, p. 239.

Chapter 2: Strangers with skins like crocodiles

1. Royal Australian Navy archives.
2. ibid.
3. This Lisbon identification is related by the Western Australian historian and writer Malcolm Uren in his book *Sailormen's Ghosts*, Melbourne, 1940, p. 24. Unfortunately, Uren does not give the source for this report.
4. *Travels of Mendes Pinto*, ed. And trs. By Rebecca D. Catz, University of Chicago Press, 1989, p.85.
5. Probably the most outspoken critic of Captain Robins' findings is Jeremy Green, archaeologist at the Western Australian Maritime Museum. In 1982 he published a paper in which he concluded that 'there is no evidence for Spanish or Portuguese association [with the cannons] or for a date any earlier than the late eighteenth century'. This sweeping pronouncement, which seems to have been accepted uncritically as gospel by most other academics, is quite unjustified by the facts.

   Green based much of his argument on an attempt to prove (not altogether convincingly) that the cannons could not have been made in Europe and therefore could not have been brought to the Kimberley coast by any Portuguese expedition. The reasoning is invalid. Green seems to have been unaware of the fact that before they sent any expedition toward Australia, the Portuguese had captured large numbers of cannons from South-East Asian foes. When they conquered the port city of Malacca on the Malay coast in 1511, for instance, the victorious Portuguese captured 300 guns of various types. Further large arsenals were seized during subsequent military campaigns on the Malay peninsula. Some of this captured ordnance would no doubt have been distributed among the captains of the Portuguese fleet.

Another very relevant factor is that in 1505 two deserters from Vasco da Gama's second Portuguese expedition were enlisted by a South Indian ruler, the Zamorin of Calicut, to set up a gun foundry. Over the next two years this foundry manufactured over 500 small cannons of European design (see Richard Hall's *Empires of the Monsoon*, p. 210). Many of these cannons were subsequently captured by Portuguese naval forces. The fact that these guns were specifically described as 'small' could be significant in the context of the Carronade Island finds, particularly as far as the undecorated cannon is concerned. And last but not least, a few years later the Portuguese had themselves established a major naval base at Goa, north of Calicut, setting up both shipbuilding and gun-manufacturing industries. The latter development is referred to in one of the footnotes to Roger Green's paper, but he does not seem to have considered the possibility that India could possibly be the origin of one or even both of the Carronade Island guns.

Green labelled the second, undecorated cannon 'a Southeast Asian lantaka'. 'Lantaka' is a generic South-East Asian term for a swivel gun, the word having originated in the Philippines (in Tagalog language) before the advent of the Spaniards, and it could be used to describe any type of South-East Asian swivel gun dating from the fifteenth century onwards. Neither of the Carronade Island cannons in fact looks anything remotely like a lantaka. Traditional Malay and Javanese lantakas tended to be highly ornate—sometimes even with muzzles grotesquely shaped to resemble the mouths of dragons or crocodiles—while even the plainer guns invariably had large, bell-shaped muzzles, usually incorporating a raised gun sight. No such feature appears on the Carronade gun. European swivel guns were fashioned on more practical lines and were able to throw a heavier projectile longer distances.

Green's assertion that the Carronade cannons must date from the end of the eighteenth century at the earliest is puzzling, in light of his own background in maritime archaeology. Cast-bronze guns like those found on Carronade Island, particularly the decorated gun, began coming into general use in Europe at the very beginning of the sixteenth century—and contrary to his claim that smaller cannons of this sort were nearly always breech-loaders, the reverse seems to have been the case.

In the early sixteenth century the manufacture of cast-metal cannons in the style of the Carronade Island guns developed rapidly in several European countries, most notably Flanders and Germany, and very soon all the best European guns were being made of bronze or brass rather than iron. The muzzle-loading design was generally preferred to breech-loading because one-piece casting greatly improved the guns' power while at the same time greatly enhancing safety: breech-loaders had a nasty tendency for the detachable breech chamber to blow out, with disastrous consequences for the gun crew.

In addition to its Portuguese crown, the decorated Carronade Island gun has a clear resemblance to other European cannons of the sixteenth century particularly cannons on display today in Lisbon. In addition, maritime archaeology has made it possible to make a direct comparison with guns carried on board ships of the Spanish Armada in 1588. Many of these vessels, including Portuguese ships, were wrecked while trying to return to Spain. Their guns have in recent times been salvaged and photographed, and some are strikingly similar in appearance to the Carronade Island cannons.

As for a cannon which Green cited in his paper as being similar in appearance to the decorated Carronade Island gun and which he described as having come to Australia from Surabaya in the 1920s, Green concedes that Surabaya was simply the place where the 'Quintus Bosz gun' was obtained by its Australian owner. He had no information on where, when or by whom it was made. (Email to the author, 9 October 2002).

6. See Colin Martin and Geoffrey Parker, *The Spanish Armada*, London, 1988, p. 216. An interesting comparison with Davey's swivel gun can be seen on p. 219, gun no. 7, a medio sacre—this is bigger, but of a similar shape.
7. Malcolm Uren, op.cit., refers to these finds on pp.24,25 and 239 of *Sailormen's Ghosts.*

8. The Royal Australian Navy seems to have either mislaid or lost the original account, but again by good fortune a copy of it was reprinted in full while it was still in the Navy's possession. (Howe's letter reproduced by Lawrence FitzGerald in his *Java la Grand—The Portuguese Discovery of Australia*, Hobart, 1984).
9. Jeremy Green, however, disagrees, maintaining that such Aboriginal genealogical traditions 'can be quite unreliable'. (Email to the author, 9 October 2002)
10. Howe used the word 'alligators', but this is clearly a mistranslation from the Aboriginal since alligators do not exist in Australia. The Aborigines could only have been speaking of crocodiles, of which there are a great number in the Kimberley region.
11. Ian Crawford, *We Won the Victory*, Fremantle, 2001, p. 78.
12. See AJ Brown, *Ill-starred Captains: Flinders and Baudin*, Adelaide, 2000, p. 358.
13. For João de Barros's account of Pacheco's voyage, see his monumental work *Asia*, edn Coimbra 1932, Dec. III, Liv. III, Cap. III, pp. 122-27.
14. Another interesting botanical feature of Napier Broome Bay's main settlement, Kalumburu, is its giant mango trees. The mango tree originated in India then spread through South-East Asia, but the word 'mango' is actually a Portuguese version of the Malay word, 'manga'.
15. Manuel de Faria e Sousa, *Asia Portuguesa*, Lisbon, 1703, Porto, 1945, Vol. II, Cap. III, p. 28-9.
16. Later, some of Barros's books were actually prohibited by the Inquisition.
17. See George Grey's *Journals of Two Expeditions of Discovery in North-West Australia*, London, 1841, Adelaide, 1964, p. 214.
18. Ian Crawford, a former Curator of Anthropology at the Western Australian Museum, spent five years in the 1960s recording Aboriginal rock art in the Kimberley. See his *Wandjinas: a Western perspective, paper delivered at a seminar on Aboriginal rock art of the Kimberley*, Perth, 1997, pp. 47-48. Also *We Won the Victory*, p. 110.
19. A contemporary picture of the combat dress of sixteenth century European soldiers can be seen in Derrick's *Images of Ireland* an Elizabethan army on the march, published in 1580. Of particular relevance is the clothing and armour of the spearmen.

20. Date quoted by Grahame Walsh, anthropologist and Aboriginal rock art authority: verbal communication to the author, 18 August 2006.
21. Tasman's ships consisted of two square-rigged *jachten* of around 100 tons carrying eighty sailors and seventeen soldiers, and the small galiot *Bracq*, with a crew of fourteen. This little ship, designed for use in shallow waters, had two masts, a gaff rig and lee boards on the sides of its hull. A second Aboriginal rock painting on Bigge Island, placed just above the figures with pipes in their mouths, depicts a small ship with a triangular sail. It has generally been assumed that this painting represents a pearling lugger, but the rig suggests it could quite possibly be the *Bracq*.
22. Fernão Lopes de Castanheda, *Historia do Discobrimento & Conquista da India pelos Portugueses*, Lisbon, 1736, Coimbra, 1929, Liv. V, Cap. V, p. 58.
23. Ian Crawford, *We Won the Victory*, p. 74.
24. ibid., p. 85.
25. DJ Mulvaney, *Encounters in Place*, Brisbane, 1989, p. 27.
26. Ian Crawford, op. cit., p. 73.
27. ibid., p. 123.
28. Fr. Anscar McPhee: letter to the author, July 2000.

Chapter 3: Braving the Coast of Danger

1. The main source of this information is the official chronicler, Barros, op. cit. Dec. II, Cap. III , pp. 188-9. Mendonça's arrival in command of a carrack is also recorded by Castanheda, op. cit., Liv. V, Cap. XV, p. 23. The Portuguese chroniclers at various times refer to the mysterious 'Island(s)' of Gold in both the singular and the plural form.
2. José Alberto Leitão Barata, *Os Senhores da Navegação*, Lisbon, 2003, pp. 82, 143.
3. The circumstances of Mendonça's arrival at Hormuz are related by Barros, op. cit. Dec. IV, Cap. II, p. 128, Castanheda, op. cit. Liv. VII, Cap. LXVII, p. 117, and Faria e Sousa. op. cit. Vol. IV, Cap. II, p. 145.
4. Some of Barros's writings were later banned by the Inquisition. There are also some reasons for doubting the veracity of the chronicler's statements about

what Mendonça had been doing in 1520–21. This evidence is discussed in a later chapter.

5. This salutary incident is related in Richard Hall's *Empires of the Monsoon*, London, 1996, p. 117.
6. See Laurence Bergreen's assessment in *Over the Edge of the World*, London, 2003, p. 24.
7. A Cortesão, *O Descobrimento da Australia em 1522*, Lisbon, 1939, p. 13.
8. These quotations are from a 1962 edition of Pigafetta's account, *Magellan's Voyage Around the World*, published by Northwestern University Press, USA, and based on a translation from the Italian first published in 1906, pp. 122-3.
9. F Peron, *A Voyage of Discovery to Terra Australis*, Melbourne, 1975 edition, originally published in English trans. London, 1809, p. 255.
10. Hawkins's comments were written in 1593. The quotations are from *The Observations of Sir Richard Hawkins*, JA Williamson (ed.), originally published in London in 1622, facsimile edition, London, 1933, pp. 40-42.
11. Useful sources of information on early charts include JH Parry's *The* Age *of Reconnaissance*, Los Angeles, 1981, pp. 100-113; Rear-Admiral GS Ritchie's introduction to *The Sea Chart*, D Howse and M Sanderson, Newton Abbot, 1973, and *Decorative Maps*, by Roderick Barron, London, 1989.
12. Cornelius Wytfliet, *Descriptionis Ptolemaicae Augmentum: siue, Occidentis Notitia Brevis Commentario*, Louvain, 1597, p. 38.
13. George Collingridge, *The Discovery of Australia*, op. cit., p. 219.
14. In Portuguese it would have been *Rio Grande* with a final 'e'. This is Australia's first *Rio Grand*, but by no means the last. We may also note here that the Vallard cartographer habitually writes what should be a final 'd' to look like an embellished 't': to simplify matters, the name will be transcribed throughout as *Rio Grand*.
15. Not to be confused with Flinders Island in Bass Strait.
16. We should note here a peculiarity which will occur throughout the Vallard map. As written in the map, this place name at first sight appears to be 'G Ferra'. In Portuguese, however, the gender of an adjective has to agree with that of the noun to which it refers—and so 'Golfo', being a masculine noun, has to be followed by the masculine adjective, 'Ferro', not the feminine form,

'Ferra'. The map contains many other similar examples. It is difficult to be sure whether this is a grammatical error on the part of the cartographer, or simply an idiosyncrasy of writing style. In this book the correct grammatical form will be assumed.

17. See JC Beaglehole, The *Journals of Captain James Cook on His Voyage of Discovery; Vol. 1, Voyage of the Endeavour 1768–1771*, Cambridge, 1955. The extracts here are dated 9 June 1770, p. 341.
18. Accounts of these Aboriginal traditions are fragmentary and confused. Possibly one of the more reliable is related in Kenneth Byron's *Lost Treasures in Australia and New Zealand*, Sydney, 1964, p. 177.
19. Beaglehole, op. cit., p. 325.

Chapter 4: Mendonça's calling card: a sinker from the bay of prawns

1. The Australian section of *The Endeavour Journal of Joseph Banks* was published as a separate edition (ed. Paul Brunton) in Sydney in 1998. This entry is on p. 34.
2. See 'Ancient Lead Weight found with Loisels Pumice near Hook Point, Fraser Island, Queensland', by WT Ward, IP Little, GM Roberts, BL Gulson, BM O'Leary and DM Price, in *Archaeology in Oceania,* vol. 34, 1999, pp. 25-30.
3. St Catherine of Alexandria should not be confused with St Catherine of Siena, now patron saint of Italy, whose feast day is 29 April. There are quite often several saints with the same name, and some are accorded more importance in some countries than in others. Another complicating factor is that in earlier times the feast days of particular saints often varied from diocese to diocese, or from one religious order to another.
4. A description of this first survey of the entrance to the Brisbane River can be found in GRC McLeod's paper 'A Short History of the Dredging of the Brisbane River, 1860 to 1910', published in the *Journal of the Royal Historical Society of Queensland, Vol, X*, No. 3, 1977–78, p. 137. McLeod notes that once European settlement began, farming and tree-felling along the upper reaches of the river caused erosion, and floods carried topsoil down to the river's entrance. Mud and silt were deposited in the channels and on the previously sandy banks, making entrance to the river more difficult. However, even before dredging

began in 1862 some coastal steamers were still able to come up the river as far as Brisbane.

5. Rex Gilroy refers to the first of these stories in his monograph 'Carving Out a New History', *Australasian Post*, 19 December 1985. The other source, Ron Heron, recounted the oral tradition in a telephone interview with the author, October 2003.
6. See Flinders' journal of the *Norfolk* voyage, 7 August 1799, in *Terra Australis* (ed. Tim Flannery), Melbourne, 2000, p. 38.

Chapter 5: In search of Botany Bay: cracking the Mendonça code

1. For a full exposition of this theory, see WAR Richardson's , *The Portuguese Discovery of Australia: fact or fiction?*, Canberra, 1989 and *Was Australia Charted Before 1606?*, Canberra, 2006.
2. *The Rotz Boke*, British Library facsimile edition, 1981 (ed. Helen Wallis), p. 59.
3. McIntyre put forward a theory in which he tried to show that the size of the eastern promontory had been exaggerated. He attempted to justify this by theorising that the Dieppe map-makers had originally drawn their map on paper pasted onto a globe, which was later cut up into segments, or 'gores', which were then pasted back together to form a flat chart. Because of the curved shape of the gores, McIntyre postulated, some extra details would have had been drawn in to fill the gaps, and he therefore felt entitled to delete these 'extra' coastlines—two of which fortuitously happened to form part of the eastern promontory.

   McIntyre's theory was ingenious but flawed. Not only did it rely on some unproven and puzzling assumptions about the technical procedures of the Dieppe cartographers, it also overlooked the fact that there was no realistic possibility that any kind of map-making globe would have been carried on a sixteenth century ship charting unknown coasts on the far side of the world. It follows from this that any original map showing the east coast of Australia could only have been drawn as a conventional plane chart. And that being the case, even if for the sake of argument one supposed that later on the Dieppe cartographers did actually go to the trouble of redrawing their map on a globe, they would have had no cause whatever to cut the globe up again in order to make a chart—for the very obvious reason they already possessed the original chart, or charts, from which the globe had been copied.

Furthermore, in spite of his best efforts McIntyre was still left with an eastern protuberance—even if a somewhat smaller one—which obstinately refused to go away. This remained an uncomfortable problem, for there is simply no such promontory on the east coast of Australia, whether *very* large or just somewhat large. Those wishing to read McIntyre's theory in full will find it set out in *The Secret Discovery of Australia*, published in 1977.

4. See *1421: The Year China Discovered the World*, by Gavin Menzies, London, 2002.
5. An error in joining portolans could also explain the missing stretch of the north Queensland coast between Repulse Bay and Shoalwater Bay: this omission occurs at about the place in the Vallard map where one would expect to find a join between two of the original charts.
6. As we will see in a later chapter, there are also references to white sand as 'snow' in other Dieppe maps—this time describing an island on Australia's tropical north-west coast with the name variously given as *Ille Nege*, or *Ye de Neige*. *Neige* is French for 'snow'—hence 'Snowy Island'. In the original Portuguese it would have been *Ilha da Neve*.
7. Although Cook likewise sighted the entrance to Sydney harbour and named it Port Jackson, he also omitted to stop and investigate more closely.
8. This replacement of some early European place names by Aboriginal ones reflects a trend which may be justified in some circumstances, but seems questionable in this case, given the historical significance of the name bestowed by Cook.
9. There is no authoritative record of this supposed discovery. It is cited by Rex Gilroy in his article 'Carving Out a New History', *Australasian Post*, December 19, 1985—but he provides no substantiation for the claim.
10. *The Endeavour Journal of Joseph Banks*, op. cit., p. 22. The Aboriginals' 'scymeters' (scimitars) were evidently boomerangs.

Chapter 6: South to Cape Beautiful and a Royal River

1. For an example of the latter version, see Pieter Schooness's *Inscriptions on Padrões, Postal Stones, Tombstones and Beacons*, published by the South African Cultural History Museum, Cape Town, 1991, p. 13.

2. In contrast, the promontory's present name commemorates an obscure London merchant, Thomas Wilson, whose only claim to fame seems to be that he happened to be a friend of the British explorers, George Bass and Matthew Flinders. Such an exceptional Australian landmark surely deserves something better. The promontory's Aboriginal name of *Wamoom* (like the indigenous people who inhabited it for thousands of years) seems to have been all but forgotten.
3. In recent times, evidence in the form of stone tools has been found indicating that Aborigines did once frequent Kangaroo Island—but that was many thousands of years ago during the last Ice Age, when Kangaroo Island was connected to the mainland by a land bridge.
4. Several books have been published to mark the 200th anniversary of the voyages of Flinders and Baudin. This quotation is from Anthony J Brown's *Ill-Starred Captains*, Adelaide, 2000, p.178.

Chapter 7: A Ring of Stone at the Bay of the Portuguese

1. For details of this theory, see *1421: The Year China Discovered the World*, by Gavin Menzies, op. cit., pp. 153-4. Menzies states (p.38) that the Chinese treasure ships were 480 feet (146 metres) long. Observers estimated the Mahogany Ship's length to be about 50 feet (approx. 15 metres).
2. 'The Bittangabee Ruin—Who Built It?', Mike Pearson, National Parks and Wildlife Service paper, 1980, p. 1.
3. 'Ben Boyd National Park Bicentennial Project: Davidson Whaling Station, Boyd's Tower, Bittangabee Ruins', by Anne Bickford, Sandy Blair and Peter Freeman, Sydney, 1988, pp. 73-92.
4. It has been suggested that a second, unstated purpose of the expedition may have been to discover new areas of farmland for Victorian graziers.
5. This entry appears on p. 245 of the microfilm version. [Mfm G27514-518]
6. For an account of the sad fate of the Tharwa/Durga language, see Diana Eades's *The Dharwal and Dhurga Languages of the New South Wales South Coast*, Canberra, 1976.

7. Michael Organ, *Ilawarra and South Coast Aborigines, 1770–1850*, Wollongong, 1990.
8. Kenneth Gordon McIntyre, *The Secret Discovery of Australia*, Sydney, 1977, 1990, p. 171. McIntyre's date was later discredited by other new facts which came to light.
9. In any case the argument is questionable. The site *was* inhabited, for several years at least, during the whaling era. The fact that it has been preserved more or less untouched since then means that a thorough archaeological investigation could, at the very least, provide useful information about life at a whaling station in the 1840s.
10. David Collins, *An Account of the English Colony in New South Wales*, London, 1798; reprint Sydney, 1975, Vol. 1, p. 408.
11. This extract is from the journals of Matthew Flinders in their abridged edition, *Terra Australis*, ed. Tim Flannery, Melbourne, 2001, pp. 10-11.
12. The diaries of Oswald Brierly, Mitchell Library, Sydney, quoted by HP Wellings in *Eden and Twofold Bay*, Sydney, 1996, p. 36.
13. AW Howitt, *The Native Tribes of South-east Australia*, London, 1904, reprint Canberra, 1996, p. 82.
14. HP Wellings, *Eden and Twofold Bay*, Sydney, 1966, p. 20.

Chapter 8: An Island of Mahogany and the Vessel of the Gods

1. Cook made similar errors when he mistook Banks Peninsula on the east coast of the South Island for an island and charted Stewart Island as a headland.
2. References to Cook's Journals are generally from the editions edited for the Hakluyt Society by JC Beaglehole, 1955–67
3. The idea is not all that new. As long ago as 1873, the former Keeper of Maps at the British Museum, Richard Henry Major, wrote of the Dieppe maps that not only were the dangers of the northeast coast of Australia distinctly shown, but that also 'the discovery of New Zealand is clearly indicated'. See Major's *Further Facts Relating to the Early Discovery of Australia*, London, 1873, p. 4.
4. This and other opinions about the helmet's age are all preserved in the Museum of New Zealand (Te Papa Tongarewa) archives.

5. The Portuguese Military Museum in Lisbon has also given an opinion, based on photographs, that the helmet is of sixteenth century origin and was manufactured in Italy. (Letter to the author, 23 January 2002)
6. This opinion from the DSIR's Industrial Processing Division is actually quoted in the museum's report, even though the report's author reaches the opposite conclusion. See Museum of New Zealand 'Records', p. 136.
7. See *Armada Guns*, Michael Lewis, London, 1961, p. 21. The bigger Spanish Armada cannons fired a 50-pound shot.
8. Professor Geoff Taylor, formerly of the Department of Earth Sciences, Australian National University, telephone conversation with the author, 1999.
9. A comprehensive description of the development of Medieval and early Renaissance military helmets is to be found in *Medieval Costume, Armour and Weapons*, by Eduard Wagner, London, 1958.
10. Andrew Sharp, *The Voyages of Abel Janszoon Tasman*, London, 1968, pp.122-3.
11. For an account of some of the Maori history of this part of New Zealand, see GL Adkin's, *The Great Harbour of Tara*, Christchurch, 1959.
12. See *Kei Puta te Wairau: A History of Marlborough Maori Times*, by WJ Elvy, Christchurch, 1957, pp. 91-92.
13. The name *Illa do Sal* appears a little further north on the map than the actual location of White Island. This is probably because the Dieppe cartographers assumed it referred to the much larger island which we have identified as Great Barrier Island.
14. Written by Moh Turei, a leader of the Ngati Porou tribe, born about 1830. Translated by Anne Salmond and Merimeri Penfold, 1991. From Anne Salmond's *Two Worlds*, Auckland, 1991, p. 62.
15. A good account of Maori life on the Tamaki isthmus is to be found in *Maori Auckland*, by David Simmons, Auckland, 1987.
16. See *Early Charts of New Zealand*, PB Maling (ed.), Wellington, 1969, plate 34.
17. Andrew Sharp, op. cit., pp. 183-4.
18. Only the letters 'MA' are visible on the stone to indicate the month of Mendonça's arrival; so theoretically it could be either March (Março in Portuguese) or May (Maio). But March would probably have been too early for a ship

sailing from India with the seasonal monsoon winds, so May seems the more logical choice.

Chapter 9: The poet-captain of Dieppe

1. A comprehensive account of the Parmentiers' voyage can be found in Jean-Michel Barrault's *Le Sacre et La Pensée*, Paris, 1989. It is available only in French. Barrault's book is based on journals from the voyage held in the Dieppe Museum. Most of the details about the Parmentiers' voyage quoted here are drawn from Barrault's account. The indication that one or more interpreters of Portuguese and Malay who sailed on the voyage were probably 'renegades' from a ship intercepted by Ango's corsairs is to be found on p. 16.
2. In *Studies from Terra Australis to Australia*, ed. John Hardy and Alan Frost, Canberra, 1989, p. 49.
3. *Voyages Avantureux du Capitaine Jan Alfonce Saintongeois*, Poitiers, 1559, f. 28r.
4. ibid., f. 35v.
5. *La Cosmographie*, published in 1544. f. 150v.
6. *Voyages Avantureux*, op. cit., f. 35r.
7. ibid., f. 36v.
8. ibid., f.35r.
9. *La Cosmographie*, op. cit., f. 147v.
10. ibid., f.150v.
11. ibid., f.158v.
12. Juan Fernandez (Robinson Crusoe's island) is actually at about 33 degrees south. Alfonse seems to be confusing it with the smaller islands of San Ambrozio and San Felix which lie on much the same longitude as Juan Fernandez, but at 26 degrees south.
13. *La Cosmographie*, op. cit., f. 158v.
14. In the Vallard map *Cap Formose* is shown lying at about 45 degrees south, and it would be considerably further south after correction. However, the latitude scale on the margin of the map is obviously incorrect. While the reference line of Capricorn is drawn at approximately the right place, on both sides of it the latitude scale is too small—an error due probably to a combination of factors

including primarily the coastline of northern Australia being confused with that of Java, a certain amount of guesswork, and the problems of trying to construct a global map using 'flat earth' geometry.

15. See *Ango et Son Epoque*, by T Calderon, Paris, 1891, p. 180.
16. *Chance Discovery of Australia and New Zealand by Portuguese and Spanish Navigators*, by Roger Hervé (trs. John Dunmore), p. 17.
17. Barrault, op. cit., p. 84.
18. Francisco da Sa's description of Java is quoted by Diogo do Couto in his *Decadas da Asia*, op. cit., Dec. IV, Liv. III, pp. 56-57.

Chapter 10: To a distant sea and 'a cape which no one has yet gone beyond'

1. See *Voyages through the Moluccan Archipelago and to New Guinea*, by DH Kolff, translated by GW Earl, London, 1840, p. 35.
2. Bernard Ryder's account, 'Wild Life in the Far North-West: Dutch cannon and Latin-speaking blacks where Dampier landed in 1688', appeared in *Walkabout*, 1 January 1936.
3. 'The stone house structures of High Cliffy Island, north-west Kimberley', Sue O'Connor, in *Australian Archeology*, 1987, No. 25, pp. 30-39.
4. A full exposition of CG von Brandenstein's theory is set out in his *Early History of Australia: the Portuguese colony in the Kimberley*, Albany WA, 1994.
5. *La Géographie*: 'La vie sauvage dans le nord-ouest de l'Australie', Paris, 1935, Vol. 63.
6. *A New Voyage Around the World*, William Dampier, London, 1703, Vol. 1, p. 464.
7. This also gives a clue to what two otherwise puzzling islands in this section of the Vallard map might represent. The larger of these has the name *Illa Tressla*. In Portuguese 'tres' is the word for 'three', so this name could mean 'Three Islands'. Only one island is shown, but it is triangular in shape and has a reef. Charts by Desceliers and Pasterot depict much the same thing. This suggests that it could be an enlargement, intended as a navigation warning, of part of the Abrolhos—most probably the dangerous Pelsaert Group where the *Batavia* and the *Zeewijk* were wrecked in later years. This group of low islands linked and

fringed by reefs is also triangular. Similarly, a smaller reef-fringed island shown further north in the Vallard map and called *Illa Ponenya* could perhaps have been originally an enlargement of one of the more northerly Abrolhos groups.

8. See *Duyfken and the First Discoveries of Australia*, by Priscilla Murdoch, Sydney, 1974, p. 7.
9. Information on Western Australian ironsand was supplied by Ron Sait, senior mining engineer, Australian Geological Survey Organisation (email to the author, 15 October 2001). The other possible identification for *Rio Mamona* is the Moore River, further south again.
10. The Portuguese adventurer and historian Antonio Galvão wrote in *The Discoveries of the World, 1555*, that what is called in Spanish the *Rio de la Plata*, in South America, was given its name by de Solis 'because he saw signs of silver, and therefore called it Rio da Prata, that is, the river of silver'. Galvão uses the Portuguese word for silver, 'prata'. This quotation is otherwise from the English translation by Richard Hakluyt, London, 1601, 1862, p. 122.
11. Cape Leeuwin was named after the Dutch ship *Leeuwin*, which sighted it in 1621.
12. *Voyages Avantureux*, op. cit., f. 36r.

Chapter 11: The ignominious voyage of Jeronimo de Figueiredo

1. *Historical Disquisition*, London 1809, p. 152.
2. *The Principal Voyages, Traffiques & Discoveries of the English Nation*, Richard Hakluyt, London, 1600, 1926, Everyman edition, Vol. 3, p. 204.
3. For most of these details, see Diogo do Couto, op. cit., Liv. VIII, Cap. XLIX, p. 279.
4. Peter Heylyn, *Cosmographie*, London, 1656, appendix p. 1093.
5. See Manuel de Faria e Sousa, op. cit., Vol. III, Cap. IV, pp. 40-41.
6. The Bible, I Kings 9:26-28.

Bibliography

Adkin, GL, *The Great Harbour of Tara*, Christchurch, 1959.

Albuquerque de, L, *Dicionario de Historia dos Descobrimentos Purtugueses*, Lisbon, 1994.

Ameal, J, *Santos Portuguesis*, Porto, 1957.

Andrade de, F, *Crónica de Dom João III*, Coimbra, 1613, Porto 1976.

Aughton, P, *Endeavour*, Gloucestershire, 1999.

Austin, KA, *The Voyage of the Investigator*, Adelaide, 1968.

Badger, G, *The Explorers of the Pacific*, Sydney, 1988.

Barata, JAL, *Os Senhores da Navegação*, Lisbon, 2003.

Banks, Sir J, *The Endeavour Journal of Joseph Banks*, Sydney, 1998.

Barbosa, D, *The Book of Duarte Barbosa*, London, 1918–21.

Barrault, JM, *Le Sacre et La Pensée: 1529 de Dieppe a Sumatra*, Paris, 1989.

Barron, R, *Decorative Maps*, London, 1989.

Barros, FE, *Tracado e Contrucão das Naus Portuguesas*, Lisbon, 1933.

Barros de, J, *Asia*, Lisbon, 1553, Coimbra, 1932, 1944.

Bawlf, S, *The Secret Voyage of Sir Francis Drake*, London, 2003.

Bergreen, L, *Over the Edge of the World: Magellan's Terrifying Circumnavigation of the Globe*, London, 2003.

Blackmore, HL, *Arms and Armour*, London, 1965.

Boxer, CR, *The Portuguese Seaborne Empire 1415–1825*, London, 1969.

Boxer, CR, *The Dutch Seaborne Empire 1600–1800*, London, 1973.

Boxer, CR, *Four Centuries of Portuguese Expansion*, Johannesburg, 1961.

Boxer, CR, *On a Portuguese Carrack's Bill of Lading in 1625*, Coimbra, 1938.

Boxer, CR, *Portuguese Society in the Tropics*, Madison, 1965.

Boxer, CR, *The Tragic History of the Sea, 1589–1622*, Cambridge, 1959.

Boxer, CR, *From Lisbon to Calicut*, Minneapolis, 1956.

Braga, JM, *Portugal's Contribution to the Discovery of Australia*, MS, 1970.

Brandenstein von, CG, *Early History of Australia: the Portuguese Colony in the Kimberley*, Albany, WA, 1994.

Brown, A, *Ill-starred Captains: Flinders and Baudin*, Adelaide, 2000.

Brown, S, *Men From Under the Sky: the Arrival of Westerners in Fiji*, Suva, 1973.

Brown, SR, *Scurvy*, Camberwell, Victoria, 2003.

Bryant, J, *Captain Matthew Flinders*, London, 1928.

Byron, KW, *Lost Treasures in Australia and New Zealand*, Sydney, 1964.

Calderon, T, *Ango et Son Epoque*, Paris, 1891.

Camões de, L, *The Lusiads*, Eng. trs. London, 1880, 1963, New York, 1966.

Campbell, J, 'Geographical Memoir of Melville Island and Port Essington', *Royal Geographical Society Journal*, vol. 4, 1834.

Castanheda de, FL, *Historia do Descobrimento & Conquista da India pelos Portugueses*, Coimbra, 1924.

Castanheda de, FL, *The First Booke of the Historie of the Discouerie and Conquest of the East Indias, Enterprised by the Portingales*, Eng. trs. London, 1582, New York, 1973.

Castlereagh, D, *The Glorious Age of Exploration*, New York, 1973.

Couto do, D, *Decadas da Asia*, Lisbon, 1736.

Clancy, R, *The Mapping of Terra Australis*, Sydney, 1995.

Clark, M, *Abel Tasman*, Melbourne, 1959.

Clark, M, *History of Australia*, Melbourne, 1993.

Collingridge, G, *The Discovery of Australia*, Sydney, 1895.

Collingridge, G, *The First Discovery of Australia and New Zealand*, Sydney, 1906.

Collingridge, V, *Captain Cook: Obsession and Betrayal in the New World*, London, 2002.

Collins, D, *An Account of the English Colony in New South Wales*, Christchurch, 1910.

Cook, J, *Captain Cook's Journal*, ed. WJL Wharton, London, 1893, Adelaide, 1968.

Cook, J, *The Journals of Captain James Cook on his Voyages of Discovery*, ed. JC Beaglehole, Cambridge, 1955–74.

Cooper, HM, *The Unknown Coast*, Adelaide, 1953.

Cortesão, A, *O Descobrimento da Australásia*, Lisbon, 1939.

Cortesão, A & Mota da, T, *Avelino, Portugaliae Monumenta Cartographica*, Lisbon, 1962.

Cortesão, A, *The First Account of the Far East in the 16th Century*, London, 1938.

Cortesão, A, *History of Portuguese Cartography*, Lisbon, 1971.

Coulton, GG, *Medieval Panorama*, Cambridge, 1938.

Crawford, IM, *The Art of the Wandjina*, Melbourne, 1968.

Crawford, IM, *We Won the Victory*, Fremantle, 2001.

Dampier, W, *A Voyage to New Holland*, Gloucester, 1981.

Danvers, FC, *The Portuguese in India*, London, 1894.

David, A (ed.), *The Charts and Coastal Views of Captain Cook's Voyages*, London, 1988–1997.

Derrick, RA, *A History of Fiji*, Suva, 1950.

Dor-Ner, Z, *Columbus and the Age of Discovery*, New York, 1991.

Duffy, AR, *European Armour in the Tower of London*, London, 1968.

Dunmore, J, *The Fateful Voyage of the St Jean Baptiste*, Christchurch, 1969.

Dunmore, J, *Visions and Realities: France in the Pacific*, Waikanae, 1997.

Dyson, J, *Columbus: For Gold, God and Glory*, Toronto, 1991.

Eades, D, *The Dharawal and Dhurga Languages of the New South Wales South Coast*, Canberra, 1976.

Edwards, H, *Australian and New Zealand Shipwrecks and Sea Tragedies*, Sydney, 1978.

Eisler, W, *The Furthest Shore: Images of Terra Australis from the Middle Ages to Captain Cook*, Cambridge, 1995.

Estensen, M, *Discovery: the Quest for the Great South Land*, Sydney, 1998.

Estensen, M, *Terra Australis Incognita*, Sydney, 1998.

Faria e Sousa de, M, *Asia Portuguesa*, Lisbon, 1703; Porto, 1945.

Faria e Sousa de, M, *A History of Portugal*, Eng. trs., London, 1698.

Fernandez-Armesto, F, *The Spanish Armada*, Oxford, 1988.

Finkel, G, *The Dutchman Bold*, Sydney, 1975.

FitzGerald, L, *Java la Grande - The Portuguese Discovery of Australia*, Hobart, 1984.

Fletcher, R, *The Quest for El Cid*, London, 1989.

Flinders, M, *Observations on the Coasts of Van Diemen's Land, on Bass's Strait and its Islands, and on part of the Coasts of New South Wales*, London, 1801; Adelaide, 1965.

Flinders, M, Flannery, T, (ed) *Terra Australis*, Melbourne, 2000.

Flinders, M, *A Voyage to Terra Australis*, London, 1814.

Flood, J, *Rock Art of the Dreamtime*, Sydney, 1997.

Forster, C, *France and Botany Bay*, Melbourne, 1996.

Franca d'O Pinto da, A, *Portuguese Influence in Indonesia*, Jakarta, 1970.

Frost, A, 'Towards Australia: the Coming of the Europeans, 1400 to 1788', *Australians to 1788,* Sydney, 1987.

Frost, A, *Voyage of the Endeavour*, Sydney, 1998.

Gaffarel, P, *Jean Ango*, Rouen, 1889.

Galvão, A, *The Discoveries of the World, from their First Original unto the Year of Our Lord 1555*, Eng. trs. London 1862.

Gill, JCH, *The Missing Coast: Queensland Takes Shape*, Brisbane, 1988.

Gilroy, R, *Mysterious Australia*, Mapleton, Queensland, 1995.

Gilroy, R, *Pyramids in the Pacific: the Unwritten History of Australia*, Katoomba, 2000.

Godden, E, *Rock Paintings of Aboriginal Australia*, Sydney, 1988.

Gould, RT, *Captain Cook*, London, 1935.

Great Britain Department of Environment, *Armours of Henry VIII*, London, 1977.

Hakluyt, R, *The Principal Navigations, Voyages, Traffiques & Discoveries of the English Nation*, London and New York, 1907.

Hakluyt, R, *Divers Voyages Touching the Discovery of America and Islands Adjacent*, London, 1850.

Hale, JR, *Age of Exploration*, Amsterdam, 1966.

Hall, R, *Empires of the Monsoon*, London, 1996.

Hargrave, L, *Lope de Vega*, Sydney, 1911.

Hart, H, *Sea Road to the Indies*, London, 1952.

Hart-Davis, D, *Armada*, London, 1988.

Hawkesworth, J, *An Account of a Voyage Round the World with a full account of the Voyage of the Endeavour in the year MDCCLXX along the East Coast of Australia, by James Cook*, compiled by D. Warrington Evans, Brisbane, 1969.

Heath, B, *Discovering the Great South Land,* Sydney, 2005.

Henderson, J, *Sent Forth a Dove: Discovery of the Duyfken*, Perth, 1999.

Hervé, R, *Chance Discovery of Australia and New Zealand by Portuguese and Spanish Navigators between 1521 and 1528*, Paris, 1982; Eng. trs. Palmerston North, NZ, 1983.

Heylyn, P, *Endeavouring a Discovery of the Unknown Parts of the World, especially of Terra Australis Incognita or the Southern Continent*, London, 1656.

Holthouse, H, *Ships in the Coral*, Sydney, 1986.

Howitt, AW, *The Native Tribes of South-east Australia*, Canberra, 1996.

Hunt, S & Carter, P, *Terre Napoleon: Australia Through French Eyes 1800–1804*, Sydney, 1999.

Hutton NM, *The Early Days of Kurnell to Cronulla*, Sydney, 1983.

Jack, RL, *Northmost Australia*, London, 1901.

Jessy, JS, *History of South-east Asia, 1500–1945*, Kuala Lumpur, 1970.

Kaplan, M, *The Portuguese*, London 1991.

Kelly, FM, *A Short History of Costume and Armour*, London, 1972.

Kenny, J, *Before the First Fleet: Europeans in Australia 1606–1777*, Kenthurst NSW, 1995.

Kennedy, J, *A History of Malaya*, London, 1970.

Klammer, G, *The Easting Pioneers: Portuguese Seafarers of the 16th Century in the Southern Indian Ocean*, Australia, 1971.

Langdon, R, *The Lost Caravel*, Sydney, 1975.

Lennon, S, *The Story of New Norcia*, New Norcia, WA, 1991.

Lewis, M, *Armada Guns*, London, 1961.

Ley, CD, *Portuguese Voyages 1498–1663*, London, 1947.

Lipscombe, T, *On Austral Shores,* Annandale NSW, 2005.

Loney, JK, *Australian Sea Mysteries*, Geelong, 1983.

Loney, JK, *Famous Wrecks*, Geelong, 1983.

Loney, JK, *Sea Adventures and Wrecks on the NSW South Coast*, Portarlington, Vic. 1977.

Loney, JK, *The Wreck of the Ly-ee-moon*, Geelong.

Loney, JK, *Wrecks in Australian Waters*, Newport, Vic. 1978.

Macarthur, A, *His Majesty's Bark Endeavour*, Sydney, 1997.

McCarthy, FD, *Australian Aboriginal Rock Art*, Sydney, 1962.

McIntyre, KG, *The Secret Discovery of Australia*, Sydney, 1977.

Mackaness, G, *George Augustus Robinson's Journey into South East Australia*, Sydney, 1941.

McKenna, M, *Looking for Blackfellas' Place*, Sydney, 2002.

McKiggan, I, 'The Portuguese Expedition to Bass Strait in AD 1522', *Journal of Australian Studies*, Melbourne, 1977, Vol. 1, pp. 2-32.

McKinnon, S, *From a Shattered Sun: Timor*, Wisconsin, 1991.

Macknight, CC, 'On the non-discovery of Australia', *Canberra Historical Journal*, 1983, no. 12, pp. 34-46.

Maffei, GP, *Historiarum Indicarum*, Rome, 1585.

Mahony, D, *Botany Bay: Environment Under Stress*, Sydney, 1979.

Maling, P (ed.), *Early Charts of New Zealand,* Wellington, 1969.

Mandeville, Sir J, *Mandeville's Travels*, Oxford, 1967.

Markham, CR, *The Voyages of Pedro Fernandez de Quiros*, London, 1904.

Martin, A, *Jean Ango—Armateur Dieppois*, Paris, 1884.

Martin, C, *Full Fathom Five: Wrecks of the Spanish Armada*, London, 1975.

Martin, C & Parker, G, *The Spanish Armada*, London, 1988.

Mawer, GA, *Ahab's Trade: The Saga of South Seas Whaling*, New York, 1999.

Menzies, G, *1421: The Year China Discovered the World*, London, 2002.

Mitchell, A, *The Fragile South Pacific*, Austin, Texas, 1990.

Moon, R, *The Kimberley: an Adventurer's Guide*, Victoria, 1997.

Mulvaney, DJ, *Encounters in Place: Outsiders and Aboriginal Australians 1606–1985*, Brisbane, 1989.

Mundle, R, *Fatal Storm: the 54th Sydney to Hobart Yacht Race*, Sydney, 1999.

Murdoch, P, *Duyfken and the First Discoveries of Australia*, Sydney, 1974.

Norman, AVB, *The Medieval Soldier*, London, 1971.

O'Connor, S, 'The Stone House Structures of High Cliffy Island, Northwest Kimberley', *Australian Archaeology*, no. 5, 1987.

Organ, M, *A Documentary History of the Illawarra and South Coast Aborigines*, Wollongong, 1990.

Osburn, R, *History of Warrnambool,* Warrnambool, 1888.

Parham, WT, *Island Volcano: White Island or Whakaari*, Auckland, 1973.

Parry, JH, *The Age of Reconnaissance*, London, 1963.

Pasco, Crawford, *Early Exploration of Australia*, reprint, Sydney, 1977.

Pearson, M, *Great Southern Land* , Canberra, 2005.

Perez, E, *Kalumburu: the Benedictine Mission*, Perth, 1977.

Pieris, PE, *Ceylon and the Portuguese*, Delhi, 1986.

Pigafetta, A, *First Voyage Around the World*, Eng. trs. Manila, 1969.

Pinto, F Mendes, *Peregrinacam*, London, 1653; Chicago, 1989.

Pinto, F Mendes, *Les Voyages Adventureux de Fernand Mendez Pinto*, Paris, 1932.

Pires, T, *The Suma Oriental*, Eng. trs. London, 1944.

Polo, M, *The Travels of Marco Polo*, Eng. trs. Geneva, 1955.

Prado y Tovar, Diego de, *New Light on the Discovery of Australia as revealed by the journal of Captain Don Diego de Prado y Tovar*, ed. HN Stevens, London, 1930.

Pritchard, WT, *Polynesian Reminiscences*, London, 1968.

Ptak, R, *Portuguese Asia*, Stuttgart, 1987.

Quiros, PF, *The Voyages of Pedro Fernandez de Quiros*, (ed. CR Markham), London, 1904.

Ravenstein, EG, *A Journal of the First Voyage of Vasco da Gama*, London, 1899.

Reed, AW, *Aboriginal Place Names*, Sydney, 1967.

Reed, AW, *Aboriginal Words of Australia*, Sydney, 1998.

Richards, JM, *Goa*, New Delhi, 1982.

Richardson, WAR, 'Jave-la-Grande: A Place Name Chart of its East Coast', *The Great Circle,* 1984, Vol. 6, no. 1.

Richardson, WAR, *The Portuguese Discovery of Australia: fact or fiction?*, Canberra, 2006.

Richardson, WAR, *Was Australia Charted Before 1606?,* Canberra, 2006.

Robertson, G, *The Discovery of Tahiti*, London, 1948.

Robertson, I, *A Traveller's History of Portugal*, London, 2002.

Robertson, W, *Historical Disquisition Concerning the Knowledge which the Ancients had of India*, London, 1791, 1809.

Rotz, J, *The Boke of Idrography*, facsimile ed. Helen Wallis, Oxford, 1981.

Russell, PE, *Prince Henry*, London, 1960.

Russell, PE, *Prince Henry the Navigator: the Rise and Fall of a Culture Hero*, Oxford, 1984.

Ryan, J, *Images of Power: Aboriginal Art of the Kimberley*, Melbourne, 1993.

Ryder, BC, 'Wild Life in the Far North-West', *Walkabout*, 1936, Vol. 2, no. 3, pp. 32-33.

Saint Augustine's Abbey, *The Book of Saints*, London, 1966.

Salmond, A, *Two Worlds: First Meetings Between Maori and Europeans 1642–1772*, Auckland, 1991.

Sanceau, E, *Indies Adventure: Alfonso de Albuquerque*, London, 1936.

Scarr, D, *Fiji: A Short History*, Sydney, 1984.

Schoonees, P, *Inscriptions on Padrões, Postal Stones, Tombstones and Beacons*, Cape Town, 1991.

Sharp, A, *The Voyages of Abel Janszoon Tasman*, London, 1968.

Sigmond, JP & Zuiderbaan, LH, *Dutch Discoveries of Australia*, Adelaide, 1979.

Simmons, D, *Maori Auckland*, Auckland, 1987.

Singer, C, *A History of Technology*, Oxford, 1984.

Solis, DG, *Armadas da Carreira da India do 1560*, Lisbon, 1938.

Spate, OHK, *Let Me Enjoy*, Canberra, 1965.

Stephens, HM, *Albuquerque*, Oxford, 1892.

Stone, GC, *A Glossary of the Construction, Decoration and Use of Arms and Armour in all Countries and in all Times*, New York, 1961.

Tasman, AJ, *Abel Janszoon Tasman's Journal of his Discovery of Van Diemen's Land and New Zealand in 1642, with Documents relating to his Exploration of Australia in 1644*, Amsterdam, 1898.

Taylor, R, *Unearthed: the Aboriginal Tasmanians of Kangaroo Island*, Adelaide, 2002.

Tench, W, *1788: A Narrative of the Expedition to Botany Bay and A Complete Account of the Settlement at Port Jackson*, ed. Tim Flannery, Melbourne, 2002.

Thorn, A, & Raymond, R, *Man on the Rim: The Peopling of the Pacific*, Sydney, 1989.

Tiley, R, *Australian Navigators*, Sydney, 2002.

Toft, K, *The Navigators: Flinders vs. Baudin*, Sydney, 2002.

Tower of London, *Exhibition of Spanish Royal Armour*, London, 1960.

Turbet, P, *The Aborigines of the Sydney District Before 1788*, Sydney, 2001.

Uren, M, *Sailormen's Ghosts*, Melbourne, 1945.

Villiers, A, *The Indian Ocean,* London, 1952.

Villiers, J, *East of Malacca*, Bangkok, 1985.

Villiers, J, *The Portuguese Presence in Southeast Asia*, Bangkok, 1985.

Villiers, J, *Motives for Portuguese Expansion in Eastern Indonesia*, Bangkok, 1985.

Wagner, E & Durdik, ZD, *Medieval Costume, Armour and Weapons*, London, 1958.

Walsh, GL, *Australia's Greatest Rock Art*, Bathurst, 1988.

Ward, R, *Australia Since the Coming of Man*, Sydney, 1982.

Ward, WT, Little, IP, Roberts, GM, Gulson, BL, O'Leary, BM and Price, DM, 'Ancient Lead Weight found with Loisels Pumice near Hook Point, Fraser Island, Queensland', *Archaeology in Oceania*, Vol. 34, 1999, pp. 25-30.

Ward, WT & Little, IP, 'Sea-rafted pumice on the Australian east coast: numerical classification and stratigraphy', *Australian Journal of Earth Sciences*, 2000, vol. 47, pp.95-109.

Water, M, *A Year with the Saints*, Victoria, 1997.

Wellings, HP, *The Imlay Brothers, Peter, Alexander and George Imlay*, Sydney, 1966.

Wellings, HP, *Shore Whaling at Twofold Bay*, Eden, NSW, 1964.

Whitehouse, EB, *The Northern Approaches: Australia in Old Maps 820 to 1770*, Brisbane, 1994.

Whitmont, D, *An Extreme Event*, Sydney, 2001.

Williams, F, *Written in Sand : Fraser Island*, Brisbane, 1982.

Williams, G & Frost, A (eds.), *Terra Australis to Australia*, Melbourne, 1988.

Wiseman, RM, *The Spanish Discovery of New Zealand in 1576*, Auckland, 1996.

Wiseman, RM, *Pre-Tasman Explorers*, Auckland, 1998.

Wood, GA, *The Discovery of Australia*, London, 1922.

Wood, GA, *The Voyage of the Endeavour*, Melbourne, 1926.

Index

About the author

A journalist by profession, Peter Trickett has written for newspapers and magazines in Australia, New Zealand and Britain, specialising in science, historical and investigative reporting. While working as a science journalist in Oxford, England, he won a national travelling fellowship for science journalism, using this award to study and write on science and environmental topics in Japan. In New Zealand, where he worked as a senior writer with the *New Zealand Herald* and the *Listener* magazine, he won awards for historical journalism. Since coming to Australia in 1982 he has worked as an editor for the Australian National University in Canberra and as a Senior Officer with the Department of Foreign Affairs and Trade, where he was a writer and editor for the Australian Government's overseas aid program. British-born, he is an Australian citizen, a member of the Independent Scholars Association of Australia and an experienced small boat sailor. This is his first book.